TASK SCHEDULING FOR PARALLEL SYSTEMS

BICENTENNIAL
1807
WILEY
2007
BICENTENNIAL

THE WILEY BICENTENNIAL—KNOWLEDGE FOR GENERATIONS

*E*ach generation has its unique needs and aspirations. When Charles Wiley first opened his small printing shop in lower Manhattan in 1807, it was a generation of boundless potential searching for an identity. And we were there, helping to define a new American literary tradition. Over half a century later, in the midst of the Second Industrial Revolution, it was a generation focused on building the future. Once again, we were there, supplying the critical scientific, technical, and engineering knowledge that helped frame the world. Throughout the 20th Century, and into the new millennium, nations began to reach out beyond their own borders and a new international community was born. Wiley was there, expanding its operations around the world to enable a global exchange of ideas, opinions, and know-how.

For 200 years, Wiley has been an integral part of each generation's journey, enabling the flow of information and understanding necessary to meet their needs and fulfill their aspirations. Today, bold new technologies are changing the way we live and learn. Wiley will be there, providing you the must-have knowledge you need to imagine new worlds, new possibilities, and new opportunities.

Generations come and go, but you can always count on Wiley to provide you the knowledge you need, when and where you need it!

WILLIAM J. PESCE
PRESIDENT AND CHIEF EXECUTIVE OFFICER

PETER BOOTH WILEY
CHAIRMAN OF THE BOARD

TASK SCHEDULING FOR PARALLEL SYSTEMS

Oliver Sinnen
Department of Electrical and Computer Engineering
The University of Aukland
New Zealand

WILEY-INTERSCIENCE
A JOHN WILEY & SONS, INC., PUBLICATION

Library of Congress Cataloging-in-Publication Data:

Sinnen, Oliver, 1971-
  Task scheduling for parallel systems / by Oliver Sinnen.
      p. cm.
  Includes bibliographical references and index.
  978-0-471-73576-2
1. Parallel processing (Electronic computers)   2.   Multitasking (Computer
science)   3.   Computer scheduling. I. Title.
  QA76.58.S572 2007
  044′.35—dc22   2006052157

Printed in the United States of America.

10 9 8 7 6 5 4 3 2 1

à Patrícia

CONTENTS

Even though the area of parallel computing has existed for many decades, programming a parallel system is still a challenging problem, much more challenging than programming a single processor system. With the current dual-core and multicore processors from IBM, AMD, Intel, and others, mainstream PCs have entered the realm of parallel systems. The investigation and understanding of the foundations of parallel computing is therefore more important than ever.

One of these foundations is task scheduling. To execute a program consisting of several tasks on a parallel system, the tasks must be arranged in space and time on the multiple processors. In other words, the tasks must be mapped to the processors and ordered for execution. This so-called task scheduling is a very complex problem and crucially determines the efficiency of the parallel system. In fact, task scheduling is an NP-hard problem; that is, an optimal solution generally cannot be found in polynomial time (unless $P = NP$). This has been motivating the development of many heuristics for its near optimal solution.

This book is devoted to task scheduling for parallel systems. Anyone who gets involved for the first time in task scheduling is overwhelmed by the enormous wealth of heuristics, models, and methods that have been contributed during the last decades. One of my main objectives for this book is to bring order into this jungle of task scheduling. However, the book does not simply categorize and order scheduling heuristics. Instead, it investigates and presents task scheduling by extracting and discussing common models, methods, and techniques, and by setting them into relation. Hence, this book is not a mere survey of scheduling algorithms, but rather an attempt at a consistent and unifying theoretical framework.

Another objective I have with this book is to go beyond the classic approach to task scheduling by studying scheduling under more advanced and accurate system models. These system models consider heterogeneity, contention for communication resources, and involvement of the processor in communication. For efficient and accurate task scheduling, a realistic system model is most crucial. This book is the first publication that discusses advanced system models for task scheduling in a comprehensive form.

Task Scheduling for Parallel Systems is targeted at practicing professionals, researchers, and students. For those who are new to task scheduling, the first chapters carefully introduce parallel systems and their programming, setting task scheduling into the context of the program parallelization process. Practitioners involved in parallel programming will gain an understanding of fundamental aspects of the parallelization process. This knowledge will help them to write more efficient code.

Compiler and parallelization tool developers will benefit from a deeper understanding of the scheduling problem, which is also a generalization of many other problems they face (e.g., loop scheduling). A chapter on graph models promotes the understanding of these relations. For task scheduling researchers, this book serves as a comprehensive reference, based on a unifying framework. The research community will especially value the later chapters on advanced scheduling and sophisticated scheduling models. Graduate students of parallel computing and compiler courses can use this book to thoroughly study task scheduling, which is supported by the exercises at the end of each chapter. The extensive index and the large number of bibliographic references make this book a valuable tool for everybody interested in task scheduling.

For a brief introduction to task scheduling and an overview of this book, including a short summary of each chapter, refer to Chapter 1.

OLIVER SINNEN
The University of Aukland
New Zealand

ACKNOWLEDGMENTS

This book has its roots in my PhD thesis, which I presented at the Instituto Superior Técnico, Technical University of Lisbon in Portugal. I therefore want to sincerely thank Prof. Leonel Sousa again for his academic advice, the fruitful discussions, and his criticism during that time. In this context I also thank Prof. Frode Sandnes, Oslo University College, Norway, who introduced me to genetic algorithms and has been an invaluable research colleague for many years.

I would like to thank The University of Auckland for the opportunity to write this book while being a lecturer in the Department of Electrical and Computer Engineering.

Cover design: Digital image of a painting created by Patrícia Raquel de Vasconcelos da Silva (2006).

OLIVER SINNEN
The University of Aukland
New Zealand

ACKNOWLEDGMENTS

CHAPTER 1

Introduction

1.1 OVERVIEW

Although computer performance has evolved exponentially in the past, there have always been applications that demand more processing power than a single state-of-the-art processor can provide. To respond to this demand, multiple processing units are employed conjointly to collaborate on the execution of one application. Computer systems that consist of multiple processing units are referred to as *parallel systems*. Their aim is to speed up the execution of an application through the collaboration of the processing units. With the introduction of dual-core and multicore processors by IBM, AMD, Intel, and others, even mainstream PCs have become parallel systems.

Even though the area of parallel computing has existed for many decades, programming a parallel system for the execution of a single application is still a challenging problem, profoundly more challenging than programming a single processor, or sequential, system. Figure 1.1 illustrates the process of parallel programming. Apart from the formulation of the application in a programming language—this is the programming for sequential systems—the application must be divided into subtasks to allow the distribution of the application's computational load among the processors. Generally, there are dependences between the tasks that impose a partial order on their execution. Adhering to this order is essential for the correct execution of the application. A crucial step of parallel programming is the allocation of the tasks to the processors and the definition of their execution order. This step, which is referred to as *scheduling*, fundamentally determines the efficiency of the application's parallelization, that is, the speedup of its execution in comparison to a single processor system.

The complexity of parallel programming motivates research into automatic parallelization techniques and tools. One particularly difficult part of automatic parallelization is the scheduling of the tasks onto the processors. Basically, one can distinguish between *dynamic* and *static scheduling*. In dynamic scheduling, the decision as to which processor executes a task and when is controlled by the runtime system. This is mostly practical for independent tasks. In contrast, static scheduling

Task Scheduling for Parallel Systems, by Oliver Sinnen
Copyright © 2007 John Wiley & Sons, Inc.

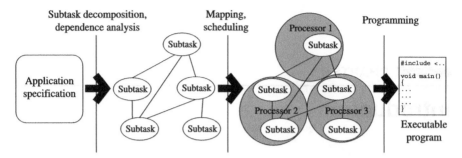

Figure 1.1. Parallel programming—process of parallelization.

means that the processor allocation, often called mapping, and the ordering of the tasks are determined at compile time. The advantage of static scheduling is that it can include the dependences and communications among the tasks in its scheduling decisions. Furthermore, since the scheduling is done at compile time, the execution is not burdened with the scheduling overhead.

In its general form (i.e., without any restrictions on the application's type or structure), static scheduling is often referred to as *task scheduling*. The applications, or parts of them, considered in task scheduling can have arbitrary task and dependence structures. They are represented as directed acyclic graphs (DAGs), called *task graphs*, where a node reflects a task and a directed edge a communication between the incident nodes. Weights associated with the nodes and edges represent the computation and communication costs, respectively. For example, consider the small program segment in Figure 1.2 and its corresponding task graph. Each line of the program is represented by one node and the edges reflect the communications among the nodes; for instance, line 2 reads the data line 1 has written into the variable *a*, hence the edge from node 1 to node 2.

The task graph represents the task and communication structure of the program, which is determined during the subtask decomposition and the dependence analysis. An edge imposes a precedence constraint between the incident nodes: the origin node must be executed before the destination node. For example, in Figure 1.2 node 1 must

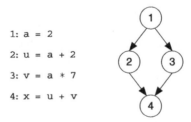

Figure 1.2. Example of task graph representing a small program segment.

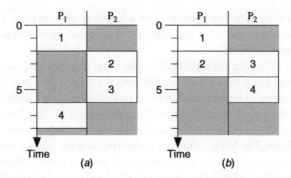

Figure 1.3. Two sample schedules of the task graph in Figure 1.2.

be executed before nodes 2 and 3 (both read the value of a written in line 1), which in turn must be executed before node 4 (line 4 adds the results of line 2 (u) and line 3 (v)).

The challenge of task scheduling is to find a spatial and temporal assignment of the nodes onto the processors of the target system, which results in the fastest possible execution, while respecting the precedence constraints expressed by the edges. As an example, consider the two schedules in Figure 1.3 of the above task graph on two processors, P_1 and P_2. For simplicity, it is here assumed that the nodes have identical weights of two time units, and all edge weights are zero. In both schedules each processor executes two nodes, yet schedule (b) is shorter than schedule (a). The reason is the precedence constraints among the nodes: in the schedule (a), the two nodes that can be executed in parallel, nodes 2 and 3, are allocated to the same processor. In schedule (b), they are allocated to different processors and executed concurrently.

What is a trivial problem in this example becomes very difficult with larger, more complex task graphs. In fact, finding a schedule of minimal length for a given task graph is, in its general form, an NP-hard problem (Ullman [192]); that is, its associated decision problem is NP-complete and an optimal solution cannot be found in polynomial time (unless NP = P). As a consequence of the NP-hardness of scheduling, an entire area emerged that deals with all aspects of task scheduling, ranging from its theoretical analysis to heuristics and approximation techniques that produce near optimal solutions.

This book is devoted to this area of task scheduling for parallel systems. Through a thorough introduction to parallel systems, their architecture, and parallel programming, task scheduling is first carefully set into the context of the parallelization process. The program representation model of task scheduling—the task graph—is studied in detail and compared with other graph-based models. This is one of the first attempts to analyze and compare the major graph models for the representation of programs.

After this ground-laying introduction, the task scheduling problem is formally defined and its theoretical background is rigorously analyzed. Throughout the entire book, this unifying theoretical framework is employed, making the study of task scheduling very consistent. For example, task scheduling without the consideration of communication costs is studied as a special case of the general problem that recognizes the costs of communication.

But the effort of having a comprehensive and easy to understand treatment of task scheduling does not stop there. After establishing the theory, the focus is on common concepts and techniques encountered in many task scheduling algorithms, rather than presenting a loose survey of algorithms. Foremost, these are the two fundamental scheduling heuristics—*list scheduling* and *clustering*—which are studied and ana-lyzed in great detail. The book continues by looking at more advanced topics of task scheduling, namely, *node insertion*, *node duplication*, and *genetic algorithms*.

While the concepts and techniques are extracted and treated separately, the frame-work is backed up with references to many proposed algorithms. This approach has several advantages: (1) common concepts and terminology simplify the understand-ing, analysis, and comparison of algorithms; (2) it is easier to evaluate the impact of a technique when it is detached from other techniques; and (3) the design of new algorithms may be inspired to use new combinations of the presented techniques.

This book also explores further aspects of the theoretical background of scheduling. One aspect is scheduling on heterogeneous processors, including the corresponding scheduling model and the adapted algorithms. Another aspect is the study of variations of the general task scheduling problem. A comprehensive survey of these variations, which again can be treated as special cases of the general problem, shows that most of them are also NP-hard problems.

The book then goes beyond the classic approach to task scheduling by studying scheduling under other, more accurate, parallel system models. Classic scheduling is based on the premise that the target system consists of a set of fully connected processors, which means each processor has a direct communication link to every other processor. Interprocessor communication is performed by a dedicated commu-nication subsystem, in a way that is completely free of contention. It follows that an unlimited number of interprocessor communications can be realized concurrently without the involvement of the processors. Qualitative analysis and recent experimen-tal evaluations show that not all of these assumptions are fulfilled for the majority of parallel systems. This issue is addressed in two steps.

In the first step, a model is investigated that extends task scheduling toward con-tention awareness. Following the spirit of the unifying scheduling framework, the investigated contention model is a general and unifying model in terms of network representation and contention awareness. It allows modeling of arbitrary heteroge-neous systems, relating to processors and communication links, and integrates the awareness of end-point and network contention. Adapting scheduling algorithms to the contention model is straightforward. Exemplarily, it is studied how list scheduling can be employed under the contention model.

In the second step, the scheduling framework is extended further to integrate involvement of the processors in communication. The resulting model inherits all abilities of the contention model and allows different types of processor involve-ment in communication. Processor involvement has a relatively strong impact on the scheduling process and therefore demands new approaches. Several approaches to handle this difficulty are analyzed and the adaptation of scheduling heuristics is discussed.

Throughout this book, numerous figures and examples illustrate the discussed concepts. Exercises at the end of each chapter deepen readers' understanding.

1.2 ORGANIZATION

This book is organized as follows.

Chapter 2 reviews the relevant background of parallel computing, divided into two parts. The first part discusses parallel computers, their architectures and their communication networks. The second part returns to parallel programming and the parallelization process, reviewing subtask decomposition and dependence analysis in detail.

Chapter 3 provides a profound analysis of the three major graph models for the representation of computer programs: dependence graph, flow graph, and task graph. It starts with the necessary concepts of graph theory and then formulates a common principle for graph models representing computer programs. While the focus is on the task graph, the broad approach of this chapter is crucial in order to establish a comprehensive understanding of the task graph, its principle, its relations to other models, and its motivations and limitations.

Chapter 4 is devoted to the fundamentals of task scheduling. It carefully introduces terminology, basic definitions, and the target system model. The scheduling problem is formulated and subsequently the NP-completeness of the associated decision problem is proved. One of the aims of this chapter is to provide the reader with a unifying and consistent conceptual framework. Consequently, task scheduling without communication costs is studied as a special case of the general problem. Again, the complexity is discussed, including the NP-completeness of this problem. The chapter then returns to the task graph model to analyze its properties in connection with task scheduling.

Chapter 5 addresses the two fundamental heuristics for scheduling—list scheduling and clustering. Both are discussed in general terms, following the expressed intention of this book to focus on common concepts and techniques. For list scheduling, a distinction is made between static and dynamic node priorities. Given a processor allocation, list scheduling can also be employed to construct a schedule. The area of clustering can be broken down into a few conceptually different approaches. Those are analyzed, followed by a discussion on how to go from clustering to scheduling.

Chapter 6 has a look at more advanced aspects of task scheduling. The first two sections deal with node insertion and node duplication. Both techniques can be employed in many scheduling heuristics. Again, for the sake of a better understanding, they are studied detached from such heuristics. The chapter then returns to more theoretical aspects of task scheduling. Integrating heterogeneous processors into scheduling can be done quite easily. A survey of variants of the general scheduling problem shows that scheduling remains NP-hard in most cases even after restricting the problem.

The last aspects to be considered in this chapter are genetic algorithms and how they can be applied to the scheduling problem.

Chapter 7 investigates how to handle contention for communication resources in task scheduling. The chapter begins with an overview of existing contention aware scheduling algorithms, followed by an outline of the approach taken in this book. Next, an enhanced topology graph is introduced, based on a thorough analysis of communication networks and routing. Contention awareness is achieved with edge scheduling, which is investigated in the third section. The next section shows how task scheduling is made contention aware by integrating edge scheduling and the topology graph into the scheduling process. Adapting algorithms for scheduling under the contention model is analyzed in the last section, with the focus on list scheduling.

Chapter 8 investigates processor involvement in communication and its integration into task scheduling. It begins by classifying interprocessor communication into three types and by analyzing their main characteristics. To integrate processor involvement into contention scheduling, the scheduling model is adapted. The new model implies changes to the existing scheduling techniques. General approaches to scheduling under the new model are investigated. Using these approaches, two scheduling heuristics are discussed for scheduling under the new model, namely, list scheduling and two-phase heuristics.

Parallel Systems and Programming

This chapter reviews parallel systems and their programming. The intention is to establish the necessary background and terminology for the following chapters. It begins with the basis of parallel computing—parallel systems—and discusses their architectures and communication networks. In this context, it also addresses programming models for parallel systems. The second part of the chapter is devoted to the parallelization process of parallel programming. A general overview presents the three components of the parallelization process: subtask decomposition, dependence analysis, and scheduling. The subsequent sections discuss subtask decomposition and dependence analysis, which build the foundation for task scheduling.

2.1 PARALLEL ARCHITECTURES

Informally, a parallel computer can be characterized as a system where multiple processing elements cooperate in executing one or more tasks. This is in contrast to the von Neumann model of a sequential computer, where only one processor executes the task. The numerous existing parallel architectures and their different approaches require some kind of classification.

2.1.1 Flynn's Taxonomy

In a frequently referenced article by Flynn [67], the design of a computer is characterized by the flow (or stream) of instructions and the flow (or stream) of data. The taxonomy classifies according to the multiplicity of the instruction and the data flows. The resulting four possible combinations are shown in Table 2.1.

The SISD (single instruction single data) architecture corresponds to the conventional sequential computer. One instruction is executed at a time on one data item. Although the combination MISD (multiple instruction single data) does not seem to be meaningful, pipeline architectures, as found in all modern processors, can be considered MISD (Cosnard and Trystram [45]).

Task Scheduling for Parallel Systems, by Oliver Sinnen
Copyright © 2007 John Wiley & Sons, Inc.

Table 2.1. Flynn's Taxonomy

	Single Data	Multiple Data
Single Instruction	SISD	SIMD
Multiple Instruction	MISD	MIMD

In SIMD (single instruction multiple data) architectures, which are also called data parallel or vector architectures, multiple processing elements (PEs) execute the same instruction on different data items. Figure 2.1(a) shows the SIMD structure with one central control unit and multiple processing elements. The central control unit issues the same instruction stream to each PE, which works on its own data set. Especially for regular computations from the area of science and engineering (e.g., signal processing), where computations can be expressed as vector and matrix operations, the SIMD architecture is well adapted.

There are only a few examples of systems that have a pure SIMD architecture, for instance, the early vector machines (e.g., the Cray-1 or the Hitachi S-3600) (van der Steen and Dongarra [193]). Today, the SIMD architecture is often encountered within a vector processor, that is, one chip consisting of the central control unit together with multiple processing elements. A parallel system can be built from multiple vector processors and examples for such systems are given later. Also, most of today's mainstream processor architectures feature an SIMD processing unit, for example, the MMX/SSE unit in the Intel Pentium processor line or the AltiVec unit in the PowerPC processor architecture.

The second parallel architecture in the taxonomy has MIMD (multiple instruction multiple data) streams, depicted in Figure 2.1(b). In contrast to the SIMD structure, every PE has its own control unit (CU). Therefore, the processor elements operate

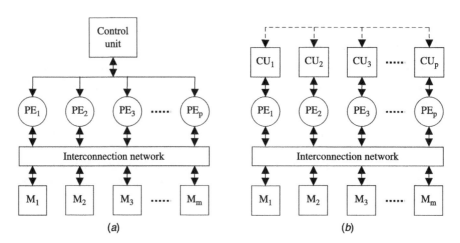

Figure 2.1. SIMD (a) and MIMD (b) architecture.

independently of each other and execute independent instructions on different data streams. A parallel execution of a global task (i.e., the collaboration of the processing elements) is achieved through synchronization and data exchange between the PEs via the interconnection network. Examples for MIMD architectures are given in the following discussion of memory architectures.

An MIMD architecture can simulate an SIMD architecture by executing the same program on all the processors, which is called SPMD (single program multiple data) mode. In general, however, executing the same program on all processors is not the same as executing the same instruction stream, since processors might execute different parts of the same program depending on their processor identification numbers. The term *processor* stands here for the combination of control unit plus processing element. From now on, this definition of a processor shall be used, if not otherwise stated.

As mentioned earlier, modern parallel systems often consist of multiple vector processors, for example, the Cray J90 or the NEC SX-6 (van der Steen and Dongarra [193]). The Earth Simulator, a NEC SX-6 based system, was the world's fastest computer in 2002–2004 [186]. Within Flynn's taxonomy, these systems can be considered to have an MIMD architecture with an SIMD subarchitecture.

2.1.2 Memory Architectures

It is generally agreed that not all aspects of parallel architectures are taken into account by Flynn's taxonomy. For both the design and the programming model of a parallel system, the memory organization is a very important issue not considered by that classification.

The memory organization of a parallel system can be divided into two aspects: the location and the access policy of the memory. Regarding the location, memory is either *centralized* or *distributed* with the processors. For both cases, systems with a common memory, distributed or not, to which all processors have full access, are called *shared-memory* machines. In systems where there is no such shared memory, processors have to use explicit means of communication like *message passing*. With these two aspects of memory organization in mind, the three most common memory organizations can be examined.

Centralized Memory   In a centralized memory multiprocessor, illustrated in Figure 2.2, memory is organized as a central resource for all processors. This typically results in a *uniform memory access* (UMA) characteristic, where the access time to any memory location is identical for every processor. Since the common memory can be accessed by all processors, these systems are called *centralized shared-memory multiprocessors* (Hennessy and Patterson [88]). Due to the UMA characteristic, systems with this architecture are also often called symmetric multiprocessors (SMP).

Distributed Memory   The alternative to a centralized architecture is an architecture where the memory is physically distributed with the processors. These systems can be further distinguished according to their memory access policy.

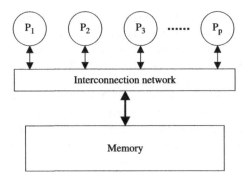

Figure 2.2. Centralized memory multiprocessor.

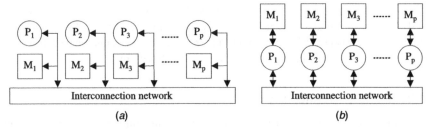

Figure 2.3. Distributed memory multiprocessors: (*a*) shared-memory and (*b*) message passing (memory access goes through processors).

A *distributed shared-memory multiprocessor* (Hennessy and Patterson [88]), as illustrated in Figure 2.3(*a*), integrates the distributed memories into a global address space. Every processor has full access to the memory; however, in general with a *nonuniform memory access* (NUMA) characteristic, as reading from or writing to, local memory is faster than from and to remote memory.

Systems without shared memory are called *distributed memory multiprocessors* or, according to the way the processors communicate, *message passing architectures*[1] (Culler and Singh [48]). Figure 2.3(*b*) displays a distributed memory multiprocessor without shared memory. The difference to the distributed shared-memory multiprocessor is that the local memories are only accessible through the respective processors.

Memory Hierarchy   Some distributed memory systems, especially large systems with shared memory, often use some kind of hierarchy for the memory organization. A common example is that a small number of processors (2–4) share one central memory—the processors and the memory comprise a computing node—and

---

[1]To distinguish the two distributed memory architectures, the supplements shared-memory or message passing shall be used. Otherwise both types are meant.

multiples of these nodes are interconnected on a higher level. Examples of systems with such a hierarchical memory architecture are the Sequent NUMA-Q, the SGI Origin 2000/3000 series, and the IBM Blue Gene/L (van der Steen and Dongarra [193]), which is the world's fastest computer at the time this is written [186].

Motivations The following paragraphs briefly look at the motivations for the different architectures.

Centralized shared-memory architectures are an intuitive extension of a single processor architecture; however, the contention for the communication resources to the central memory significantly limits the scalability of these machines. Bus-based systems therefore have a small number of processors (usually ≤ 8), for example, systems with $\times 86$ processors, and only more sophisticated interconnection networks allow these systems to scale up to 64 processors. For example, in the Sun Enterprise series (van der Steen and Dongarra [193]), systems with low model numbers (e.g., 3000) are connected by a bus, whereas the high-end model, the Sun Enterprise 10000, uses a crossbar (see Section 2.2) for up to 64 processors.

In contrast, distributed memory architectures with message passing allow a much simpler system design, but their programming becomes more complicated. In fact, commodity PCs can be connected via a commodity network (e.g., Ethernet) to build a so-called cluster of workstations or PCs (Patterson [147], Sterling et al. [181]). The big advantage of distributed memory systems is their much better scalability. Hence, it is no surprise that the massively parallel processors (MPPs) are distributed memory systems using message passing with up to hundred thousand processors (e.g., Thinking Machines CM-5, Intel Paragon, and IBM Blue Gene/L).

Distributed shared-memory architectures try to integrate both approaches. They provide the ease of the shared-memory programming paradigm and benefit from the scalability of distributed memory systems, for instance, Cray T3D/T3E, the SGI Origin 2000/3000 series, or the HP SuperDome series (van der Steen and Dongarra [193]). Yet, shared-memory programming of these architectures can have limited efficiency in as much as the heterogeneous access times to memory are often hidden from the programmer.

2.1.3 Programming Paradigms and Models

Shared-Memory Versus Message Passing Programming The programming paradigms for parallel systems have a strong correspondence to the memory access policies of multiprocessors. Fundamentally, one can distinguish between shared-memory and message passing programming. In the former paradigm, every processor has full access to the shared memory, and communication between the parallel processors is done implicitly via the memory. Only concurrent access to the same memory location needs explicit synchronization of the processors. In message passing programming, every exchange of data among processors must be explicitly expressed with send and receive commands.

It must be noted that the employed programming paradigm does not always correspond to the underlying memory organization of the target system. Message passing can be utilized on both shared-memory and message passing architectures. In a shared-memory system, the passing of a message is often implemented as a simple memory copy. Even distributed shared memory can be emulated on message passing machines with an additional software layer (e.g., survey by Protić et al. [154]).

Parallel Random Access Machine (PRAM)   PRAM (Fortune and Wyllie [68]) is a popular machine model for algorithm design and complexity analysis. Essentially, the simple model assumes an ideal centralized shared-memory machine with synchronously working processors. PRAMs can be further classified according to how one memory cell can be accessed: only exclusively by one processor or concurrently by various processors. Memory access to different cells by different processors can always be performed concurrently.

The advantage of PRAM is its simplicity and its similarity to the sequential von Neumann model. Yet, owing to the increasing gap between processing and communication speed, it has become more and more unrealistic.

LogP   With the proposal of the LogP model, Culler et al. [46, 47] recognized the fact that the widely used PRAM model is unrealistic due to its assumption of cost-free interprocessor communication, especially for distributed systems. The LogP model gained its name from the parameters that are used to describe a parallel system:

- L: Upper bound on the *latency*, or delay, incurred in communicating a message from a source to a destination processor.
- o: *Overhead*—time during which a processor is engaged in sending or receiving a message; during this time the processor cannot perform other operations.
- g: *Gap*—minimal time between consecutive message transmissions or between consecutive message receptions; the reciprocal of g corresponds to the per-processor bandwidth.
- P: Number of processors.

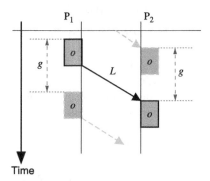

Figure 2.4.   Interprocessor communication in LogP.

Furthermore, the structure of the processor network is not described by LogP, but its capacity is limited to $\lceil L/g \rceil$ simultaneous message transfers between all processors. An implicit parameter is the message size M, which consists of one or a small number of words. Based on this parameter, an interprocessor message transfer in the LogP model proceeds as illustrated in Figure 2.4. In contrast to PRAM, LogP is an asynchronous model.

2.2 COMMUNICATION NETWORKS

Fast communication is crucial for an efficient parallel system. A determining aspect of the communication behavior is the network and its topology. In the previous section, some kind of interconnection network for communication among the units of the parallel system was supposed. This section reviews the principal network types, of which each offers a different trade-off between cost and performance.

Initially, interconnection networks can be classified into *static* and *dynamic networks*. Static networks have fixed connections between the units of the system with point-to-point communication links. In a dynamic network, the connections between units of the parallel system are established dynamically through switches when requested. Based on this difference, static, and dynamic networks are sometimes referred to as *direct* and *indirect* networks, respectively (Grama et al. [82], Quinn [156]).

2.2.1 Static Networks

The essential characteristic of a static network is its topology, as the interconnections between the units of the parallel system are fixed. Most static networks are processor networks used in distributed memory systems, where every processor has its own local memory.

Processor network topologies are usually represented as undirected graphs[2]: a vertex represents a processor, together with its local memory and a switch, and an undirected edge represents a communication link between two processors (Cosnard and Trystram [45], Culler and Singh [48], Grama et al. [82]). Figure 2.5(*b*) depicts an example for a network graph consisting of four processors. Figure 2.5(*a*) illustrates the implicit association of memory and a switch with each processor.

Once a topology is modeled as an undirected graph, terminology from graph theory can be utilized for its characterization. The *degree* of a vertex is defined as the number of its incident edges, denoted by δ. The eccentricity of a vertex is the largest distance, in terms of the number of edges, from that vertex to any other vertex. Furthermore, the *diameter* of an undirected graph, denoted by D, is defined as the maximum eccentricity of all vertices of the graph. Another notable indicator for a network is its *bisection width*. It is defined as the minimum number of edges that have to be removed to

---

[2]Basic graph concepts are introduced in Section 3.1 and the undirected graph model of topologies will be defined more formally in Section 7.1. For the current purpose, this informal definition suffices.

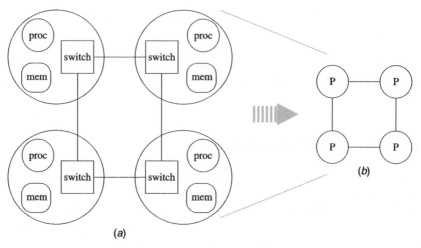

(a)

(b)

Figure 2.5. An undirected graph representing a processor network (*b*); (*a*) an illustration that the switch and the memory associated with each processor are implicit in the common representation of processor networks.

partition the network into two equal halves. In other words, it is the number of edges that cross a cut of a network into two equal halves.

To achieve a network with a small maximum communication time, the goal is to have a small diameter D. At the same time, the mean degree $\bar{\delta}$ of the network should be small, since it determines the hardware costs. Last but not least, the network should have a large bisection width, because it lowers the contention for the communication links. However, the bisection width is also a measure for the network costs, as it provides a lower bound on the area or volume of its packaging (Grama et al. [82]).

Fully Connected Networks   A network in which every processor has a direct link to any other processor, as depicted in Figure 2.6, is called fully connected. It has

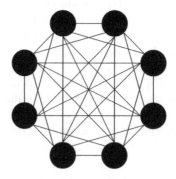

Figure 2.6. An 8-processor fully connected network.

the nice property that it is *nonblocking*; that is, the communication of two processors does not block the connection of any other two processors in the network. All other static networks studied in this section are blocking networks.

Obviously, the degree of each vertex is $\delta = p - 1$, for a network of p processors, the diameter is $D = 1$ and the bisection width is $p^2/4$. Fully connected networks are extreme in both ways: they provide the smallest possible diameter and the highest possible degree and bisection width. In practice, the quadratic growth of the number of links—p processors are fully connected by $p(p - 1)/2$ links—renders a fully connected network unemployable for medium to large numbers of processors. However, the fully connected network is important in terms of theoretical work as it is often used as a model.

Meshes   In practical terms, the most popular class of static networks are meshes. Processors are arranged in a linear order in one or more dimensions, whereby only neighboring processors are interconnected by communication links. According to their dimension, meshes can be grouped into linear networks or rings (1D), grids (2D), and tori (3D). Figure 2.7 visualizes meshes up to three dimensions.

A distinction can be made between acyclic and cyclic meshes. For the latter, the links are wrapped around at the end of the respective dimension to its beginning. The few additional links in comparison with acyclic meshes reduce the diameter significantly, namely, by half in each dimension. Thus, in practice, meshes are usually cyclic.

Table 2.2 summarizes the properties of acyclic and cyclic meshes up to three dimensions. In general, an n-dimensional cyclic mesh with q_k processors on each

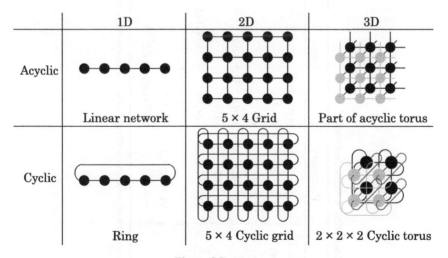

	1D	2D	3D
Acyclic	Linear network	5 × 4 Grid	Part of acyclic torus
Cyclic	Ring	5 × 4 Cyclic grid	2 × 2 × 2 Cyclic torus

Figure 2.7. Meshes.

Table 2.2. Properties of Static Networks[a]

Network	Processors	Links	Mean Degree $\bar{\delta}$	Diameter D	Bisection Width
Fully connected	p	$p(p-1)/2$	$p-1$	1	$p^2/4$
Linear network	p	$p-1$	$2-2/p$	$p-1$	1
Ring	p	p	2	$\lfloor p/2 \rfloor$	2
Grid	$q_1 \times q_2$	$q_1(q_2-1)+q_2(q_1-1)$	$4-2/q_1-2/q_2$	q_1+q_2-2	$\min(q_1,q_2)$
Cyclic grid	$q_1 \times q_2$	$2q_1q_2$	4	$\lfloor q_1/2 \rfloor + \lfloor q_2/2 \rfloor$	$\min(2q_1,2q_2)$
Torus	$q_1 \times q_2 \times q_3$	$q_2q_3(q_1-1)+q_1q_3(q_2-1)$ $+q_1q_2(q_3-1)$	$6-2/q_1-2/q_2$ $-2/q_3$	$q_1+q_2+q_3-3$	$\min(q_1q_2,q_2q_3,q_1q_3)$
Cyclic torus	$q_1 \times q_2 \times q_3$	$3q_1q_2q_3$	6	$\lfloor q_1/2 \rfloor + \lfloor q_2/2 \rfloor + \lfloor q_3/2 \rfloor$	$\min(2q_1q_2,2q_2q_3,2q_1q_3)$
Hypercube	$p=2^d$	$(p/2)\log p = d2^{d-1}$	$\log p = d$	$\log p = d$	$p/2 = 2^{d-1}$

[a] q_k denotes the number of processors in dimension k; d is the dimension of the hypercube.

dimension k (with $k = 1, 2, \ldots, n$) has the following diameter and mean degree:

$$D = \sum_{k=1}^{n} \left\lfloor \frac{q_k}{2} \right\rfloor \quad \text{and} \quad \bar{\delta} = 2n. \tag{2.1}$$

Its bisection width is

$$\min_{i} \frac{2 \prod_{k=1}^{n} q_k}{q_i}, \tag{2.2}$$

assuming the number of processors in the dimension q_i, through which the bisection cut is made, is even.

The mesh topologies can be very attractive for scientific and engineering computations, since they correspond to data structures like vectors and matrices, which are heavily used for this kind of computations. In the ideal case, the data structures can be distributed uniformly among the processors.

Mesh networks can be found in a large variety of systems. The scalable coherent interface (SCI) standard specifies a ring-based network. A two-dimensional grid is used in the Intel Paragon. Three-dimensional tori are often employed in MPP systems, for example, the Cray T3E or the IBM Blue Gene/L.

Hypercubes   A hypercube is a network with the interesting property that the degree equals the diameter, $\delta = D = d$, where d is the dimension of the hypercube. The number of processors is given by $p = 2^d$. A hypercube of dimension d can be recursively constructed from two $(d - 1)$-cubes, in which each processor of one cube is connected to the processor at the same position in the other cube. Figure 2.8 shows how this is performed for a 4D cube using two 3D cubes. This construction procedure also leads intuitively to the bisection width of $p/2 = 2^{d-1}$, which is the number of edges that connect the two identical $(d - 1)$-cubes.

Hypercubes have the nice property that the diameter and the degree only grow logarithmically with the number of processors. Furthermore, routing can be implemented with very little effort using Gray codes to denote the processors (Cosnard and Trystram [45]). However, these theoretical advantages are opposed by practical shortcomings. The number of processors must be a power of 2. Moreover, building a large hypercube in hardware is difficult, as the links differ much in length. An example of systems employing a hypercube is the SGI Origin 2000/3000 series.

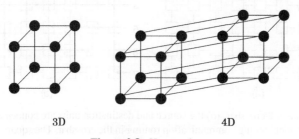

3D   4D

Figure 2.8. Hypercubes.

Summary   Table 2.2 summarizes the properties of the static networks analyzed so far, including acyclic and cyclic meshes with up to three dimensions.

Many other static network topologies have been proposed in the past: most notably, topologies that are constructed modularly by substituting the nodes of one topology with subtopologies. A detailed study of more sophisticated topologies and further reading can be found in Cosnard and Trystram [45], Culler and Singh [48], Grama et al. [82], and Quinn [156].

2.2.2 Dynamic Networks

In contrast to static networks, dynamic networks establish the connection between the units of the parallel system on demand. The connections are made through internal network switches that are not associated with any processor—hence the alternative name of indirect network. This is in contrast to static networks, where every switch is implicitly associated with one processor; see Figure 2.5. A dynamic network is constituted by a set of switches, which route the communication through the network from its source to its destination.

Crossbars   The most powerful dynamic network is a crossbar. It employs a grid of nm switches to connect n units on one side with m units on the other side, as illustrated in Figure 2.9. Through this simple approach, it is a *nonblocking* network; that is, a connection between unit i on the left and unit j at the top does not block the connection of any other unit on the left with any other unit at the top. An example of this nonblocking behavior is shown in Figure 2.9(*b*). From the static networks considered in Section 2.2.1, only the fully connected network has the same nonblocking property. However, it also comes at a similar cost, as the number of switches in a crossbar has the same complexity as the number of links in a fully connected network, namely, $O(p^2)$

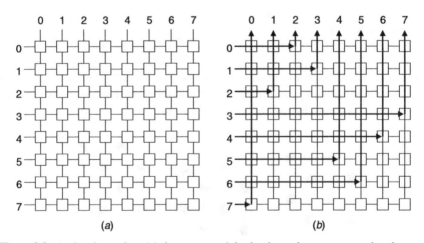

Figure 2.9. An 8×8 crossbar; (*a*) the source and destination units are connected to the numbered lines; (*b*) nonblocking communication routing in the crossbar. The squares represent the switches.

(assuming $n = m = p$). Even though the diameter grows linearly with the number of processors, $D = 2p$, this is often not an issue, since crossbars are usually implemented in a very compact form. In any case, the quadratically growing number of switches is more limiting to the scalability. On the upside, the bisection width grows linearly with the number of processors.

In conclusion, crossbars are not very scalable costwise and are only used for small to medium sized networks. They are found in larger shared-memory architectures (Section 2.1.2), for example, Sun Enterprise 10000, Sun Fire E25K, and on the node level in the SGI Origin 2000/3000 series and the HP SuperDome series.

Multistage Networks   In comparison to crossbars, multistage networks employ a reduced number of switches. This makes them more scalable in terms of cost but also makes them *blocking* networks; that is, certain combinations of source–destination connections block each other. Yet, the aim of their design is to keep the number of conflicts low in practice.

A common multistage network is the *butterfly network*. Figure 2.10 visualizes a three-dimensional butterfly network with 8 source units and 8 destination units. Every communication from a source unit (left side of Figure 2.10(a)) travels through switches (squares) until it reaches its destination unit (right side of Figure 2.10(b)). So this network consists of three stages. As noted earlier, conflicts can arise between certain combinations of source–destination connections. For example, the source–destination connections 2–2 and 6–3 conflict, since both utilize the same link between the second and the third stage (Figure 2.10(b)).

Butterfly networks are equivalent to *omega networks* and one can be turned into the other by relabeling/reordering the switches (Cosnard and Trystram [45]). The omega network uses the so-called perfect shuffle pattern (Cosnard and Trystram [45]) to interconnect the switches. An example for an omega network is found in the IBM RS/6000 SP-SMP.

The number of switches in a butterfly or omega network is $(p/2) \log p$ (switches-per-stage × stages), or $2^{d-1}d$ with $p = 2^d$, which is significantly less compared to the p^2 switches in a crossbar. Furthermore, the diameter grows only logarithmically with the number of the processors, $D = \log p + 1 = d + 1$, and the

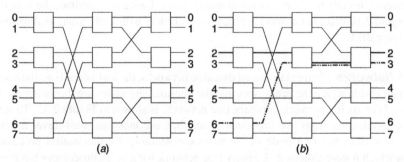

Figure 2.10. A 3D butterfly network with 3 stages; (a) the source and destination units are connected to the numbered lines; (b) example for blocking communication conflict.

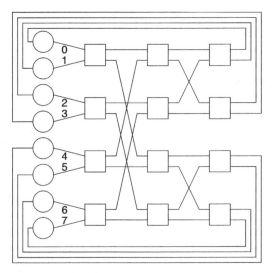

Figure 2.11. A 3D butterfly connecting 8 processors by wrapping around the destination to the source.

bisection width is $p/2$ (in Figure 2.10 the bisection cut is done horizontally between lines 3 and 4).

Dynamic networks are often employed in larger scale centralized shared-memory architectures (Section 2.1.2) as a connection between the processors on the one side and the memory banks on the other (e.g., Convex Exemplar, Sun Enterprise 10000).

Dynamic networks are also utilized to interconnect processors, or processing nodes consisting of several processors, for example, in the IBM RS/6000 SP-SMP, the HP SuperDome series, or the Sun Fire E25K. As in the static networks discussed in Section 2.2.1, memory is usually associated with each processor or processing node, consisting of several processors. For crossbars and multistage networks, the processor network is built by wrapping around the connection lines on the destination side back to the source side (Hennessy and Patterson [87]). In other words, destination line i is connected directly to the processor at source line i. Using this method, the butterfly in Figure 2.10 can interconnect 8 processors (circles) with each other, as illustrated in Figure 2.11.

Tree Networks In tree structured dynamic networks, the leaf nodes (i.e., the nodes without children) are the processors and the intermediate nodes are the switches. An example of such a network, a binary tree network, is shown in Figure 2.12. To send a communication from one processor to another, the message must travel upward until it reaches the root node of the subtree containing the destination processor, from which it descends to it. A binary tree network for $p = 2^d$ processors has $p - 1$ switches. Hence, it has less switches than the discussed multistage networks, while it has also a logarithmically growing diameter, $D = 2 \log p$. The big shortcoming of

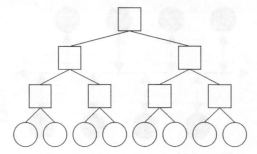

Figure 2.12. Binary tree network with 8 processors (circles).

the tree is its bisection width of 1. The links at higher levels are easily congested with communications. All communications between the processors on the left side and the processors on the right side of a tree go through the root node and its links. Fat trees alleviate this problem by having more, parallel links in higher levels of the tree. In other words, the bandwidth between the switches is increased on the way to the root.

Trees are especially interesting for certain types of communication, for example, a broadcast, where one message is sent to all processors. Once the message is at the root, all processors receive the message at the same time, with a logarithmic delay in terms of the number of processors. For this reason, the IBM Blue Gene/L employs, in addition to its torus network, a tree network.

Summary Table 2.3 summarizes the properties of the dynamic networks considered so far. The expressions are given for processor networks; that is, only processors are connected to the network, not separate memory banks. For the crossbar and the butterfly/omega network, this implies that the destination lines are wrapped around to the processors which are connected at the source lines, as shown in Figure 2.11 for an 8-processor butterfly. As a consequence, the expressions for the number of links, the diameter, and the bisection width include those links that connect the processors with the dynamic network. This makes these expressions more comparable to each other

Table 2.3. Properties of Dynamic Networks Connecting p Processors

Network	Processors	Switches[a]	Links	Maximum Links per Switch	Diameter D	Bisection Width
Crossbar	p	p^2	p^2	4	$2p$	$3p/2$
Butterfly/ omega	$p = 2^d$	$(p/2)\log p$ $= d2^{d-1}$	$p(\log p + 1)$ $= 2^d(d+1)$	4	$\log p + 1$ $= d + 1$	$p/2$
Binary tree	$p = 2^d$	$p - 1$	$p - 1$	3	$2\log p = 2d$	1

[a] Switches—the number of switches not associated with a processor.

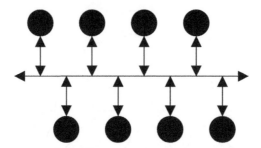

Figure 2.13. An 8-processor bus topology.

and to static networks. Instead of the mean degree $\bar{\delta}$ (Section 2.2.1), the maximum number of links per switch is given.

Bus   One particular dynamic network, the bus, deserves special attention, since it is found in many small-sized parallel systems and also as a building block in larger scale computers. Conceptually, it is quite different from the dynamic networks analyzed so far. The units of the parallel system are grouped around the bus, which is shared among them for their communications, as shown in Figure 2.13. A controller assigns the bus to a unit on request and during the assigned time the unit can use the bus exclusively.

In terms of performance scalability, it is situated on the opposite side of the spectrum, when compared with the crossbar. Already one communication between two processors blocks the bus for any other connection. As a consequence, the bus is usually time-shared among the connected units in practice. While its simplicity makes the bus very attractive, it very quickly becomes the performance bottleneck of a parallel system. Consequently, the number of connected units is small (usually ≤ 8).

2.3   PARALLELIZATION

The objective of a parallel system is the fast execution of applications, faster than they can be executed on single processor systems. For this aim, the application's formulation as a program must be in a form that allows it to benefit from the multiple processing units. Unfortunately, parallel programming (i.e., the formulation of an application as a program for a parallel system) is significantly more complex than sequential programming. The additional procedures involved in parallel programming are often called the *parallelization* of the program, especially if the parallel version of the program is derived from an existing sequential implementation.

Figure 2.15 illustrates the process of parallel programming and for comparison Figure 2.14 shows that of sequential programming, where the program is directly created from the application specification. For its parallelization, the application must be divided into subtasks, or simply tasks, which in general are not independent. Precedence constraints, caused by the dependence of a task on the result of another task,

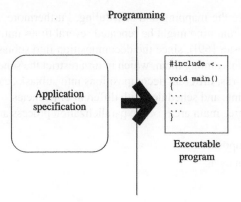

Figure 2.14. Sequential programming.

impose a partial order of execution among the tasks. Thus, a correct execution of the application demands a dependence analysis, on whose basis the tasks must be ordered for execution. Given the division of the application into tasks and the analysis of their dependence relations, the final step of parallelization consists of mapping the tasks onto the processors and the determination of their execution order, which corresponds to the assignment of start times to the tasks. The assignment of start times is called *scheduling* and can only be accomplished with an existing mapping. Therefore, generally both are meant, the spatial (i.e., mapping) and the temporal assignment of the tasks to the processors, when referring to scheduling. Based on the schedule of the tasks, the parallel program can be created. This process of parallelization is depicted in Figure 2.15 from left to right: beginning with the specification of the application, the program undergoes subtask decomposition, dependence analysis, and scheduling. At the end of the parallelization process, the program code must be generated as in sequential programming.

In many practical approaches to parallel programming, the steps of parallelization are not clearly separable. For example, tasks are often grouped into processes or threads (the so-called orchestration (Culler and Singh [48]) or agglomeration

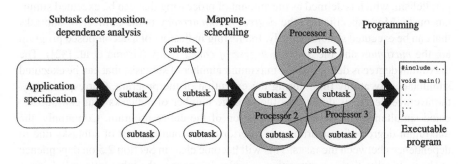

Figure 2.15. Parallel programming—process of parallelization.

(Foster [69])) before the mapping and scheduling. Furthermore, the parallelization part of parallel programming might be repeated several times until a satisfying solution is obtained (Foster [69]), since the decomposition into subtasks determines the precedence constraints among them, which in turn restrict the scheduling on the processors. In other words, different decompositions into subtasks can lead to different precedence constraints and schedules with different efficiencies.

Summarizing, three main areas of the parallelization process are identified:

- Subtask decomposition
- Dependence analysis
- Scheduling

Subtask decomposition and dependence analysis build the foundation for scheduling. The next two sections focus on these areas in order to lay the groundwork for the treatment of scheduling. Examples used in the following sections help one to better comprehend the aspects of parallelization introduced here.

2.4   SUBTASK DECOMPOSITION

In general, the decomposition of a program into subtasks cannot be examined isolated from other aspects of the parallelization process. First, the decomposition is driven by the type of computation, and the programming techniques and language used for the program formulation. Second, the subsequent techniques employed in the parallelization process have a strong influence on the decomposition process. Finally, the architecture of the target system and the corresponding parallel programming paradigm affect the decomposition. Before these three aspects are discussed, the next two subsections look at metrics that characterize a decomposition and common decomposition techniques.

2.4.1   Concurrency and Granularity

The decomposition of an application into subtasks determines its concurrency or parallelism, which is defined as the amount of processing that can be executed simultaneously. More specifically, the *degree of concurrency* is the number of subtasks that can be executed simultaneously. Interesting indicators of a decomposed program are the *maximum* and the *average degree of concurrency* (Grama et al. [82]). The maximum degree is defined as the maximum number of subtasks that can be executed simultaneously at any time during the execution of the program. Correspondingly, the average degree is defined as the average number of subtasks that can be executed simultaneously during the execution of the entire program. Commonly, the maximum degree of concurrency is inferior to the total number of subtasks due to dependences between the tasks. This will become clear in Section 2.5 on dependence analysis. Ideally, the decomposition completely exposes the inherent parallelism or concurrency of an application.

Another notion widely used in the context of parallelism is that of *granularity*. By saying, for example, "a program is coarse grained," it is meant that the distinguishable chunks of the program, in other words the subtasks, are large considering the size of the entire program. Typically, the granularity of a program is informally differentiated between *small/fine*, *medium*, and *coarse grained*. Granularity can also be understood as the relation of the data transfer to the computational work of a program object. For instance, a small amount of computational work in comparison to the data transfered is considered to be small grained. Although all this is a rather informal definition of granularity, its utilization in the literature is quite usual. With the aid of a graph model, however, granularity can be defined formally, which is done in Section 4.4.3. Until then, the notion of granularity will be used in the usual sense of comparing the computational size of objects.

2.4.2 Decomposition Techniques

Grama et al. [82] describe several decomposition techniques: data, recursive, exploratory, and speculative decomposition. While this is not a complete list of decomposition techniques, it comprises commonly used techniques. Data decomposition and recursive decomposition are general-purpose techniques, while exploratory decomposition and speculative decomposition are more specialized and only apply to certain classes of problems. Their structural approach is similar to recursive decomposition. In practice, it also happens that these techniques are combined in what might be called hybrid decomposition.

Data Decomposition In data decomposition, the focus is on the data rather than on the different operations applied to the data. This technique is powerful and commonly used for programs that operate on large data structures, as often found in scientific computing. In terms of Flynn's taxonomy (Section 2.1), this method corresponds to SIMD. The data is partitioned into N usually equally sized parts, whereby N depends on the desired degree of concurrency. If there are no dependences among the operations performed on the different parts, these N parts can be mapped onto N processors and executed in parallel. In other words, the degree of concurrency is N.

Figure 2.16 illustrates the example of a data decomposition applied to the matrix–vector multiplication $A \cdot b = y$, with $N = 4$. The four tasks that result from this decomposition are independent; for example, task 1 calculates $y[i] = \sum_{j=1}^{n} A[i, j]b[j]$ for $1 \leq i \leq n/N$ and task 2 for $n/N + 1 \leq i \leq 2n/N$.

Data decomposition can be differentiated further by the way the data is partitioned. The partitioning can be applied to the output data, input data, intermediate data, or a mixture of them. In the earlier matrix–vector multiplication example, the output data is completely partitioned and the input data is partially partitioned (matrix A is partitioned, vector b is not).

Recursive Decomposition This decomposition technique uses the divide-and-conquer strategy. A problem is solved by first dividing it into a set of independent subproblems. Done recursively, a tree structure of subproblems is created as depicted

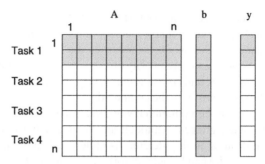

Figure 2.16. Data composition of matrix–vector multiplication $A \cdot b = y$. Data is partitioned into $N = 4$ parts corresponding to 4 tasks; the gray-shaded part shows the data accessed by task 1 (Grama et al. [82]).

in Figure 2.17. The leaves of the tree are basic subproblems that cannot be divided further into smaller subproblems. Each node that is not a leaf node combines the solutions of its subproblems. The result is finally computed in the root of the tree, when its subproblems are combined. Each node of the tree represents one task, and the edges between the nodes reflect their dependences (task dependence and graphs for dependence representation are discussed in Section 2.5 and Chapter 3, respectively). Since all leaves are independent from each other, the maximum degree of concurrency is the number of leaves.

The divide-and-conquer structure can be found in many applications and algorithms, for example, in Mergesort (Cormen et al. [42]).

Exploratory Decomposition This technique has the same tree-like structure as the recursive decomposition. Exploratory decomposition is employed when the underlying computation corresponds to a search of a space for solutions. For such a problem, the search space is represented by a tree, where each node represents a point in the search space and each edge a path to the neighboring points. To decompose

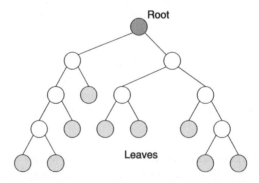

Figure 2.17. Tree created by a recursive decomposition.

such a problem for parallel execution, an initial search tree is generated until there is a sufficiently large number of leaf nodes that can be executed in parallel (i.e., until the required degree of concurrency is met). Each of these leaf nodes corresponds to an independent task, which searches for a solution in the subtree induced below this node. The overall computation can finish as soon as one of the parallel tasks finds a solution. This fact is the main difference between exploratory decomposition and data decomposition, where all subproblems must be solved in their entirety.

Speculative Decomposition   Speculative decomposition takes the idea of exploratory decomposition even one step further. In applications where the choice for a certain computation branch depends on the result of an earlier step, a speedup can be achieved by speculatively executing all (or a subset of all) branches. At the time the data for making the choice is finally available, only the result of the correct branch is chosen; the others are discarded. For example, an if statement in a programming language has two lines of execution—one if the evaluated expression is true, another if it is false. In speculative decomposition, each of the two branches is made a task and executed concurrently. At the time the expression of the if statement can finally be evaluated, the result of the correct branch is chosen and the result of the other branch is discarded.

2.4.3 Computation Type and Program Formulation

Essential for the subtask decomposition is the *granularity*, the *dependence structure*, and the *regularity* inherent in an application and its description as a program. The computation type of an application and its expression as a program may already give a strong indication for its decomposition into subtasks.

Computation Type   Most applications belonging to one area might employ certain types of computation and data structures. For instance, an application of signal processing is likely to perform linear algebra on matrices. In this case, data decomposition is probably the right technique. Other application areas might typically use other types of data structures (e.g., lists and graphs) and computations (e.g., matching and search algorithms). For these application areas, recursive or exploratory decomposition can be a good candidate. Furthermore, depending on the computation type, an application might consist of several distinct steps or one iterative block of computation. In the former case, each step can be represented by one task, hence using coarse grained parallelism, while in the latter case the iterative computation suggests a regular, small grained decomposition, potentially based on data decomposition.

The type of the application also limits the inherent degree of concurrency. For example, the matrix–vector multiplication in Figure 2.16 performs n^2 multiplications and $n(n-1)$ additions. All multiplications can be performed concurrently, but the additions must wait for the multiplications to finish. Hence, the degree of concurrency is bounded by n^2. Also, it is often pointless to partition an application into subtasks that strongly depend on each other's results, preventing their parallel execution due to the small degree of concurrency. As will be shown in the next section, some dependence relations are inherent in an application, while others arise from its actual implementation as a program.

Thus, the application area often defines the type of computation, which in turn largely determines the inherent granularity, dependence structure, and regularity of an application. Yet, the parallelization process often initiates from the application's description as a (sequential) program.

Program Formulation   One application can be expressed by many, sometimes substantially, different programs. A regular computation expressed in a loop supports its division into iterations (e.g., using data decomposition)—that is, a fine grained parallelism—while the same computation expressed as a recursive function might be easier divided into a few large tasks based on recursive decomposition. Also, a program, for example, written with a few large functions, is surely a candidate for a coarse grained decomposition, while a program with many small functions is probably better handled in a medium grained approach. This also depends strongly on the parallelization techniques that can be applied to the utilized programming paradigm and language.

2.4.4   Parallelization Techniques

Different parallelization techniques require or favor different subtask decompositions. This is illustrated by the following two examples.

When talking about parallelization techniques, one should not forget that in most cases a program is parallelized manually by its designer. A popular approach for parallel programming is message passing, where communication between the processors of a parallel system is performed by the exchange of messages (Section 2.1.3). It is typically the designer who decides which parts of the program are to be executed concurrently, thereby defining the subtasks of the program. The designer's job is then to insert the message passing directives for the synchronization and communication between those tasks. Such an approach normally favors a coarse grained decomposition.

Parallelizing compilers, on the other hand, focus on the parallelization of loops in programs. They therefore support regular decompositions with rather small or medium granularity.

Furthermore, it is important to determine how dependence relations between tasks can be avoided or eliminated by a transformation of the program.

2.4.5   Target Parallel System

Another dominant factor in the decomposition process is the target parallel system. At the beginning of this chapter different parallel architectures and systems were discussed. Some of those systems are appropriate for small grained parallelism, while others are better suited for coarse grained tasks. The number of processors, the communication speed and the communication overhead are, among others, important aspects for an inclination toward a certain form of parallelism.

A parallel system, whose interprocessor communication is based on message passing, typically performs interprocessor communication with a relatively expensive overhead. Consequently, subtasks should be quite large so that the overhead becomes small when compared with the computation time of the tasks; otherwise the speedup of the parallel program is degraded.

A massively parallel system, with hundreds of processors, needs a high degree of parallelism in order to involve all processors in the execution of the program, which obviously cannot be achieved with only a few large tasks.

These two examples demonstrate the influence of the system architecture on the subtask decomposition of a program. Also, the programming paradigm can have a similar effect. A centralized shared-memory system, which inherently supports fine grained parallelism, can also be programmed with a message passing paradigm. Yet, when doing so, the same applies as for a message passing system; hence, coarser granularity is indicated.

Earlier discussion showed that the decomposition of a program is a complex task that is strongly interwoven with other parts of the parallelization process. As was mentioned at the beginning of this section, it is not possible to treat subtask decomposition separately from its context. Nevertheless, this section gave an overview of the general subject and techniques and referred to the associated problems. In the next section, subtask decomposition will be seen in practice for simple program examples.

2.5 DEPENDENCE ANALYSIS

Dependence analysis (Banerjee et al. [17], Wolfe [204]) distinguishes between two kinds of dependence: *data dependence* and *control dependence*. The latter represents the dependence relations evoked by the control structure of the program, which is established by conditional statements like `if...else`. Section 2.5.3 analyzes this kind of dependence. The discussion starts with data dependence, which reflects the dependence relations between program parts, or tasks, caused by data transfer.

2.5.1 Data Dependence

The best way to build an understanding for dependence is to start with a simple example. Consider the following equation:

$$x = a * 7 + (a + 2). \tag{2.3}$$

Assume now that Eq. (2.3) corresponds to the initial application specification of the parallelization process as described in Section 2.3 (see Figure 2.15). The first step is to divide the equation into subtasks. This can be done by considering every operation as one task. For example, suppose the value 2 is assigned to a, a program can be written from Eq. (2.3) with four tasks as presented in Example 1.

Example 1   Program for $x = a * 7 + (a + 2)$

```
1: a = 2
2: u = a + 2
3: v = a * 7
4: x = u + v
```

Every line of this program is a statement of the form *variable = expression*. In the dependence analysis literature, statements are the entities between which dependence relations are defined, especially since classical dependence analysis is strongly linked to cyclic computations (i.e., loops), which will be covered in Section 2.5.2. In scheduling, the notion of a task is preferred, as it reflects a more flexible concept in terms of the size of the represented computation. A task can range from a small statement to basic blocks, loops, and sequences of these. To emphasize the fact that the ideas of this section have general validity, a statement will be referred to as a task. The concepts of dependence presented here are unaffected by the choice of notation.

The *variable* of each statement *variable = expression* holds the result of the task and thereby acts as its output, while the variables in the *expression* are the input of the task. In Example 1, the output variable of task 1 (a) is an input variable of tasks 2 and 3, and the output variables of task 2 (u) and task 3 (v) are the input variables of task 4. In other words, tasks 2 and 3 depend on the outcome of task 1, and task 4 on the outcomes of tasks 2 and 3. It is said that tasks 2 and 3 are *flow dependent* on task 1, caused by variable a. Correspondingly, task 4 is flow dependent on tasks 2 and 3, caused by the variables u and v. Consequently, only tasks 2 and 3 are not dependent on each other and can therefore be executed in parallel.

Flow dependence is not the only type of dependence encountered in programs. Consider the equation in Example 2 and its corresponding formulation as a program.

Example 2   Program for $x = a * 7 + (a * 5 + 2)$

```
1: a = 2
2: v = a * 5
3: u = v + 2
4: v = a * 7
5: x = u + v
```

While task 3 is flow dependent on task 2 (caused by variable v), there is a new form of dependence between tasks 3 and 4. In flow dependence, a write occurs before a read, but task 3 reads variable v before task 4 writes it. The supposed input of task 3 is the value computed by task 2 and not that of task 4, but both use variable v for their output. In order to calculate the correct value of x, task 3 must be executed before task 4. Task 4 is said to be *antidependent* on task 3.

Another type of dependence is output dependence. Consider the program extract in Example 3.

Example 3   Output Dependence
. . .
2: o = 10
. . .
4: o = 7
. . .

Suppose the variable o is only utilized in the two tasks 2 and 4 of the program. After a sequential execution, the value of o is 7. In a parallel execution, however, the value of o is determined by the execution order of the two tasks: if task 4 is executed before task 2, the final value of o is 10! Obviously, the sequential and the parallel versions should produce the same result. Thus, task 4 is said to be *output dependent* on task 2—for a correct semantic of the program task 2 must be executed before task 4.

The antidependence in Example 2 and the output dependence of Example 3 can be resolved by modifying the programs. For instance, if Example 2 is rewritten as displayed in Example 4, its antidependence is eliminated—task 4 is no longer antidependent on task 3.

Example 4   Eliminated Antidependence from Program in Example 2
1: a = 2
2: v = a * 5
3: u = v + 2
4: w = a * 7
5: x = u + w

The dependence was eliminated by substituting variable v in tasks 4 and 5 with the new variable w. Likewise, the output dependence of Example 3 is eliminated by the utilization of an additional variable.

Real Dependence   In general, antidependence and output dependence are caused by variable (i.e., memory) reuse or reassignment. For this reason, they are sometimes called *false dependence* as flow dependence is called *real dependence* (Wolfe [204]). Flow dependence is inherent in the computation and cannot be eliminated by renaming variables.

Nevertheless, the concepts of output and antidependence are important in dependence analysis of concrete programs, for example, in a parallelizing compiler (Banerjee et al. [13], Polychronopoulos et al. [153]). There, the program is a concrete implementation of an application specification and might contain these types of false dependence. The compiler must either eliminate the dependence relations or arrange the tasks in precedence order.

Dependence in Scheduling As will be seen later in Section 3.5, the graph representation of programs used in task scheduling only addresses real data dependence. Other dependence types are bound to a given implementation and are not inherent in the represented computation. Thus, a graph represents either an application specification with the inherent real data dependence relations or a concrete program, where all false dependence relations have been eliminated. Based on this convention, every data dependence in a graph is equivalent to a data transfer, called communication, between the respective tasks.

2.5.2 Data Dependence in Loops

Data dependence in loops (i.e., iterative (or cyclic) computations) is conceptually identical to the data dependence in linear programs as described previously. For the parallelization of a program, loops are very attractive, since they typically consume the lion's share of the total execution time of the program. They are different from linear programs in two aspects: (1) loops form a regular computation—the same statements are executed multiple times; and (2) loops often contain array variables. Both aspects together have an impact on the dependence analysis of loops. The following discussion of data dependence in loops is restricted to flow dependence, yet the definitions and conclusions are valid for all types of data dependence.

Single Loops Example 5 presents a loop over the index variable i with a loop body, or loop kernel, consisting of the two statements, or tasks, S and T.

Example 5   Single Loop
```
for i = 2 to 100 do
   S:  A(i) = B(i+2) + 7
   T:  B(i*2+1) = A(i-1) + C(i+1)
end for
```

During the execution of the loop, i takes $2, 3, \ldots, 100$ as its values. For each value, an instance of the loop body is executed, and this instance is called an iteration of the loop. In each iteration, instances of the tasks S and T are executed and the instances corresponding to the iteration for $i = j$ are denoted by $S(j)$ and $T(j)$. Unlike in the previous discussion where the variables used in the examples are scalars, the variables in the tasks S and T are elements of the disjoint arrays A, B, and C. Thus, data dependences arise from the reference to the same array element of different task instances. The two tasks read from and write to elements of the arrays A and B, potentially creating a dependence relation. Elements of C are only read in task T, therefore not causing any dependence relation.

Within one iteration, that is, between the instances $S(j)$ and $T(k)$, with $j = k$, it can easily be verified that there is no dependence: for any value of $i = j$, $S(j)$ and $T(j)$ access different elements of A and B. $S(j)$ writes to $A(j)$ and $T(j)$ reads from

$A(j-1)$, but i never equals $i-1$ for $i = 2, \ldots, 100$. Likewise for array B, since $i + 2 \neq i * 2 + 1$ for $i = 2, \ldots, 100$. If such a dependence exists, it is referred to as *intraiteration dependence*.

Dependence does arise between instances of different iterations. Example 6 shows the first four iterations, similar to the linear programs studied in the previous section.

Example 6   First Iterations of Example 5

```
S(2):  A(2)  = B(4)  + 7
T(2):  B(5)  = A(1)  + C(3)

S(3):  A(3)  = B(5)  + 7
T(3):  B(7)  = A(2)  + C(4)

S(4):  A(4)  = B(6)  + 7
T(4):  B(9)  = A(3)  + C(5)

S(5):  A(5)  = B(7)  + 7
T(5):  B(11) = A(4)  + C(6)
...
```

Thus, it is easy to spot that instance $T(3)$ is flow dependent on $S(2)$: the output variable of task $S(2)$, $A(2)$, is the input of $T(3)$. The same can be observed for the relation between $T(4)$ and $S(3)$, and between $T(5)$ and $S(4)$, caused by other elements of array A. The dependence *distance*, that is, the number of iterations between the two task instances forming a dependence relation, is 1. In general, it can be stated that the instance $T(i + 1)$ is dependent on the instance $S(i)$. Such a dependence is called *uniform* since its distance is constant. Analogous to the intraiteration dependence, this type is referred to as *interiteration dependence*. For flow dependence relations, intra- and interiteration dependences correspond to intra- and interiteration communications, respectively. Note that a task can even depend on itself, under the condition that the dependence distance is greater than 0. Another logical step is to consider intraiteration dependence as a special case of interiteration dependence, namely, with distance 0. In fact, dependence analysis for loops, for example, as presented by Banerjee et al. [17], does not distinguish between intra- and interiteration dependence. However, it will prove useful for the treatment of graph representations in Chapter 3.

In Example 6 one observes further dependence relations. The instance $S(3)$ is flow dependent on $T(2)$, caused by the array element $B(5)$, and $S(5)$ is flow dependent on $T(3)$, caused by $B(7)$. These are two examples of the dependence pattern caused by the output variable $B(i * 2 + 1)$ of T and the input variable $B(i + 2)$ of S. Other dependence pairs of this pattern are: $S(7)$ depends on $T(4)$, $S(9)$ depends on $T(5), \ldots, S(99)$ depends on $T(50)$. Since the distance of these dependence relations is not constant—the minimum distance is 1 and the maximum is 49—it is said to be *nonuniform*.

Double Loops—Loop Nests   The next logical step in dependence analysis of loops is to extend the described concepts to double and multiple nested loops. Example 7 displays a double loop over the indices i and j, containing the two statements, or tasks, S and T in the kernel.

Example 7   Double Loop

```
for i = 0 to 5 do
  for j = 0 to 5 do
    S:  A(i+1,j)   = B(i,j) + C(i,j)
    T:  B(i+1,j+1) = A(i,j) + 1
  end for
end for
```

Suitable for the double loop, the arrays used in the tasks are two dimensional, whereby the index variable i of the outer loop is used only in the subscripts of the arrays' first dimension and the index variable j only in the subscripts of the second dimension. While this is common practice in nested loops, it is neither a guaranteed nor a necessary condition for the dependence analysis in nested loops. The arrays, for example, might only have one dimension, and the subscripts might be functions of more than one index variable. Relevant for a dependence relation is only the reference of two different tasks to the same array element.

What the index variable is to the single loop is now, in a straightforward generalization, an index vector of two dimensions. An instance of the double loop kernel (i.e., an iteration) is determined by the two corresponding values of the index variables i and j. Also, an instance of one of the tasks S and T is denoted by $S(i,j)$ and $T(i,j)$, respectively. The extension to a more general nest of loops follows a similar pattern—every loop simply contributes one dimension to the index vector. In the same way, the index variable of a single loop can be treated as an index vector of one dimension.

By examining the tasks of the loop in Example 7, it becomes apparent that instance $S(i+1, j+1)$ depends on instance $T(i, j)$, caused by the references to the elements of array B, and instance $T(i+1, j)$ depends on $S(i, j)$, caused by the references to the elements of array A. As a logical consequence of the generalization from the index variable to an index vector, the dependence distance is also expressed as a *distance vector*. For the identified dependence relations in Example 7, the distance vectors are $(1, 1)$, for $S(i+1, j+1)$ depending on $T(i, j)$, and $(1, 0)$, for $T(i+1, j)$ depending on $S(i, j)$. So there are two uniform dependences, as the distance vector is constant for every dependence.

The determination of the dependence relations and the distance vectors for the loops in Example 5 and Example 7 are relatively simple. In real programs, however, various circumstances can make dependence analysis more complicated and time consuming. Sometimes it might even be impossible to determine the dependence relation of a program: for example, when a subscript of an array, which is read and written in various tasks, is a function of an input variable of the program. In that case, the dependence relations can only be established at runtime. A conservative approach, in the sense that the discovered dependence structure includes all constraints of the true structure, is then to assume dependence between all respective tasks.

Loop Stride The dependence analysis presented so far is based on the assumption that the loop stride (i.e., increment or decrement of the index variable per iteration) is 1. In order to handle different strides, as they happen to appear in real loops, the loop must be normalized to a stride of 1—at least during the analysis phase. The normalization involves a transformation of all expressions where the original index variable is included (Banerjee et al. [17]).

Dependence Tests The area of dependence analysis in loops mainly concentrates on array subscripts that are linear functions of the index variables. But even with this restriction, the determination of a dependence can be quite time consuming, as exact solutions are based on integer programming. This, together with the fact that it is sometimes sufficient to know whether there is a dependence or not, led to the utilization of approximation methods. It suggests that these approximation methods often do not determine the distance vector; at most, they compute a direction vector, which is the vector of the signs of the distance vector's components. Some loop transformations only require knowledge of such a direction vector. Many so-called dependence test algorithms have been proposed in the past; consult, for example, Banerjee et al. [15, 16], Blume and Eigenmann [23], Eisenbeis and Sogno [58], Petersen and Padua [148], Pugh [155], Wolfe [202], and Yazici and Terziologu [212].

2.5.3 Control Dependence

In contrast to data dependence, control dependence is not caused by the transfer of data among tasks. Control dependence relations describe the control structure of a program (Banerjee [14], Towle [189]). Consider the sequence of statements in Example 8.

Example 8   Control Dependence
1: **if** u = 0 **then**
2:     v = w
3: **else**
4:     v = w + 1
5:     x = x - 1
6: **end if**

The tasks in lines 2, 4, and 5 are *control dependent* on the outcome of the if statement in line 1. In other words, tasks 2, 4, and 5 should not be executed until the statement of line 1 is evaluated.

Control dependence can be transformed into data dependence (Banerjee et al. [17]), and in this way the representation and analysis techniques for data dependence can be applied. The transformation proceeds by replacing the if statement with a

Boolean variable, say, *b*, to which the result of the if statement's argument evaluation is assigned. The control dependent tasks are rewritten, as shown in Example 9, so that they are only executed when *b* is true and false, respectively.

Example 9   Control Dependence of Example 8 Transformed to Data Dependence
```
1: b = [u = 0]
2: v = w when b
3: v = w + 1 when not b
4: x = x - 1 when not b
```

The operator when indicates that the statement to its left is only executed when its argument to the right is true. Resulting from the transformation, tasks 2–4 are now flow dependent on task 1, caused by the variable *b*.

Transforming control dependence into data dependence is described by Allen and Kennedy [11] and by Banerjee [14]. As mentioned earlier, the transformation of control dependence into data dependence permits one to unify the treatment of both dependence types. The graph representation of programs can benefit from this and only reflect data dependence relations, as described in Section 3.2.

2.6 CONCLUDING REMARKS

In this chapter, parallel systems and their programming were reviewed. This review focused on the parallel architectures and the communication networks, both of which determine the communication behavior of a parallel system. In order to produce accurate and efficient schedules, a good understanding of the communication behavior is crucial. Scheduling is a crucial part of the parallelization process in parallel programming. The process as a whole was studied and the two steps that precede scheduling—subtask decomposition and dependence analysis—were analyzed in detail. Altogether, this chapter established the foundation, background, and terminology for the following chapters.

Naturally, the discussion of parallel architectures and their networks in Sections 2.1 and 2.2 cannot be complete. For more details and further reading the reader should refer to Cosnard and Trystram [45], Culler and Singh [48], Grama et al. [82], Hamacher et al. [84], Hennessy and Patterson [88], and Kung [109]. Many of the system examples in Section 2.1 are taken from the *Overview of Recent Supercomputers* [193], published yearly since 1996 by van der Steen and Dongarra on the site of the TOP500 Supercomputer Sites [186].

Decomposition techniques are studied in greater detail in Grama et al. [82], on which Section 2.4.2 is based.

The dependence analysis discussed in Section 2.5 is based primarily on the publications by Banerjee et al. [17] and by Wolfe [204]. More on dependence and its analysis, especially in loops, can be found, apart from the references given in the text, in the literature by Allen and Kennedy [12], Banerjee et al. [15, 16], Blume et al. [24], Polychronopoulos [152], and Wolfe [202, 203].

Plenty of general books on the aspects of parallel computing and programming have been published: for example, Culler and Singh [48], El-Rewini and Lewis [64, 122, 123], Foster [69], Grama et al. [82], Hwang and Briggs [95], Kumar et al. [108], Leighton [118], Parhami [143], Polychronopoulos [152], Quinn [156], and Wilkinson and Allen [200].

Before starting the analysis of scheduling in Chapter 4, Chapter 3 will introduce graph models for the representation of parallel programs. Task scheduling is based on such models.

2.7 EXERCISES

2.1 Figure 2.1(*a*) in Section 2.1.1 shows a schematic diagram of an SIMD system. This can be interpreted as a parallel system consisting of one vector processor. Draw the schematic diagram of a system consisting of multiple vector processors, that is, a multiple SIMD system.

2.2 From a memory architecture point of view (Section 2.1.2), there are two fundamentally different approaches to the design of a parallel system: shared-memory architecture and message passing architecture.

(a) Describe the difference between the two architectures.

(b) Which architecture has a cluster of workstations interconnected through a LAN (local area network)?

(c) Which architecture has a single workstation equipped with four processors?

(d) On which of these two systems can the message passing programming model be used? Which of these two systems can share the memory in a global address space?

2.3 In Section 2.1.2 the memory architectures of parallel systems are studied. Conceptually, the distributed shared-memory architecture is the most interesting one, but also the most complex. In a system with such an architecture:

(a) What is meant by local and remote memory access? What is the usual major difference between them?

(b) Is such a system likely to be a UMA or a NUMA system?

(c) What is the danger of NUMA systems for their efficiency?

(d) Why are memory hierarchies used in distributed shared-memory systems?

(e) Can the message passing programming model be used on this system?

2.4 LogP is an advanced model for parallel programming (Section 2.1.3).

(a) Why was the LogP model introduced?

(b) What does "LogP" stand for?

2.5 The communication network is an essential part of a parallel system. Many different network topologies are discussed in Section 2.2. For a network that connects $p = 128$ processors, calculate the number of links, the diameter, and the bisection width for the following topologies:

(a) Fully connected, cyclic grid (2D mesh), cyclic torus (3D mesh), and hypercube.

(b) Crossbar, butterfly, and binary tree. Also, calculate the number of switches for each network.

2.6   A butterfly network is a blocking network (Section 2.2.2), which means that certain combinations of source–destination connections are mutually exclusive. Figure 2.10 depicts a 3D butterfly network with eight source and destination lines. Which of the following connection pairs block each other: 2–6 and 7–1, 3–4 and 2–7, 5–0 and 0–5, 3–5 and 6–7, 1–1 and 3–0, 4–2 and 1–3?

2.7   Parallel programming implies the parallelization of a problem (Section 2.3). Describe the three main areas of the parallelization process.

2.8   Four quite common decomposition techniques are studied in Section 2.4.2: data decomposition, recursive decomposition, exploratory decomposition, and speculative decomposition. Which of these techniques seems indicated for parallelizing the following algorithms and applications: Quicksort (Cormen et al. [42]), chess game, fast fourier transform (FFT) (Cormen et al. [42]), and path finding algorithm. Briefly describe how you would decompose these problems.

2.9   In Section 2.5.1, three different types of data dependence are discussed: flow dependence, antidependence, and output dependence. Consider the following fragment of a program:

```
1: b = a * 3
2: a = b * 7
3: c = a + b - d
4: a = c - 11
5: d = a - b + c
```

Identify all dependences and classify their types.

2.10   Exercise 2.9 asks you to identify and classify the data dependences in a small code fragment. In loops, data dependence usually arises through the access of array elements (Section 2.5.2). Consider the following loop:

```
for i = 3 to 50 do
    S:  A(i)   = B(2*i) + 1
    T:  B(i+1) = A(i-1) + A(i+1)
    U:  C(2*i) = A(i-2) + E(i+2)
    V:  D(i)   = C(2*i+1) + B(i+1)
end for
```

(a) Without considering the subscripts of the array variables, which of the array variables $A, B, C, D,$ and E are potentially involved in the dependences among the tasks?

(b) Identify all *intra*iteration dependences and their types.

(c) Identify all *inter*iteration dependences and their types. What is the distance of each dependence? For nonuniform dependences, give the minimum and the maximum distance.

2.11 Dependence analysis of nested loops is a straightforward generalization of the dependence analysis of single loops. Essentially, the index variable becomes an index vector and the dependence distance becomes a dependence vector (Section 2.5.2). Consider the following example of a double loop:

```
for i = 1 to 10 do
  for j = 1 to 10 do
    S:  A(i+1,j-1) = B(i,j) + E(i+1,j+1)
    T:  B(i,j-1) = A(i,j) + 1
    U:  C(2*i+1,j) = A(i+1,j-1) + 4
    V:  D(i,j) = C(i,j-1) - 1
  end for
end for
```

(a) Identify all dependences and their types. What is the distance vector of each dependence? For nonuniform dependences, give the minimum and the maximum distance for each component of the vector.

(b) Can the loop order be swapped, that is, the first loop iterates over *j* and the second over *i*?

2.12 Control dependence can be converted into data dependence, using a when operator as shown in Section 2.5.3. Convert the control dependences in the following code fragment into data dependences.

```
1: if u > 5 then
2:   v = u - 5
3: else if u < 2 then
4:   v = u + 2
5: else
6:   w = u - 2
7: end if
```

CHAPTER 3

Graph Representations

Graphs are deployed in many areas of parallel and distributed computing. Section 2.2 includes one example, where graphs are used for the representation of communication networks in parallel systems. Graphs are also employed for the representation of the task, communication, and dependence structure of programs. A first example is given in Figure 2.15, where the subtasks are drawn as ovals, and lines between the ovals represent the dependence relations between the respective tasks. This is the typical form of the pictorial representation of graphs (Berge [21]). Indeed, graphs are widely used for the representation of a parallel program's structure.

The task graph, often simply called DAG for directed acyclic graph, is the primary graph model used for program representation in task scheduling. Section 3.5 presents the task graph model and thoroughly discusses its properties and the concepts connected with it, as it is used heavily in the following chapters.

However, it is deemed essential to establish a broader background of graph models in order to have a general understanding of the task graph's bases, its relations to other major models, and its benefits as well as its drawbacks. Therefore, the next section will define basic graph concepts and Section 3.2 studies the graph as a model for program representation in general. It follows a discussion of the major graph models, namely, the dependence graph (Section 3.3), the flow graph (Section 3.4), and the task graph (Section 3.5).

3.1 BASIC GRAPH CONCEPTS

In order to discuss graph models, it is necessary to define graphs and establish some terminology. The initial three definitions of graph, path, and cycle are based on the notations given by Cormen et al. [42].

Definition 3.1 (Graph) *A graph G is a pair $(\mathbf{V}, \mathbf{E})$, where $\mathbf{V}$ and $\mathbf{E}$ are finite sets. An element v of $\mathbf{V}$ is called vertex and an element e of $\mathbf{E}$ is called edge. An edge is a pair of vertices (u, v), $u, v \in \mathbf{V}$, and by convention the notation e_{uv} is used for an edge between the vertices u and v.*

Task Scheduling for Parallel Systems, by Oliver Sinnen
Copyright © 2007 John Wiley & Sons, Inc.

In a directed graph, *an edge* e_{uv} *has a distinguished direction, from vertex u to vertex v, hence* $e_{uv} \neq e_{vu}$, *and such an edge shall be referred to as a* directed edge. Self-loops—*edges from a vertex to itself (i.e.,* e_{uv} *with* $u = v$)—*are possible. In an* undirected graph, e_{uv} *and* e_{vu} *are considered to be the same* undirected edge, *thus* $e_{uv} = e_{vu}$, *and since self-loops are forbidden* $u \neq v$ *for any edge* e_{uv}.

It is said edge e_{uv} *is* incident on *the vertices u and v, and if* e_{uv} *is a directed edge, it is said that* e_{uv} *leaves vertex u and enters vertex v. Similarly, if* $e_{uv} \in \mathbf{E}$, *vertex v is* adjacent *to vertex u and in an undirected graph, but not in a directed graph, vertex u is adjacent to vertex v (i.e., in an undirected graph the adjacency relation is symmetric). The set* $\{v \in \mathbf{V} : e_{uv} \in \mathbf{E}\}$ *of all vertices v adjacent to u is denoted by* **adj**(u). *For the directed edge* e_{uv}, *u is its* origin *vertex and v its* destination *vertex.*

The degree *of a vertex is the number of edges incident on it. In a directed graph, it can be further distinguished between the* out-degree *(i.e., the number of edges leaving the vertex) and the* in-degree *(i.e., the number of edges entering the vertex).*

Figure 3.1 shows the pictorial representation of two graphs: Figure 3.1(a) an undirected graph and Figure 3.1(b) a directed graph. Vertices are represented by circles, undirected edges by lines, and directed edges by arrows. Both graphs are composed of the four vertices u, v, w, x. In the undirected graph, vertex v has a degree of 2, since edges e_{vx} and e_{vw} are incident on it. Thus, vertices x and w are adjacent to v, and, as the graph is undirected, v is also adjacent to them. In the directed graph, v has a degree of 3 caused by the two entering edges e_{uv} and e_{xv} (in-degree = 2) and the leaving edge e_{vw} (out-degree = 1). Vertex x is adjacent to vertex w, caused by edge e_{wx}, whose origin is w and whose destination is x. The edges e_{xw} and e_{wx} are identical in the undirected graph but are distinct in the directed graph. One of the edges leaving vertex u also enters it and thereby builds the self-loop e_{uu}.

Definition 3.2 (Path) *A path p in a graph* $G = (\mathbf{V}, \mathbf{E})$ *from a vertex* v_0 *to a vertex* v_k *is a sequence* $\langle v_0, v_1, v_2, \ldots, v_k \rangle$ *of vertices such that they are connected by the edges* $e_{v_{i-1}v_i} \in \mathbf{E}$, *for* $i = 1, 2, \ldots, k$. *A path p contains the vertices* $v_0, v_1, v_2, \ldots, v_k$ *and the edges* $e_{01}, e_{12}, e_{23}, \ldots, e_{(k-1)k}$ *(* e_{ij} *is short for* $e_{v_iv_j}$ *) and the fact that a vertex* v_i

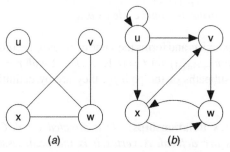

(a)   (b)

Figure 3.1. Pictorial representation of two sample graphs: (a) undirected graph and (b) directed graph. Both graphs consist of vertices u, v, w, x and various edges; vertex u has a self-loop (b).

or an edge e_{ij} is a member of the path p is denoted by $v_i \in p$ and $e_{ij} \in p$, respectively. Consequently, the path p from a vertex v_0 to a vertex v_k is also defined by the sequence $\langle e_{01}, e_{12}, e_{23}, \ldots, e_{(k-1)k} \rangle$ of edges. In this text, sometimes $p(v_0 \rightarrow v_k)$ is written to indicate the path goes from vertex v_0 to v_k. A path is simple *if all vertices are distinct. The* length *of a path is the number of edges in the path. A* subpath *of a path $p = \langle v_0, v_1, v_2, \ldots, v_k \rangle$ is a contiguous subsequence of vertices $\langle v_i, v_{i+1}, \ldots, v_j \rangle$ with $0 \le i \le j \le k$. Two paths $p_1 = \langle v_0, v_1, \ldots, v_i \rangle$ and $p_2 = \langle u_0, u_1, \ldots, u_j \rangle$ can be* concatenated *to build a new path $p = \langle v_0, v_1, \ldots, v_i, u_1, \ldots, u_j \rangle$, if $v_i = u_0$.*

In the sample undirected graph of Figure 3.1, the sequence $\langle u, w, x \rangle$ of vertices forms a simple path as does $\langle u, x, v, w \rangle$ in the directed graph, but the sequence $\langle w, x, v, w, u \rangle$ is a path in the undirected graph, which is not simple. Path $\langle u, x, v \rangle$ of length 2 in the directed graph is a subpath of the path $\langle u, x, v, w \rangle$ of length 3.

Definition 3.3 (Cycle)   *A path $p = \langle v_0, v_1, v_2, \ldots, v_k \rangle$ forms a* cycle *if $v_0 = v_k$ and the path contains at least one edge. The cycle is* simple *if, in addition, the vertices $v_1, v_2, \ldots, v_k$ are distinct.*

The same cycle is formed by two paths $p = \langle v_0, v_1, \ldots, v_{k-1}, v_0 \rangle$ and $p' = \langle v_0', v_1', \ldots, v_{k-1}', v_0' \rangle$, if an integer r with $1 \le r \le k - 1$ exists so that $v_i' = v_{(i+r) \bmod k}$ for $i = 0, 1, \ldots, k - 1$. A subpath of a cycle p is a contiguous subsequence of vertices of any of the paths, including p, forming the same cycle as p. A graph with no cycles is acyclic.

An example for a simple cycle in Figure 3.1 is the path $\langle v, w, x, v \rangle$ in both sample graphs, while $\langle v, w, x, w, x, v \rangle$ establishes a cycle in the directed graph, which is not simple. The two paths $\langle v, w, x, v \rangle$ and $\langle w, x, v, w \rangle$ describe the same cycle ($r = 2$) and $\langle v, w \rangle$ is a subpath of this cycle.

The following lemma can be established about subpaths in cycles.

Lemma 3.1 (Subpaths in Cycle)   *Let $p_c = \langle v_0, v_1, \ldots, v_{k-1}, v_0 \rangle$ be a cycle in a directed graph $G = (\mathbf{V}, \mathbf{E})$. For any pair of vertices v_i and v_j, $0 \le i < j \le k - 1$, two subpaths $p(v_i \rightarrow v_j) = \langle v_i, v_{i+1}, \ldots, v_{j-1}, v_j \rangle$ and $p(v_j \rightarrow v_i) = \langle v_j, v_{j+1}, \ldots, v_{k-1}, v_0, \ldots, v_{i-1}, v_i \rangle$ of the cycle exist. The path $p_r = \langle v_0', v_1', \ldots, v_{k-1}', v_0' \rangle = \langle v_i, v_{i+1}, \ldots, v_{j-1}, v_j, v_{j+1}, \ldots, v_{k-1}, v_0, \ldots, v_{i-1}, v_i \rangle$, concatenated from these two paths, forms the same cycle as p_c.*

Proof.   The path p_r exists and forms the same cycle as p_c according to the definition of a cycle, since $v_l' = v_{(l+r) \bmod k}$ for $l = 0, 1, \ldots, k - 1$ with $r = i$. Since $p(v_i \rightarrow v_j)$ and $p(v_j \rightarrow v_i)$ are subpaths of the path p_r, they are by definition subpaths of the cycle p_c. $\square$

Definition 3.4 (Vertex Relationships)   *In a directed graph $G = (\mathbf{V}, \mathbf{E})$, the following relationships are defined. A vertex u is the* predecessor *of vertex v and correspondingly v is the* successor *of u, if and only if edge $e_{uv} \in \mathbf{E}, u, v \in V$. The vertex v is the successor of vertex u if it is adjacent to vertex u. The set $\{x \in \mathbf{V} : e_{xv} \in \mathbf{E}\}$*

of all predecessors of v is denoted by **pred**(v) *and the set* $\{x \in \mathbf{V} : e_{vx} \in \mathbf{E}\}$ *of all successors of v, is denoted by* **succ**(v). *A vertex w is called* ancestor *of vertex v if there is a path $p(w \to v)$ from w to v, and the set of all ancestors of v is denoted by* **ance**$(v) = \{x \in \mathbf{V} : \exists p(x \to v) \in G\}$. *Logically, a vertex w is called* descendant *of vertex v if there is a path $p(v \to w)$, and the set of all descendants of v is denoted by* **desc**$(v) = \{x \in \mathbf{V} : \exists p(v \to x) \in G\}$. *In the latter case it is sometimes said that vertex w is* reachable *from v.*

Obviously, all predecessors are also ancestors and all successors are also descendants. Alternative notations are sometimes used when appropriate, for example, child or parent, given the analogy to the pedigree.

These are some examples from the directed graph in Figure 3.1(b): vertex x is a predecessor of v, which is a successor of u; hence, the set of predecessors of v is **pred**$(v) = \{u, x\}$. The set of successors of w only comprises the single vertex x, **succ**$(w) = \{x\}$. Finally, vertex u is an ancestor of w and v is a descendent of w, but also its ancestor.

As seen in the examples above, a vertex relation is not necessarily unique in a cyclic graph.

Lemma 3.2 (Vertex Relationships in Cycle) *A vertex v belonging to a cycle path p_c is for any vertex of the cycle, $u \in p_c, u \neq v$, an ancestor and a descendant at the same time.*

Proof. The lemma follows directly from Lemma 3.1 and the notion of ancestor and descendant, since in a cycle p_c, for every two vertices u and v, $u, v \in p_c, u \neq v$, there are two subpaths $p(u \to v)$ and $p(v \to u)$. $\qquad\square$

Based on the notions of predecessors and successors, two vertex types can be further distinguished in a directed graph.

Definition 3.5 (Source Vertex and Sink Vertex) *In a directed graph $G = (\mathbf{V}, \mathbf{E})$, a vertex $v \in \mathbf{V}$ having no predecessors,* **pred**$(v) = \emptyset$, *is named* source *vertex and a vertex $u \in \mathbf{V}$ having no successors,* **succ**$(v) = \emptyset$, *is named* sink *vertex.*

Alternative notations for source vertex and sink vertex are entry vertex and exit vertex, respectively. The set of source vertices in a directed graph G is denoted by **source**$(G) = \{v \in \mathbf{V} : \textbf{pred}(v) = \emptyset\}$ and the set of sink vertices by **sink**$(G) = \{v \in \mathbf{V} : \textbf{succ}(v) = \emptyset\}$.

3.1.1 Computer Representation of Graphs

For the complexity analysis of graph-based algorithms, in time and space, it is essential to consider the computer representation of graphs. There are two standard ways to represent a graph $G = (\mathbf{V}, \mathbf{E})$: as a collection of adjacency lists or as an adjacency matrix (Cormen et al. [42]).

Vertex	List
u	→w
v	→x →w
w	→u →x →v
x	→v →w

Vertex	List
u	→u →v →x
v	→w
w	→x
x	→v →w

(a)   (b)

Figure 3.2. The adjacency list representations of the two graphs in Figure 3.1: (a) for the undirected graph and (b) for the directed graph.

Adjacency List Representation

A graph can be represented as an array of $|V|$ adjacency lists, one for each vertex in V. The adjacency list belonging to vertex $u \in V$ contains pointers to all vertices v that are adjacent to u; hence, there is an edge $e_{uv} \in E$. In other words, in vertex u's adjacency list the elements of **adj**(u) are stored in arbitrary order. Figure 3.2 shows the adjacency list representations of the two sample graphs in Figure 3.1; in Figure 3.2(a) the one for the undirected graph and in Figure 3.2(b) the one for the directed graph.

For the directed graph, the sum of the lengths of the adjacency lists is $|E|$, because for each edge e_{uv} the destination vertex v appears once in the list of vertex u. For an undirected graph every edge e_{uv} appears twice, once in the list of u and once in the list of v, due to the symmetry of the undirected edge; thus, the sum of the lengths of the adjacency list is $2|E|$. Clearly, this representation form describes a graph G completely, as there is a list for every vertex and at least one entry for every edge. The amount of memory required for a graph, directed or undirected, is consequently $O(V + E)$.

In the previous asymptotic notation of the complexity, a common notational convention was adopted. The sign $|\ |$ for the cardinality (or size) of sets was omitted and it was written $O(V + E)$ instead of $O(|V| + |E|)$. This shall be used in all asymptotic notations, but only there, since it makes them more readable and is nonambiguous.

The adjacency list representation has the disadvantage that there is no quicker way to determine if an edge e_{uv} is part of a graph G than to search in u's adjacency list.

Adjacency Matrix Representation

The alternative representation of a graph as an adjacency matrix overcomes this shortcoming. A graph is represented by a $|V| \times |V|$ matrix A and it is assumed that the vertices are indexed $1, 2, \ldots, |V|$ in some arbitrary manner. Each element a_{ij} of the matrix A has one of two possible values: 1 if the edge $e_{ij} \in E$ and 0 otherwise. Figure 3.3 depicts the two adjacency matrices for the graphs of Figure 3.1 with the vertices u, v, w, x numbered $1, 2, 3, 4$, respectively.

Owing to the symmetry of undirected edges, the matrix of an undirected graph is symmetric along its main diagonal, which can be observed in the matrix of Figure 3.3(a). As the matrix is of size $|V| \times |V|$, the memory requirement of the adjacency matrix representation is $O(V^2)$.

	u	v	w	x
u	0	0	1	0
v	0	0	1	1
w	1	1	0	1
x	0	1	1	0

(a)

	u	v	w	x
u	1	1	0	1
v	0	0	1	0
w	0	0	0	1
x	0	1	1	0

(b)

Figure 3.3. The adjacency matrix representations of the two graphs in Figure 3.1: (*a*) for the undirected graph and (*b*) for the directed graph.

For many algorithms the adjacency list is the preferred representation form, because it provides a compact way to represent sparse graphs—those for which $|E|$ is much less than $|V|^2$. For dense graphs, or when the fast determination of the existence of an edge is crucial, the adjacency matrix is preferred.

Maximum Number of Edges   From the above considerations it is easy to state the maximum number of edges in a graph.

Directed Graph   The maximum number of edges in a directed graph $G = (V, E)$ is $|V|^2$, that is, the number of elements in the adjacency matrix, as every vertex may have an edge to every other vertex including itself.

$$|E| \leq |V|^2. \tag{3.1}$$

Undirected Graph   The symmetry of an undirected edge reduces the maximum number of edges by more than half, compared to a directed graph, as already shown by the symmetric adjacency matrix. The maximum number of edges of an undirected graph $G = (V, E)$ is limited by

$$|E| \leq \sum_{i=1}^{|V|-1} i = \tfrac{1}{2}|V|(|V| - 1). \tag{3.2}$$

The maximum number of edges in a directed *acyclic* graph is identical to that in an undirected graph. It follows that in both graph forms—directed and undirected—the number of edges is $O(V^2)$. In describing the running time of a graph-based algorithm in the following, the size of the input shall be measured in terms of the number of vertices $|V|$ and the number of edges $|E|$. Substituting the size of E by $O(V^2)$ would be too rough an approximation, as the number of edges of a graph varies largely among graphs.

3.1.2 Elementary Graph Algorithms

This section concludes by reviewing some of the well-known elementary graph algorithms, which built the foundations of many others (Cormen et al. [42]).

Two simple search algorithms, complementary in their approach, can be considered the most important graph algorithms, the *breadth first search* (BFS) and the *depth first search* (DFS). Both can be applied to a directed or undirected graph $G = (\mathbf{V}, \mathbf{E})$.

BFS (*Breadth First Search*)   The BFS searches a graph G beginning with a specified start vertex s. It examines all vertices adjacent to s and then continues with the vertices adjacent to these vertices and so on until all vertices reachable from s have been considered. Breadth first search is so named because it first discovers the vertices at distance 1 of s, then at distance 2, and so on. The *distance* between two vertices u and v is defined as the minimum number of edges that must be traversed to reach v from u; or expressed in another way, distance is the length of the *shortest path* from u to v. When during the BFS the predecessor, from which a vertex is discovered, is stored, upon termination a shortest path between s and any vertex v reachable from s is given by the recursive list of predecessors of v. Algorithm 1 briefly outlines BFS, which uses a FIFO (first in first out) queue as its main data structure.

Algorithm 1   BFS(G, s)
    Put s into a FIFO queue Q; mark s as discovered
    while $Q \neq \emptyset$ **do**
      Let u be the first vertex of Q; remove u from Q
      for each $v \in \mathbf{adj}(u)$ **do**
        if v not discovered **then**
          Put v into Q and mark v as discovered
        end if
      end for
    end while

DFS (*Depth First Search*)   The DFS searches a graph G using the depth first approach. It moves deeper into the graph before finishing the examination of all vertices adjacent to a vertex u. Only when all vertices reachable from a vertex v adjacent to u have been examined, does DFS return to examine the remaining undiscovered vertices adjacent to u. Consequently, DFS is readily expressed as a recursive algorithm, using a subroutine, here named DFS-Visit, which is recursively called for any undiscovered adjacent vertex as soon as it is found. DFS-Visit searches the subgraph spanned by all reachable vertices from its parameter vertex u and their adjacent edges. Algorithm 3 outlines the DFS-Visit subroutine and Algorithm 2 shows the main routine of DFS, which calls DFS-Visit for every undiscovered vertex $v \in G$.

Both the BFS and the DFS have a runtime complexity of $O(\mathbf{V} + \mathbf{E})$, since for each vertex ($O(\mathbf{V})$), every adjacent vertex is examined once and the total number of

Algorithm 2   DFS(G)
> **for** each $v \in \mathbf{V}$ **do**
>> **if** v not discovered **then**
>>> DFS-Visit(v)
>>
>> **end if**
>
> **end for**

Algorithm 3   DFS-Visit(u)
> **for** each $v \in \mathbf{adj}(u)$ **do**
>> **if** v not discovered **then**
>>> Mark v as discovered
>>> DFS-Visit(v)
>>
>> **end if**
>
> **end for**
> Mark u as finished

adjacent vertices is $O(\mathbf{E})$ (see also Section 3.1.1). For more details and an in-depth analysis of the properties of the two algorithms please refer to Cormen et al. [42].

Topological Order   Now an important concept for directed acyclic graphs is considered—the topological order of their vertices (Cormen et al. [42]). Directed acyclic graphs build an essential class of graphs for task scheduling, because they are utilized for the representation of programs (Section 3.5) in scheduling algorithms. The topological order is defined as follows.

Definition 3.6 (Topological Order)   *A topological order of a directed acyclic graph $G = (\mathbf{P}, \mathbf{E})$ is a linear ordering of all its vertices such that if $\mathbf{E}$ contains an edge e_{uv}, then u appears before v in the ordering.*

To illustrate Definition 3.6, consider the small directed acyclic graph in Figure 3.4. The topological order of the graph's vertices can be interpreted as a horizontal arrangement of the vertices (i.e., a linear order), in such a way that all edges are directed from left to right. This arrangement of the graph in Figure 3.4 is depicted in Figure 3.5.

The acyclic property is crucial for the topological order; otherwise no such order exists.

Lemma 3.3 (Topological Order and Directed Graphs)   *A directed graph $G = (\mathbf{P}, \mathbf{E})$ is acyclic if and only if there exists a topological order of its vertices.*

Proof. $\Rightarrow$: Suppose no topological order exists for G. Thus, for any ordering of the vertices of G, there is at least one edge e_{vu} with u appearing before v in the list. Consequently, there must be a path $p(u \rightarrow v)$ from u to v; otherwise v and all of its ancestors that lie between u and v in the ordering could be inserted just before u,

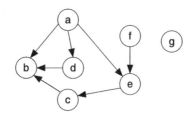

Figure 3.4. A directed acyclic graph.

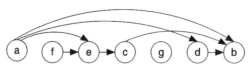

Figure 3.5. The directed acyclic graph of Figure 3.4 arranged in topological order; note that all directed edges go from left to right.

making the edge e_{vu} comply with the topology order without making a new edge violating it. With the path $p(u \rightarrow v)$, however, G is cyclic, since edge e_{vu} builds a cycle together with the path $p(u \rightarrow v)$.

$\Leftarrow$: Suppose G is cyclic; hence, it has at least one simple cycle $p_c = \langle v_0, v_1, \ldots, v_{k-1}, v_0 \rangle$. Consider the distinct vertices $v_0, v_1, \ldots, v_{k-1}$ of p_c. In a topological order, vertex v_i must appear before v_{i+1}, $0 \leq i < k - 1$, imposed by the edge $e_{i,i+1} \in p_c$. That implies v_0 comes before v_{k-1}, but then edge $e_{k-1,0}$ does not comply with the condition of the topological order. Consequently, no topology order exists for the vertices of p_c. Since no topology order can be found for the vertices of p_c, no topology order exists for G. $\square$

The vertices of a directed acyclic graph can be sorted into topological order by a simple DFS-based algorithm, which is outlined in Algorithm 4 (Cormen et al. [42]). As soon as a vertex is marked finished in the DFS (see DFS-Visit, Algorithm 3), it is inserted onto the front of a list and upon termination of DFS, the list holds the topologically ordered vertices.

Algorithm 4   Topological-Sort(G)

    Execute DFS(G) with the following addition:

     Insert each vertex of G onto the front of a list L as soon as it is marked finished

     Return L

The correctness of Algorithm 4 can be verified by the following considerations. It is ensured by the main part of DFS (Algorithm 2) that every vertex is discovered; therefore, every vertex is eventually marked as finished and inserted onto the front of the list. When a vertex u is marked as finished, none of the vertices already in the list can have an edge to u. If there were such a vertex v in the list L with $e_{vu} \in \mathbf{E}$, u would be adjacent to v and DFS-Visit would have been recursively called for u before v was marked finished. That is, u would have finished before v, which is a contradiction.

Given that Topological-Sort(G) is based on DFS, its runtime complexity is $O(\mathbf{V} + \mathbf{E})$.

3.2 GRAPH AS A PROGRAM MODEL

Sinnen and Sousa [173] classify some of the well-known graph theoretic models for the representation of parallel computations. They extracted several characteristics that are shared among most of the graph models. Premised on their findings, a general graph model is defined and its properties are analyzed in the next paragraphs.

Definition 3.7 (Graph Model) *In a graph theoretic abstraction, a program consists of two kinds of activity—computation and communication. The computation is associated with the vertices of a graph and the communication with its edges. A vertex is called* node *and the computation associated with it* task. *A task can range from an atomic instruction/operation (i.e., an instruction that cannot be divided into smaller instructions) to threads or compound statements such as loops, basic blocks, and sequences of these. All instructions or operations of one task are executed in sequential order; there is no parallelism within a task. A node is at any time involved in only one activity—computation or communication.*

Note that the granularity of the general graph model is not restricted in any way, leaving this to the individual graph models considered in the next sections. For instance, in the program examples of Section 2.5, one simple mathematical operation (e.g., summation or multiplication) is considered a task. An essential property of the nodes is their strictness.

Definition 3.8 (Node Strictness) *The nodes are strict with respect to both their inputs and their outputs: that is, a node cannot begin execution until all its inputs have arrived, and no output is available until the computation has finished and at that time all outputs are available for communication simultaneously.*

Branching None of the graph models covered in this chapter incorporates a branching concept. In other words, there is no node type deciding which of the successors participates in the program execution. An example is a branch node that represents an `if...else` statement, shown in Figure 3.6: the `if` node has two edges—one representing the *true* and the other the *false* branch.

In any execution of the program represented by the entire graph, the `if` node only leads to the subsequent execution of one of the two subgraphs spanned by the branches. This concept is used in control flow graphs (CFGs) (Wolfe [204]) for modeling control dependence and its flow. Not allowing branching has two consequences: (1) all nodes

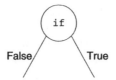

Figure 3.6. An if...else statement represented as a branch node.

of a graph participate in the execution of the program; and (2) conditional statements cannot directly be reflected by the graph structure. Thus, either the control structure of a program must be encapsulated within a node, which potentially leads to larger nodes (see granularity in Section 2.4.1), or it must be transformed into data dependence structure as addressed in Section 2.5.3.

A graph adhering to the above definitions reflects with its node and edge structure a program's decomposition into tasks and their communication structure. As mentioned earlier, such a graph is not capable of representing the control dependence, but only the data dependence structure. In contrast, data dependence is very well described by the structure of such a graph. One of the most prominent graph models is the dependence graph that exposes all types of data dependence, studied in Section 2.5.1.

3.2.1 Computation and Communication Costs

Many utilization areas of program graph models require knowledge about the computation and communication costs of the nodes and edges, respectively, unless it can be assumed they are uniform. Commonly, these costs are measured in time—the time a computation or communication takes on the respective target system—and are incorporated into the graph model by assigning weights to the graph elements. In terms of the general graph model of Definition 3.7 this can be defined as follows.

Definition 3.9 (Computation and Communication Costs)   *Let $G = (\mathbf{V}, \mathbf{E}, w, c)$ be a graph representing a program $\mathcal{P}$ according to Definition 3.7. The nonnegative weight $w(n)$ associated with node $n \in \mathbf{V}$ represents its computation cost and the nonnegative weight $c(e)$ associated with edge $e \in \mathbf{E}$ represents its communication cost.*

Examples for graph models employing such weights are some flow graphs (Section 3.4) and the task graph (Section 3.5). Further weights might be associated with the graph elements; for example, in the flow graph each edge has a weight representing its delay.

3.2.2 Comparison Criteria

Having defined a common principle for graph models implies that there are characteristics that allow one to differentiate them. Sinnen and Sousa [174] present several criteria for analyzing and distinguishing the various models. These criteria are based on similar aspects as considered during the discussion of subtask decomposition in Section 2.4. This is not surprising given the fact that the creation of a graph involves the decomposition of the program into tasks and the knowledge of their communication relations. The three main criteria are:

- *Computation Type.* What type of computation can be efficiently represented with the graph model? Graph models can be distinguished regarding the modeled granularity and the ability to reflect iterative computations and regularity.
- *Parallel Architectures.* For which parallel system architecture is the graph model usually employed? As mentioned in Chapter 2, generic parallel systems can be classified according to their instruction and data stream type, their memory architecture, and their programming model.
- *Algorithms and Techniques.* What are the algorithms and techniques that can be applied to the graph model? Typically, graphs are used for dependence analysis, program transformations, mapping, and scheduling.

Apart from these main topics, practical issues like the computer representation and the size of the graphs are also discussed by Sinnen and Sousa [174].

During the following review of the individual graph models, these criteria will be helpful when explaining the motivation for the respective graph model and discussing its distinctive features. The first graph to be examined is the dependence graph, which is intuitively derived from the dependence analysis in Section 2.5.

3.3 DEPENDENCE GRAPH (DG)

Recall the discussion of dependence in Section 2.5, where directed dependence relations are analyzed between tasks. Intuitively, one can represent the programs' tasks as nodes and the dependence relations as edges, corresponding to the graph model defined in the previous section. Consider the program in Example 1 of Section 2.5, which contains flow dependence relations. Representing every task, in this case every statement, by a node and every dependence relation by an edge directed from the cause to the dependent task, a graph reflecting the dependence structure is obtained. Figure 3.7 repeats the program code of Example 1 and depicts the corresponding dependence graph. From this graph one intuitively perceives that tasks 2 and 3 can be executed concurrently.

Next, the dependence graph is defined formally.

Definition 3.10 (Dependence Graph (DG)) *A dependence graph is a directed graph $DG = (\mathbf{V}, \mathbf{E})$ representing a program $\mathcal{P}$ according to the graph model of Definition 3.7. The nodes in $\mathbf{V}$ represent the tasks of $\mathcal{P}$ and the edges in $\mathbf{E}$ the dependence relations between the tasks. An edge $e_{ij} \in \mathbf{E}$ from node n_i to n_j, $n_i, n_j \in \mathbf{V}$, represents the dependence of node n_j on node n_i.*

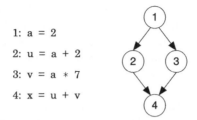

1: a = 2

2: u = a + 2

3: v = a * 7

4: x = u + v

Figure 3.7. The code from Example 1 and the corresponding dependence graph.

A dependence graph DG reflects the dependence structure of a program $\mathcal{P}$, which imposes a partial order of the nodes (i.e., the tasks) in an execution. Every edge $e_{ij} \in DG$ constrains the precedence of n_i over n_j in the execution order of the nodes. Graphically, one can imagine that the nodes have to be executed following the edges—the execution of a program progresses with the edge direction, never against it. Note that the node strictness (Definition 3.8) accurately reflects the nature of dependence: a node can only start if all of its entering dependence relations are fulfilled.

Definition 3.11 (Feasible Program)   *A program $\mathcal{P}$ is feasible if and only if a task order can be found that complies with the precedence constraints of the dependence relations.*

Clearly, the feasibility of a computation is indicated by its dependence graph.

Lemma 3.4 (DG of Feasible Program Is Acyclic)   *A program $\mathcal{P}$ is feasible if and only if its dependence graph DG is acyclic.*

Proof. This lemma can easily be proved by contradiction using Lemma 3.3 (topological order and directed graphs).

$\Rightarrow$: Suppose DG is cyclic. According to Lemma 3.3, no topological order exists for the nodes of a cyclic graph. But then, for any order of the nodes, at least two nodes are not in precedence order. As the nodes represent the tasks of $\mathcal{P}$, also no precedence order exists for them—$\mathcal{P}$ is not feasible.

$\Leftarrow$: Suppose $\mathcal{P}$ is not feasible, thus no order of the tasks can be found that complies with the precedence constraints. Consequently, no topological order exists for DG. Then, however, DG is cyclic according to Lemma 3.3. $\square$

According to the graph model of Definition 3.7, the communication of a program is represented by the edges of a graph. A further level of abstraction is introduced by the DG, as only the dependence relations caused by communication are considered. From this it is clear that the dependence reflected by a DG is the data dependence of a program. As discussed in Section 2.5.1, there are three types of data dependence: flow dependence, antidependence, and output dependence. All these types can be represented and distinguished by the DG, albeit often the mere fact that a dependence

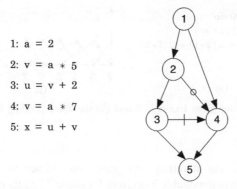

```
1: a = 2
2: v = a * 5
3: u = v + 2
4: v = a * 7
5: x = u + v
```

Figure 3.8. The code from Example 2 and the corresponding dependence graph. A line through the arrow of an edge denotes antidependence and a circle through the arrow denotes output dependence; a plain arrow stands for flow dependence.

exists is relevant. For distinction in a pictorial representation, the edge of an antide-pendence is sometimes drawn as an arrow with a line crossing through and the edge of an output dependence as an arrow with a circle through it (Banerjee [15], Wolfe [204]). A flow dependence edge is denoted by a plain arrow. Figure 3.8 shows a DG that uses this kind of notation for the program of Example 2: task 4 is antidependent on task 3 and also output dependent on task 2. All other dependence relations are flow, or real, dependences.

3.3.1 Iteration Dependence Graph

A typical utilization area for dependence graphs is in compilers. The DG gives a compiler a way to capture the precedence constraints that prevent it from reordering operations in the program. Parallelizing compilers usually focus on the paralleliza-tion of loops, as they commonly accommodate the largest share of computational load. A logical specialization of the DG is therefore the iteration dependence graph, reflecting dependence relations in loops. The theoretical background for the transition from a simple DG to an iteration dependence graph is given in Section 2.5.2, where dependence relations in loops were analyzed. Reconsider the loop in Example 5, here shown in Figure 3.9(*a*). A graphical representation of the dependence relations—the iteration dependence graph—is given in Figure 3.9(*b*).

Each instance of the statements S and T is modeled as a task and thus a node of the graph, while the edges represent the dependence relations between these instances. In Figure 3.9(*b*), a node is drawn as a dot in the coordinate system spanned by the statements and the iterations of the loop. The dependence distance vector corresponds to the spatial vector drawn in the graph illustration, whereby the vertical dimension is for the distinction between the tasks. The short arrows going bottom–up reflect the uniform dependence and the arrows going top–down reflect the nonuniform depen-dence between instances of the tasks S and T (see also Section 2.5.2). Due to the nonuniform dependence between instances of tasks S and T, the graph is irregular.

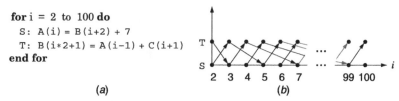

```
for i = 2 to 100 do
  S: A(i) = B(i+2) + 7
  T: B(i*2+1) = A(i-1) + C(i+1)
end for
```

(a)   (b)

Figure 3.9. (*a*) The loop from Example 5 and (*b*) the corresponding iteration dependence graph.

In a loop nest, the iteration DG gains one dimension for every loop. Figure 3.10(*a*) displays the double loop from Example 7 and its iteration DG. Here, the approach differs from the one in Figure 3.9, since each node of the graph encapsulates all statements of one iteration: that is, the entire loop kernel is one task. A more precise decomposition of the loop body into a task for every statement, as done with the iteration DG in Figure 3.9, would add one more dimension to the graph. It depends on the context in which the graph is generated if this is necessary and/or desirable. Note that if the entire loop body is modeled as one task, the graph is only able to reflect interiteration dependence. Certain parallelization techniques, for example, software pipelining (e.g., Aiken and Nicolau [8]), are based on both intra- and interiteration dependence.

In Section 2.5.2, the distance vectors $(1, 1)$ and $(1, 0)$ were identified for the dependence relations in the code of Figure 3.10(*a*): $S(i + 1, j + 1)$ depends on $T(i, j)$, and $T(i + 1, j)$ depends on $S(i, j)$. As with the single loop of Figure 3.9(*a*), the arrows in the iteration DG correspond directly to the dependence distance vectors; but recall that the two statements of the kernel are merged into one task, that is, one node in the graph. The uniform dependence relations of the loop create a regular iteration DG.

For the construction of iteration dependence graphs, it is obvious that knowledge of the distance vectors of all dependence relations is crucial. In other words, if the distance vector cannot be determined for every single dependence relation, the complete

```
for i = 0 to 5 do
  for j = 0 to 5 do
    S: A(i+1,j) = B(i,j) + C(i,j)
    T: B(i+1,j+1) = A(i,j) + 1
  end for
end for
```

(a)   (b)

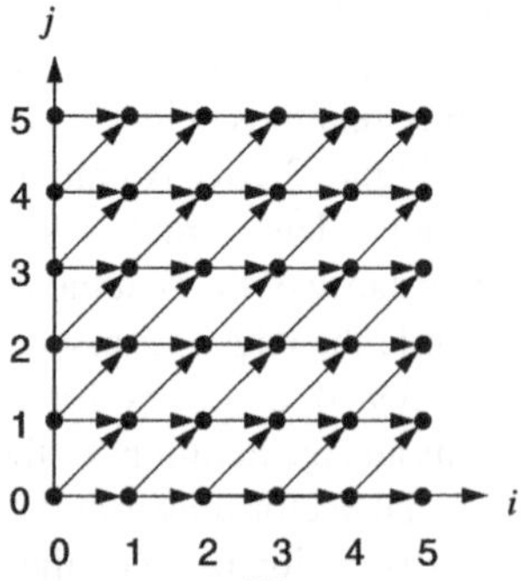

Figure 3.10. (*a*) The double loop from Example 7 and (*b*) the corresponding iteration dependence graph.

iteration DG cannot be constructed. This fact is noteworthy, especially in conjunction with the observation made at the end of Section 2.5.2: the known techniques of dependence analysis can only compute—in reasonable time—the distance vectors for dependence relations based on simple, albeit common, types of subscript functions. Not to forget that it is sometimes even impossible to determine the dependence relations, for example, when array subscripts are functions of input variables of the program. The only solution is then to create a dependence edge for every possible dependence relation in order to obtain a conservative dependence graph, that is, a graph that includes all edges that the true dependence graph would have.

3.3.2 Summary

As the name indicates, the dependence graph is a representation of the dependence structure of a program. In its general form, as in Definition 3.10, the DG is not bound to any particular computation type or granularity. This general theoretical form is widely employed, whenever it is necessary to reason about the dependence structure of a program, the feasibility of computations, or the validity of program transformations (see the literature referenced in Section 2.5.1). In practical usage, the DG is limited to a coarse grained or partial representation of a program, because its size (and with it its computer representation) grows with the number of tasks.

The dependence structure of cyclic computations is represented by the iteration dependence graph, where every node of the graph is associated with coordinates of the iteration space spanned by the index variables of the loop nest. For dependence analysis of loops and, above all, for loop transformations, the iteration DG is a valuable instrument. Banerjee et al. [17] survey many of the existing techniques. Although several of them do not require the explicit construction of the graph, the underlying theory of these techniques is based on the dependence graph. Granularity in loops is typically small to medium, which is therefore the common granularity of iteration dependence graphs. The size issue of the general DG is evaded by benefitting from the regular iterative structure of the computation. A compact computer representation, given that the dependence distances are uniform, comprises only the tasks of the loop kernel, the intraiteration dependence relations, and the distance vectors of the interiteration dependence relations.

Iteration DGs are employed not only in parallelizing compilers but also in VLSI array processor design (Kung [109]). There, the iteration DG, constructed from an initial application specification—typically signal processing—serves as an intermediate representation for the mapping and scheduling of the application to array processors. Those graphs possess two distinctive attributes: (1) a node represents one iteration, not only a part of it; and (2) the dependence relations are local, that is, the absolute values of the distance vectors' elements are typically 1 or 0 (Karp et al. [101]). The mapping and scheduling process transforms the iteration DG into a flow graph, which is treated in the next section.

While the dependence graph has no affinity for any particular parallel architecture, the iteration dependence graph, owing to its regular iterative structure, is usually employed for homogeneous systems, with SIMD or SPMD streams.

3.4 FLOW GRAPH (FG)

The iteration dependence graph, as discussed in the previous section, is not the only graph model for the representation of cyclic computations. Intuitively, a model for such computations can benefit from the inherent regular structure, as insinuated during the discussion of the computer representation of an iteration dependence graph in the previous section. It seems to be sufficient for a model to reflect the tasks of one iteration with their intra- and interiteration communications.[1]

The flow graph (FG) achieves a concise representation of the structure of an iterative computation by introducing timing information into the graph, which allows one to distinguish between intra- and interiteration communication. This is illustrated with Example 10.

Example 10   Flow Graph
```
for i = 1 to N do
   S:  A(i)   = C(i) * D(i+2)
   T:  B(i+1) = A(i) + 10
   U:  C(i+2) = A(i) + B(i)
end for
```

Analyzing the data dependence of its loop (Section 2.5.2), the following communications can be identified. Considering the three statements S, T, and U as tasks, there are two intraiteration communications—the output of S is the input of T and U, caused by $A(i)$—and two interiteration communications—the output of $T(i)$ is the input of $U(i + 1)$ (array B) and the output of $U(i)$ is the input of $S(i + 2)$ (array C). The flow graph for this computation is then as depicted in Figure 3.11.

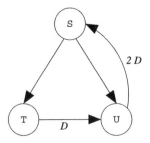

Figure 3.11. Flow graph of code in Example 10.

---

[1] Due to the close connection between dependence and communication, many concepts of dependence can be directly applied to the treatment of communication. Therefore, those concepts will be employed when appropriate for communication without further formalization, unless it is deemed ambiguous.

The graph consists of three nodes for the tasks S, T, and U, and four edges for the communications between these tasks. The edges, which reflect the communications between nodes, are drawn as arrows as usual, whereby an interiteration communication is distinguished by a label showing its communication distance, that is, dependence distance (see Section 2.5.2). As is often done in the literature (e.g., Parhi [145], Yang and Fu [208]), the communication distance, which is the delay of the communication in terms of iterations, is expressed as multiples of D, where D stands for one iteration. So edge e_{TU} has a delay of D for the communication distance of one iteration, and edge e_{US} has a delay of $2D$, that is, two iterations. Before examination of this graph model's properties, it is defined formally.

Definition 3.12 (Flow Graph) *A flow graph is a directed graph $FG = (\mathbf{V}, \mathbf{E}, D)$ representing a program $\mathcal{P}$ of an iterative computation according to the graph model of Definition 3.7. The nodes in $\mathbf{V}$ represent the tasks of $\mathcal{P}$ and the edges in $\mathbf{E}$ the communications between the tasks. An edge $e_{ij} \in \mathbf{E}$ from node n_i to n_j, $n_i, n_j \in \mathbf{V}$, represents a communication from node n_i to node n_j. Each edge $e \in \mathbf{E}$ is associated with a nonnegative integer weight $D(e)$, representing a delay count.*

During execution of the program $\mathcal{P}$, every task represented by the nodes of FG is executed once in each iteration. Only when all nodes have finished their execution, can a new iteration begin. In distinction from the data-driven execution model outlined later, this form of execution is called iteration-driven. No communications or dependences exist in $\mathcal{P}$ other than those reflected by the edges of FG. The delay $D(e)$ associated with each edge reflects the number of iterations a communication is delayed between its output from the origin node until its input into the destination node. Communication between nodes within one iteration (i.e., intraiteration communication) has consequently the delay value 0. In the sample graph of Figure 3.11, the edges e_{SU} and e_{ST} have zero delay, which by convention is omitted in the illustration.

It is possible for a flow graph to have parallel edges, that is, distinct edges that have the same origin and destination node. Graphs with this property are sometimes called multigraphs (Cormen et al. [42]). Parallel edges in the flow graph must differ, however, on their delay value, by which they can thus be distinguished. Figure 3.12 exhibits an example program whose flow graph contains parallel edges. Two communications go from task S to task T: one intraiteration (via $A(i)$) and one interiteration (via $A(i-1)$) with a delay of 1.

```
for i = 1 to N do
    S: A(i) = C(i) * D(i)
    T: B(i) = A(i) + A(i-1)
end for
```

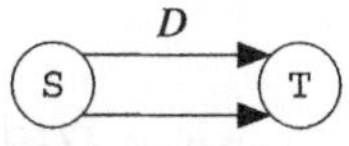

Figure 3.12. An iterative computation whose flow graph has parallel edges.

The flow graph *FG* does not contain information about the number of iterations. This also means that the number of iterations does not have to be known when constructing the graph, as is the case, for instance, with a loop whose index bound is a variable (*N* in Example 10).

It should be mentioned, that all communications must be uniform, in order to represent their communication distance as one integer value. Nonuniform communication, as found in the loop of Figure 3.9, cannot be expressed as one integer value. Yet, a flow graph can still be constructed in such cases using a conservative approximation of the distance, for example, "+1" to denote an unknown dependence distance greater than or equal to one. For certain loop parallelization algorithms, such information can be sufficient. If the accurate representation of the dependence structure is indispensable, a flow graph can only be used with uniform communication. This establishes an important limitation of the flow graph, which relates to the representable computation types.

In comparison to the DG or the iteration DG (Section 3.3), the flow graph has two essential characteristics. First, nodes in the flow graph represent tasks that can be executed several times, and not instances of tasks as in the DG, where each represented node is executed exactly once. Second, the flow graph is, in general, not acyclic. Cycles can arise in connection with interiteration dependence, but in contrast to the DG, the represented computation is still feasible.

When reading the code of Example 10, one concludes that it is a feasible program, but in fact its FG in Figure 3.11 contains two cycles—$\langle S, U, S \rangle$ and $\langle S, T, U, S \rangle$. The flow graph breaks a limitation imposed on the dependence graph: in contrast to the DG, a flow graph can include cycles. As expressed by Lemma 3.4, a DG must be acyclic in order to represent a feasible computation. So, how can a flow graph, based on the same general graph model of Definition 3.7, be a valid representation of a feasible program? According to Definition 3.8 (node strictness), the nodes must be strict regarding their input and output. Even though an FG models the data flow among nodes, and not as the DG the dependence relations, communications among nodes create (flow) data dependence relations (see Section 2.5.1). In other words, the edges implicitly represent dependence relations; thus, a communication cycle would lead to a dependence cycle.

The reason an FG remains feasible, despite the cycles, lies in the introduction of delays on the edges, which prevent cycles in the graph from turning into dependence cycles. In the flow graph, every cycle must contain at least one delayed edge, breaking the dependence chain. Seen the other way around, paths in flow graphs are only closed to cycles by interiteration communications, which by definition are delayed.

Lemma 3.5 (Feasible Flow Graph) *A flow graph $FG = (\mathbf{V}, \mathbf{E}, D)$ represents a feasible iterative computation $\mathcal{P}$ if and only if any cycle p_c in FG contains at least one edge e with a nonzero delay $D(e) \neq 0$:*

$$\forall p_c \in FG \, \exists e \in \mathbf{E} : e \in p_c \land D(e) \neq 0. \tag{3.3}$$

Proof. ⇒: Suppose *FG* contains at least one cycle whose edges all have zero delay. This corresponds to a cycle in the *DG*, which, however, must be acyclic to represent a feasible program (Lemma 3.4).

⇐: The program's feasibility is shown by demonstrating that the *DG* corresponding to the flow graph has no cycles (Lemma 3.4). One node in *DG* represents one *instance* of a node in *FG*; that is, *DG* consists of *i*-times the nodes of *FG*, therefore $|\mathbf{V}_{DG}| = i \times |\mathbf{V}_{FG}|$, with $\mathbf{V}_{DG}$ and $\mathbf{V}_{FG}$ being the node sets of *DG* and *FG*, respectively, and *i* the number of iterations. (Figure 3.13 shows the *DG* of the flow graph in Figure 3.11 assuming three iterations—$N = 3$ in the underlying program of Example 10—and each iteration comprises the nodes S, T, and U.) Communication edges in *FG* with zero delay $\{e \in \mathbf{E}_{FG} : D(e) = 0\}$, correspond to dependence edges in *DG* between the nodes of the same iteration; thus, there are $i \times |\{e \in \mathbf{E}_{FG} : D(e) = 0\}|$ edges of this type. (These are three times the edges e_{SU} and e_{ST} in Figure 3.13.) As every cycle in *FG* contains at least one nonzero edge, there cannot be a cycle in *DG* among the nodes of one iteration. Edges with nonzero delay are directed from an earlier to a later iteration, never the other way around, because the weight is nonnegative, $D(e) \geq 0 \forall e \in \mathbf{V}_{FG}$. (The graph of Figure 3.13 has three such edges: $e_{T(1)U(2)}$, $e_{T(2)U(3)}$, and $e_{U(1)S(3)}$.) This means that the entering edges of a node always have their origin nodes in the same or a previous iteration, while leaving edges have their destination node in the same or a subsequent iteration. But then there is also no dependence cycle in *DG* across iterations, as no path can return to its origin node. □

The technique employed in the above proof—creating a dependence graph from a flow graph—is called unrolling or unfolding (Parhi [144]). With the inverse technique, called projection, a flow graph can be obtained from an iteration DG (Kung [109]). Both techniques, which show the close relation of the graph models, will be considered in the following section.

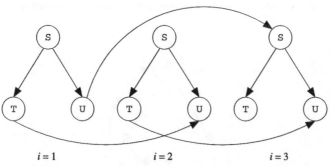

$i = 1$ \qquad\qquad $i = 2$ \qquad\qquad $i = 3$

Figure 3.13. Dependence graph, created by unrolling the flow graph of Figure 3.11 for three iterations ($N = 3$ in Example 10).

3.4.1 Data-Driven Execution Model

A communication edge of a flow graph can also be interpreted as a communication between two nodes via an intermediate queue. The communication data is written by the origin node of the communication edge into the queue from where it is read by the destination node. A delay arises when other data items, often designated tokens (Kung [109]), already populate the queue at the time a new item is placed in the queue, since these items are read before the new one, due to the FIFO (first in first out) characteristic of the queue. When the execution of the flow graph starts, the queues of the edges with a nonzero delay contain initial data items, or tokens, whose number is given by the delay count $D(e)$ of the edges. For the code of Example 10, these initial data items correspond to the initial values of the array elements of B and C. This view of a delayed communication also clarifies how nodes remain strict regarding their input and output (Definition 3.8), while at the same time cycles are feasible: on delayed edges, input is provided to a node by the queue, allowing the node to start execution independently of the status of the edge's origin node in the current iteration. As an example, consider Figure 3.14, where the same flow graph of Figure 3.11 is depicted, except now with the queue and token based interpretation. The places in the queues (which are shown in finite number in this figure) are illustrated by lines through the edges' arrows and the initial tokens by dots, namely, for one token on edge e_{TU} and two on e_{US}.

With the introduction of queues and tokens, the state of the computation is at any time reflected by the distribution of the tokens among the queues of the edges. A node is enabled to start execution, to "fire," if each input edge contains a positive number of tokens and each output edge has at least one space in the queue—that is, the node is strict (Definition 3.8). Before starting execution, the node removes a token from every input edge and puts a token on every output edge after finishing.

The token model can be considered a data-driven execution model, as the flow of the tokens invokes the execution of a node without the need for synchronization. Communication between nodes is performed asynchronously with a self-timed execution of the nodes triggered by the flow of the tokens. Moreover, the strict distinction between iterations is also not necessary due to the self-timed execution.

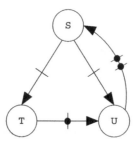

Figure 3.14. The flow graph of Figure 3.11 with queues and tokens (initial data items) on the edges; a line through an arrow denotes a place in a queue and a dot denotes an initial data item.

One of the earliest flow graphs for parallel computations is the computation graph (CG) (Karp and Miller [100], Reiter [162]). Each edge has a FIFO queue without restriction of its length and a weight is associated with each edge indicating the initial number of data words in the queue. The token concept is also used in data flow languages (Ackerman [1], Davis and Keller [53]), a graphical approach to software development. Data flow program graphs allow the programming of data flow computers, since both share the self-timed and data-driven nature. The data flow graph (DFG) (Kung [109], Parhi [145]) benefits particularly from the hardware-oriented view of the data-driven execution model, as it is used in VLSI array processor design. Enhanced with two additional weight types, compared to the simple flow graph, the DFG also reflects the computation time of the nodes and the capacity of the FIFO queues. Restricting the queue capacity can lead to deadlocks, which is of real concern in VLSI array processors. DFGs are used for the mapping of parallel computations on VLSI wavefront array processors (Kung [109]).

The data-driven execution model is a valuable mechanism whenever an accurate view of the data flow in space and time is required, as, for example, in hardware-oriented parallelization. For many purposes, however, such a detailed view of the data flow is not necessary and the iteration-driven execution model is appropriate and sufficient. In any case, the flow graph defined in Definition 3.12 is suitable for a data-driven view as soon as the edge delays $D(e)$ are considered initial tokens (Kung [109]), as described earlier. The iteration-driven execution model simply presupposes a queue mechanism for the delayed communication that is sufficiently large, and therefore no further consideration is needed.

3.4.2 Summary

The flow graph is an efficient representation of iterative computations, owing to the fact that repetitive parts of computation and communication are modeled only once. On the other side, the flow graph is limited to cyclic computations with usually uniform communication relations (nonuniform communication relations can only be approximated). The granularity of the flow graph is basically inherited from the iterative computation type and is hence fine to medium grained.

Many parallelization techniques for iterative programs are based on the program's representation as a flow graph with or without computation and/or communication costs: unfolding, retiming, software pipelining, mapping, and scheduling, just to name a few. A flow graph with computation and communication costs is simply defined as $FG = (\mathbf{V}, \mathbf{E}, D, w, c)$ (see Definitions 3.9 and 3.12). The flow graphs used by Parhi and Messerschmitt [144, 146] for some of these techniques are called iterative data-flow programs and include computation costs, that is, node weights (w). Yang and Fu [208] employ a graph called the iterative task graph (ITG), featuring both computation (w) and communications costs (c), which is called a communication sensitive data flow graph (CS-DFG) by Tongsima et al. [185]. The signal flow graph (SFG) (Kung [109], Parhi [145]), the counterpart of the DFG (see earlier data-driven execution model), is used for synchronous and uniform computations mapped and scheduled on VLSI systolic array processors, for which consequently neither a data-driven execution model nor the modeling of the computation of communication costs is necessary.

3.5 TASK GRAPH (DAG)

This section is dedicated to the task graph, which is used for task scheduling. With the understanding of graph models built in the previous sections, it is straightforward to comprehend and analyze the task graph. During the following discussion, the relationships between the graph models will be developed, including transformation techniques.

In the literature, the task graph is often simply referred to as DAG, which merely describes the graph theoretic properties of the model, namely, that it is a directed acyclic graph. In order to avoid ambiguity with other directed acyclic graphs, the name *task graph* is used in this text.

The discussion starts with the definition of the task graph.

Definition 3.13 (Task Graph (DAG)) *A task graph is a directed acyclic graph $G = (\mathbf{V}, \mathbf{E}, w, c)$ representing a program $\mathcal{P}$ according to the graph model of Definition 3.7. The nodes in $\mathbf{V}$ represent the tasks of $\mathcal{P}$ and the edges in $\mathbf{E}$ the communications between the tasks. An edge $e_{ij} \in \mathbf{E}$ from node n_i to $n_j, n_i, n_j \in \mathbf{V}$, represents the communication from node n_i to node n_j. The positive weight $w(n)$ associated with node $n \in \mathbf{V}$ represents its computation cost and the nonnegative weight $c(e_{ij})$ associated with edge $e_{ij} \in \mathbf{E}$ represents its communication cost.*

During the execution of the program $\mathcal{P}$, every represented task (i.e., node) is executed exactly once. It is presupposed that the program $\mathcal{P}$ is a feasible program, hence the acyclic property of the graph (see Lemma 3.4). No communications or dependences exist in $\mathcal{P}$ other than those reflected by the edges of G. The task graph features node and edge weights representing the costs of computation and communication, respectively, with the notable difference that the node weight is here defined as positive rather than nonnegative as in Definition 3.9. This small difference is important for certain properties of the task graph, as will be seen in Section 4.4.

Figure 3.15 illustrates a sample task graph for a small fictitious program; the nodes are named by the letters from a to k, while the node and edge weights are noted beside the respective graph elements. This sample graph will be employed as an example throughout the following chapters.

Some of the early scheduling algorithms and those used under restricted conditions (see Chapter 4) employ simplified task graphs, inasmuch as communication and sometimes computation costs are neglected. In that sense, the here defined task graph is a general model, which can be used in these algorithms by either ignoring the computation and/or communication costs or setting them to zero.

The edges of the task graph only reflect flow data dependence (remember, communications establish flow dependence relations). If other dependence relations existed in

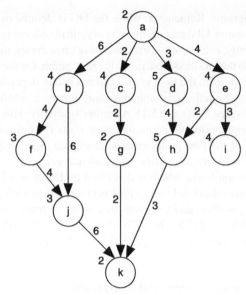

Figure 3.15. The task graph for a fictitious program. Nodes are named by letters *a–k*; node and edge weights are noted beside them.

a preliminary version of the program $\mathcal{P}$, they have been eliminated before the construction of the task graph G (the elimination of output and antidependence is explained in Section 2.5.1). For some practical purposes, the difference between the three data dependence types (i.e., flow dependence, antidependence, and output dependence) is not relevant. In such cases, a task graph can be defined, whose edges can represent all types of dependences. The only difference from a DG is then the computation and communication costs of the task graph. However, the definition of "communication costs" for output dependence and antidependence becomes very problematic. Moreover, output dependence and antidependence are provoked by variable reuse. In a distributed system without shared memory, this problem does not exist when the respective tasks are executed on different processors. For these reasons and in order to have one general model, the defined task graph here only represents flow dependences.

The task graph is a general graph model not bound to a particular computation type; that is, it can reflect iterative and noniterative computations. Still, like the DG, it is not adapted to cyclic computations in any way, making the graph size, in terms of nodes, proportional to the number of iterations. That is why the task graph is commonly used for coarse grained, noniterative computations—hence its name task graph. It shares another limitation of the DG with respect to the number of iterations: if the number depends on the input of the program, the task graph cannot be constructed for the general case. In return, the task graph is not limited to uniform interiteration communications as is the (approximation-free) flow graph.

From the definition of the task graph, several similarities with the dependence graph and the flow graph can be observed. The task graph inherits the topology of the

DG for feasible programs. Remember, while the DG is defined as a directed graph (Definition 3.10), Lemma 3.4 demands that it is acyclic in order to represent a feasible program. Consequently, every node is executed only once during the execution of $\mathcal{P}$, which corresponds to the DG model but not to the flow graph. On the other hand, edges reflect communication, as in the flow graph, and not solely dependence relations, as in the DG. Obviously, the reflected communications represent implicitly the real data dependence relations (see Section 2.5.1). A further similarity with the flow graph is the inclusion of computation and communication costs (Definition 3.9), which are contained in several of the flow graph models mentioned (Section 3.4).

The three considered graph models—dependence graph, flow graph, and task graph—share a common basis, which is described by Definition 3.7. As mentioned earlier, there are many additional similarities between the models and it is thus not surprising that it is possible, and common, to convert or transform one model into another (Sinnen and Sousa [173, 174]). The next section will outline some of these techniques, which also helps to gain a general view of graph models.

3.5.1   Graph Transformations and Conversions

In the subsequent discussion, a rough distinction will be made between transformations and conversions. Those techniques that require the creation or removal of nodes shall be called transformations and those preserving the original nodes shall be called conversions.

The discussion starts with the conversions, which are sometimes little more than a different interpretation of the given graph model.

Task Graph to DG   The conversion of the task graph into a DG is very simple, given that the dependence relations are implicitly expressed through the communication edges. Hence, this conversion is rather an interpretation of the task graph as a DG by ignoring the node and edge weights. It should be noted that the type of data dependence reflected by the edges is *flow*, or real, dependence. Other dependence relations do not exist by definition of the task graph. For this reason, the opposite conversion, from DG to task graph, is in general not valid, because dependence graphs can comprise all types of dependence.

Extracting Iterative Kernel—Flow Graph to Task Graph   Even though the task graph is not designed specifically for the representation of cyclic computations like the flow graph, it proves useful for the representation of the kernel of the iterative computation—the loop body. If one iteration consists of more than one task, surely a task graph can be constructed to represent the iteration's tasks and their communications. The same task graph can also be obtained by converting the flow graph that represents the entire iterative computation. The simple conversion is performed by removing all edges from the graph that have nonzero delays, that is, $\{e \in \mathbf{E} : D(e) \neq 0\}$: in other words, all edges that represent interiteration communications. Obtained are the nodes connected only by the intraiteration communication

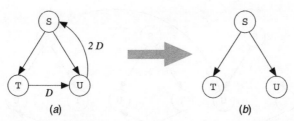

Figure 3.16. (*a*) The flow graph of Figure 3.11 is converted into the task graph of the loop kernel (*b*) by removing the delayed edges.

edges, which is the task graph of the iterative kernel. As an example, this conversion is performed with the sample flow graph of Figure 3.11 of the previous section, shown here in Figure 3.16(*a*). By removing the nonzero communication edges, the graph illustrated in Figure 3.16(*b*) is obtained. The correctness of the task graph is verified by analyzing the code of the represented program of Example 10: the output of S ($A(i)$) is the input of T and U creating the two intraiteration communications.

Using a task graph for modeling the iterative kernel is successfully employed for the scheduling of cyclic computations (Sandnes and Megson [164], Sandnes and Sinnen [166], Yang and Fu [208]). The major advantage is that once the iterative computation is represented by a task graph, task scheduling, as discussed in the following chapters, can be utilized. Note that the number of iterations does not need to be known for this technique.

Unrolling—Flow Graph to Task Graph An alternative technique to create a task graph from a flow graph is unrolling or unfolding, which was already used for the proof of Lemma 3.5. Unrolling, according to the above loose definition, is a transformation since it creates a new graph. In contrast to the previous conversion, not only is the loop kernel represented by the task graph but the entire iterative computation is. It follows that the number of iterations must be known at the time of construction—a condition that is not always fulfilled at compile time, since the number of iterations can depend on the input of the computation. Recall that the flow graph is constructed independently of the number of iterations (Section 3.4).

To construct the unrolled task graph, a graph is created consisting of the respective nodes and edges of the loop-body task graph for every iteration. A node is identified by its name in the flow graph and the iteration index to which it belongs. Those nodes and edges already reflect all task instances of the iterative computation and their intraiteration communications. The remaining interiteration communication edges are added to the graph by considering each node of the new graph and adding edges for its leaving interiteration communications, as long as the iteration index of the destination node is part of the computation. If the considered node n is of iteration i, the edge e is created between n and the respective destination node of iteration $i + D(e)$, unless $i + D(e) > N$ for $1 \leq i \leq N$, with N being the number of iterations. The node and edge weights representing the computation and communication costs, respectively, are set to the values of the corresponding flow graph elements.

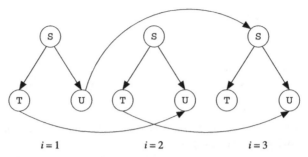

Figure 3.17. The unrolled flow graph of Figure 3.16(*a*) for three iterations.

Figure 3.17 shows the unrolling example of Section 3.4, where the flow graph of Figure 3.16(*a*) is unrolled for three iterations (i.e., $N = 3$) in the underlying code of Example 10. One clearly sees the three distinct kernel task graphs for the three iterations and the three interiteration communications. There is no leaving edge from node U in the second iteration, since the potential destination node (in iteration 4) is not part of the computation. The same holds for the interiteration communications of the nodes of the third iteration.

A graph constructed from a flow graph without the attribution of weights to the nodes and edges can be interpreted as the dependence graph—reflecting only flow data dependences—of the iterative computation. This close relationship between task graph and DG was examined earlier. More precisely, the DG obtained by unrolling a flow graph is the iteration DG of the computation, comprising only uniform dependence relations.

Sometimes the unrolling is done only for a fraction of the total number of iterations. This partial unrolling, which is sufficient for some purposes (Sandnes and Megson [164], Yang and Fu [208]), has two advantages: (1) the total number of iterations does not need to be known and (2) the size of the unrolled graph, in terms of the number of nodes, is not proportional to the number of iterations. However, the resulting graph remains a flow graph; it is not a task graph. For this reason, partial unrolling is at times employed as a prestage to the extraction of the iterative kernel as described earlier (e.g., Yang and Fu [208]).

Projection—Iteration DG to Flow Graph   The countertechnique to unrolling is the projection of an iteration DG to a flow graph (Kung [109]). Multiple equal nodes of the iteration DG (i.e., nodes that represent the same type of task) are projected into one node of the flow graph. An illustrative example is the projection of a two-dimensional iteration DG into a flow graph, reducing the iteration DG by one dimension. Figure 3.18 shows such a projection for the two-dimensional iteration DG of Figure 3.10 along the *i*-axis, resulting in the depicted flow graph.

Essentially, the projection is performed by merging all nodes along the projection direction into one node and by transforming the communication edges into new edges with delays. A delay substitutes the spatial component of the distance vector of the

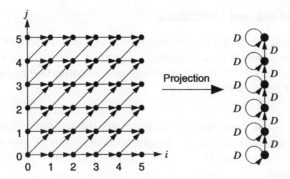

Figure 3.18. The iteration DG of Figure 3.10 is projected along the i-axis into a flow graph.

edge that is parallel to the direction of the projection. In other words, a spatial dimension of the distance vectors is transformed into a temporal one. In the example, the i-dimension is transformed into the temporal dimension of the delays.

The projection in the above example is linear along one of the axes of the iteration DG. In general, the projection is not required to be along an axis, in fact, it is not even required to be linear (Kung [109]). However, an inherent iterative structure must be present in the DG; otherwise the computation cannot be described as a flow graph, so normally a general DG cannot be transformed into a flow graph.

A common application of projection is in VLSI array processor design (Kung [109]), where the iteration DG often serves as an initial model to obtain the flow graph, which is a description of the application closer to the hardware level.

The conversions and transformations demonstrate the close relationships of the various graph models. To conclude and summarize the discussion of these techniques, the relationships are illustrated in Figure 3.19. Shown are the three major graph models—DG, flow graph, and task graph—linked by the conversions and transformations discussed in this section.

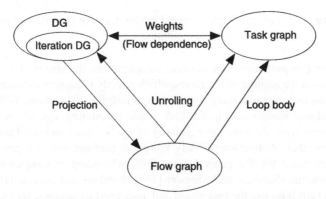

Figure 3.19. The three major graph models and the relationships among them.

3.5.2 Motivations and Limitations

This subsection discusses the motivations for the adoption of the task graph model for task scheduling, which is accompanied by a critical discussion of its limitations. Basically, this goal is achieved by summarizing and comparing the principal properties of the presented graph models.

Motivations So far, this chapter has demonstrated that the abstraction of a program as a graph can very well capture the dependence and communication structure of a program. The task graph model has several properties that make it particularly suitable for task scheduling.

- *General.* It is desirable that the graph model is as general as possible, in terms of the computation types, and the task graph is such a model. In comparison, the flow graph is restricted to iterative computations with uniform communications (otherwise it is not accurate).
- *Simple.* The task graph's focus on communications, which correspond to real data dependence, permits task scheduling to concentrate on these essential precedence constraints. Other dependences reflected in the DG can be eliminated and are not inherent to the represented computation.
- *Modeling of Computation and Communication Costs.* Scheduling algorithms for modern parallel systems must be aware of the computation and communication costs. It is therefore crucial that they are represented in the graph model; hence, a DG does not suffice.
- *Close Relationship to Other Models.* The previous section outlined the various relationships between the discussed models. In connection with a conversion or transformation, techniques and algorithms based on the task graph can be employed for other models.

Limitations The task graph as a general model does not provide any mechanism to efficiently represent an iterative computation.

- *Iterative Computations.* For iterative computations, the size of the task graph depends on the number of iterations, which directly influences the memory consumption and the processing time of task scheduling algorithms. With the loss of regularity information in the task graph, scheduling algorithms also cannot benefit from the inherent regularity of cyclic computations. Furthermore, if the number of iterations is only known at runtime, the task graph cannot be constructed for the general case. Still, scheduling techniques for cyclic computations (Sandnes and Megson [164], Sandnes and Sinnen [166], Yang and Fu [208]) do use the task graph and associated techniques, for example, to represent the iterative kernel.

Another limitation is not a particular limitation of the task graph, but of all models covered in this chapter. In fact, it was already introduced during the definition of the general graph model in Section 3.2.

- *Static Model.* The graph models according to Definition 3.7, to which the task graph model belongs, do not exhibit conditional statements of the code; that is, there is no branching. These control dependences are either transformed into data dependences or encapsulated within a node (see Section 3.2).

3.5.3 Summary

The task graph is the graph model of choice for task scheduling. It clearly exhibits the task and communication structure of a program, while also reflecting the computation and communication costs. Its properties were summarized in the previous section, when analyzing the motivations for the model's choice and its limitations. The general nature of the task graph, together with the typically coarse granularity of the represented tasks, indicates the adequacy of employing the task graph for distributed parallel systems with SPMD or MIMD streams.

Chapter 4 returns to the task graph model after establishing a basic understanding of the task scheduling problem on parallel systems. It is there that the task graph model and its properties are examined further in the context of task scheduling.

Also, Chapter 4 addresses the computation and communication costs associated with the nodes and the edges of the task graph, respectively. Until now, the node and edge weights were introduced only as abstract notations of computation and communication costs. Evidently, such costs are related to the target parallel system on which the program represented by the task graph is executed. Thus, it is necessary to define the target parallel system model, before the concept of costs in the task graph can be substantiated.

3.6 CONCLUDING REMARKS

This chapter presented and analyzed in depth the three major graph models for the representation of computer programs: dependence graph, flow graph, and task graph. It started with the fundamental concepts of graph theory and then formulated a common foundation for graph models representing computer programs. With this background, the three models and their properties were discussed.

While this chapter serves as an introduction to graph models for program representation, its objective was to introduce and analyze the task graph model of task scheduling. This broad approach was deemed crucial in order to establish a comprehensive understanding of the task graph. It was shown that the task graph model inherits the structure of the dependence graph and many of its properties. Furthermore, the flow graph, as a concise representation of iterative computation, was related to the task graph by means of transformations and conversions, discussed in

Section 3.5.1. At the end, the motivations for the task graph model were presented and its limitations were analyzed.

The chapter's focus on the task graph resulted in the omission of several other, less common, graph models, some of which are analyzed by Sinnen and Sousa [173]:

- *Task Interaction Graph* (*TIG*) (Stone [182]). An undirected graph model only capturing the communication relations between entire processes as opposed to tasks. The TIG is used for mapping of processes onto parallel processors (e.g., Sadayappan et al. [163], Stone [182]).
- *Temporal Communication Graph* (*TCG*) (Lo [129]). A model that integrates the task and process oriented view of a parallel program. The TCG is based on the space–time diagram introduced by Lamport [116] and can be interpreted for mapping and scheduling algorithms as a TIG as well as a task graph (Lo et al. [130]).
- *Control Flow Graph* (*CFG*) *and Control Dependence Graph* (*CDG*) (Allen and Kennedy [12], Wolfe [204]). These two closely related directed graph models represent the control flow and control dependence of a program. As a consequence, they reflect conditional execution paths (branching, see Section 3.2), as opposed to all other graph models discussed in this chapter. Both are used in compilers to analyze and handle the control flow of a program.

For the fundamental problem of integrating iterative and noniterative computations into one efficient graph model, the idea of a hierarchical graph (Sinnen and Sousa [175]) emerged, where a node of a higher level can represent an entire subgraph.

With the end of this chapter, the foundation for the discussion of task scheduling in the next chapter has been laid.

3.7 EXERCISES

3.1 Section 3.1 reviews basic graph concepts for undirected and directed graphs. Consider the following directed graph $G = (\mathbf{V}, \mathbf{E})$ consisting of 6 vertices and 9 edges:

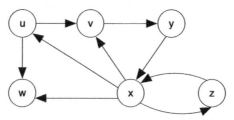

(a) Which vertex has the highest degree?

(b) What is the average in-degree of the vertices? What is the average out-degree?

   (c) Is vertex y an ancestor or a descendant of vertex x?

   (d) Is vertex v reachable from vertex w? Is vertex w reachable from vertex v?

3.2 Paths and cycles are important and powerful concepts in graphs (Section 3.1). Consider again the directed graph of Exercise 3.1.

   (a) How many simple cycles has the graph? Specify them.

   (b) What is the length of the longest simple path, in terms of the number of edges, in this graph? Specify a path with such a length.

3.3 Commonly, there are two different approaches to the representation of a graph in a computer (Section 3.1.1). Consider again the directed graph of Exercise 3.1.

   (a) Give an adjacency matrix representation of this graph.

   (b) Give an adjacency list representation of this graph.

3.4 The topological order is an important concept for directed acyclic graphs (Section 3.1.2). Find a topological order for the following directed acyclic graph $G = (\mathbf{V}, \mathbf{E})$ consisting of 10 vertices:

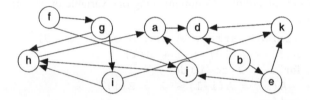

Which vertices are source (entry) vertices and which are sink (exit) vertices?

3.5 Construct the dependence graph for the code fragment of Exercise 2.9.

3.6 Construct the iteration dependence graph for the double loop of Exercise 2.11.

3.7 Construct the flow graph (Section 3.4) for the following iterative computation:

```
for i = 1 to n do
   S:  A(i) = B(i) - 1
   T:  B(i+1) = (A(i-1)-A(i))/2
   U:  C(2 * i) = B(i-1) + E(i+2)
   V:  D(i) = C(2 * i-2) + B(i+1)
end for
```

3.8 Construct a task graph (Section 3.5) for the code below. Each line shall be represented by one task, named by its line number, and the costs shall be assumed as follows:

   • *Computation*. Assignment alone: 1 unit; add/subtract operation: 2 units; multiply operation: 3 units; divide operation: 4 units.

   • *Communication*. Communicating a variable with a small letter and with a capital letter costs 1 unit and 2 units, respectively (imagine variables with capital letters to have higher precision).

```
1: a = 56
2: b = a * 10 + 2
3: C = (b - 2) / 3
4: D = 91.125
5: E = D * a
6: F = D * b + 1
7: g = 11 + a
8: H = (E + F) * g
```

3.9 Construct a task graph (Section 3.5) for the code below. Take alternatively a coarse grained or a fine grained approach and make clear which part of the code is represented by which node of the task graph. It might be useful to convert control dependence into data dependence (Section 3.2). The costs shall be assumed as follows:

- *Computation.* Assignment alone: 1 unit; add/subtract operation: 2 units; multiply operation: 3 units; function call: 25 units; variable comparison: 3 units (conditional execution: 1 unit).

- *Communication.* Communicating one variable or one array element costs 1 unit.

```
1: {Input: arrays A and B of size 10}
2: A(1) = A(10) + 2 * B(1)
3: for i = 2 to 10 do
4:    A(i) = A(i-1) + 2 * B(i)
5: end for
6: x = function(A)
7: y = function(B)
8: if x > y then
9:    result = x - y
10: else
11:    result = y - x
12: end if
```

3.10 Given is the following flow graph $FG = (\mathbf{V}, \mathbf{E}, D)$ consisting of 6 nodes:

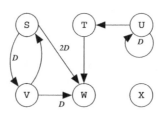

Extract the task graph representing the iterative kernel (Section 3.5.1). To simplify, assume that all computation and communication costs are of unit size and can therefore be neglected.

3.11 Unrolling is a technique used to transform generally cyclic flow graphs into acyclic task graphs (Section 3.5.1). Consider again the flow graph of Exercise 3.10. Unroll this flow graph into a task graph for 4 iterations. To simplify, assume that all computation and communication costs are of unit size and can therefore be neglected. Name the nodes of the task graph using their original flow graph name and a suffix indicating the iteration number (e.g., S.2 for node S of iteration 2).

3.12 As discussed in Section 3.5.1, unrolling has two disadvantages: (1) the number of iterations must be known and (2) the size of the resulting task graph grows with the number of iterations. Partial unrolling is a technique that avoids both problems by only unrolling the graph for a small, fixed number of iterations, which is lower than the total number of iterations. Consider again the flow graph of Exercise 3.10.

   (a) Partially unroll this flow graph for 2 iterations; that is, the total number of iterations N is higher than 2. It can be assumed that the number of total iterations N is always an even number. Remember, the resulting graph remains a flow graph.

   (b) Extract the task graph representing the iterative kernel (Section 3.5.1) of the unrolled flow graph (see also Exercise 3.10).

3.13 The purpose of Exercises 3.5 – 3.9 was to construct a graph from a given code fragment. In this exercise it is the other way around: the task is to write a simple code fragment that corresponds to a given graph. The code should only consist of array variables, assignments, and add operations, while the number of iterations is given as N. Initialization of variables can be ignored. Write a simple code fragment whose flow graph is identical to:

   (a) The flow graph of Exercise 3.10.

   (b) The partially unrolled flow graph of Exercise 3.10.

3.14 In Exercise 3.11 the purpose was to unroll a flow graph into a task graph. This exercise is about reconstructing the original flow graph from an unrolled task graph (costs are neglected). Let the acyclic directed graph below be an unrolled flow graph for 4 iterations.

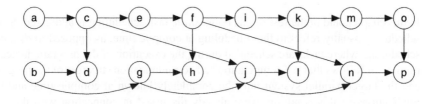

Reconstruct the original flow graph.

Task Scheduling

This chapter introduces the general problem of task scheduling for parallel systems. It formulates the static task scheduling problem and discusses its theoretic background. Hence, the chapter's aim is to introduce the reader to task scheduling and to lay the groundwork for the following chapters. This is done by providing a uniform and consistent conceptual framework.

The scheduling problem defined in this chapter is based on a strongly idealized model of the target parallel system. It has been shown that this model, referred to as the *classic model*, is often not sufficient to obtain schedules with high accuracy and short execution times. More sophisticated models are therefore discussed in Chapters 7 and 8. Yet, most algorithms found in the literature have been proposed for the classic model. In order to build a comprehensive understanding of task scheduling, it is thus indispensable to analyze and discuss this model and its algorithms. What is more, most of the techniques and algorithms can be employed for the more advanced models studied in later chapters.

The chapter starts with terminology and basic definitions for task scheduling and formulates the scheduling problem. Its associated decision problem is NP-complete and a corresponding proof is provided. Subsequently, task scheduling without communication costs is studied as a special case of the general problem. Again, the complexity is discussed, including the NP-completeness of this problem. The chapter then returns to the task graph model to analyze its properties in connection with task scheduling.

4.1 FUNDAMENTALS

In this chapter, and in the rest of this book, *static* task scheduling is addressed. Static scheduling usually refers to the scheduling at compile time, as opposed to *dynamic* scheduling, where tasks are scheduled during the execution of the program, hence at runtime. It follows that the computation and communication structure of the program and the target parallel system must be completely known at compile time and the implications of this condition were already discussed in connection with the task graph model in the previous chapter (Section 3.5.2).

Task Scheduling for Parallel Systems, by Oliver Sinnen
Copyright © 2007 John Wiley & Sons, Inc.

The discussion in this chapter is oriented toward what can be considered the general scheduling problem: no restrictions are imposed on the input task graph—it may have arbitrary computation and communication costs as well as an arbitrary structure—and the number of processors is limited.

Another important, less general scheduling problem does not consider the costs of communication. Historically, the investigation of scheduling without communication delays preceded the consideration of communication costs in scheduling. Nevertheless, this chapter will start with the general problem that includes communication delays. This has the advantage that scheduling without communication delays can easily be treated afterwards as a special case of the general problem. This improves the understanding and interrelation of the task scheduling problems and also reduces the number of necessary definitions. Furthermore, the more advanced task scheduling discussed in later chapters is based on scheduling with communication delays.

Several other scheduling problems arise by restricting the task graph, for example, by having unit computation or communication costs, or by employing an unlimited number of processors. Again, these problems can be treated as special cases of the general problem and they are discussed later in Section 6.4.

Basics With the preparations of the previous chapter, the notion of a schedule and associated conditions can be defined right away. A system of tasks together with their precedence constraints is already defined by the task graph model.

In Section 2.3, when the parallelization process was outlined, the scheduling problem was introduced as the spatial and temporal assignment of tasks to processors. The spatial assignment, or mapping, is the allocation of tasks to the processors.

Definition 4.1 (Processor Allocation) *A processor allocation $\mathcal{A}$ of the task graph $G = (\mathbf{V}, \mathbf{E}, w, c)$ (Definition 3.13) on a finite set $\mathbf{P}$ of processors is the processor allocation function proc: $\mathbf{V} \to \mathbf{P}$ of the nodes of G to the processors of $\mathbf{P}$.*

The temporal assignment is the attribution of a start time to each task. Even though scheduling refers only to the attribution of start times, it presupposes the allocation of the tasks to processors and therefore commonly both are defined by a schedule.

Definition 4.2 (Schedule) *A schedule S of the task graph $G = (\mathbf{V}, \mathbf{E}, w, c)$ on a finite set $\mathbf{P}$ of processors is the function pair $(t_s, proc)$, where*

- *$t_s: \mathbf{V} \to \mathbb{Q}_0^+$ is the start time function of the nodes of G.*
- *proc: $\mathbf{V} \to \mathbf{P}$ is the processor allocation function of the nodes of G to the processors of $\mathbf{P}$.*

So the two functions t_s and *proc* describe the spatial and temporal assignment of tasks, represented by the nodes of the task graph, to the processors of a target parallel system, represented by the set $\mathbf{P}$. The node $n \in \mathbf{V}$ is scheduled to start execution at

$t_s(n)$ on processor $proc(n) = P$, $P \in \mathbf{P}$, which is denoted by $t_s(n, P)$; hence,

$$t_s(n, P) \Leftrightarrow t_s(n), proc(n) = P, \ P \in \mathbf{P}. \tag{4.1}$$

4.2 WITH COMMUNICATION COSTS

Definition 4.2 describes a schedule, but it does not ensure the compliance with the precedence constraints of the task graph. Conditions for this will be established shortly, but first the model of the target parallel system must be defined.

Definition 4.3 (Target Parallel System—Classic Model) *A target parallel system* $\mathbf{P}$ *consists of a set of identical processors connected by a communication network. This system has the following properties:*

1. Dedicated System. *The parallel system is dedicated to the execution of the scheduled task graph. No other program or task is executed on the system while the scheduled task graph is executed.*

2. Dedicated Processor. *A processor $P \in \mathbf{P}$ can execute only one task at a time and the execution is not preemptive.*

3. Cost-Free Local Communication. *The cost of communication between tasks executed on the same processor, local communication, is negligible and therefore considered zero. This assumption is based on the observation that for many parallel systems remote communication (i.e., interprocessor communication) is one or more orders of magnitude more expensive than local communication (i.e., intraprocessor communication).*

4. Communication Subsystem. *Interprocessor communication is performed by a dedicated communication subsystem. The processors are not involved in communication.*

5. Concurrent Communication. *Interprocessor communication in the system is performed concurrently; there is no contention for communication resources.*

6. Fully Connected. *The communication network is fully connected. Every processor can communicate directly with every other processor via a dedicated identical communication link.*

What kind of parallel systems does this model represent (Section 2.1)? Given the identical processors and the fully connected network of identical communication links, the system is completely homogeneous. The characteristic of expensive remote communication in comparison with local communication is typical for NUMA or message passing architectures, where the memory is distributed across the processors. Other parallel systems are also reflected by this model; essential is the characteristic of expensive remote communication. For example, on a UMA system remote communication can become expensive due to cache effects. Two tasks running on the same processor may communicate via the processor's cache, while they have to communicate through the main memory when they run on different processors. Another cause for expensive remote communication in a UMA system can be the employment of the message passing programming paradigm.

Based on this model, the meaning of the computation and communication costs of the nodes and edges in a task graph, respectively, can be defined. The weights of the task graph elements were introduced as abstract costs in Definition 3.13.

Definition 4.4 (Computation and Communication Costs) *Let* $\mathbf{P}$ *be a parallel system. The computation and communication costs of a task graph* $G = (\mathbf{V}, \mathbf{E}, w, c)$ *expressed as weights of the nodes and edges, respectively, are defined as follows:*

- $w: \mathbf{V} \to \mathbb{Q}^+$ *is the computation cost function of the nodes* $n \in \mathbf{V}$. *The computation cost* $w(n)$ *of node* n *is the time the task represented by* n *occupies a processor of* $\mathbf{P}$ *for its execution.*
- $c: \mathbf{E} \to \mathbb{Q}_0^+$ *is the communication cost function of the edge* $e \in \mathbf{E}$. *The communication cost* $c(e)$ *of edge* e *is the time the communication represented by* e *takes from an origin processor in* $\mathbf{P}$ *until it completely arrives at a different destination processor in* $\mathbf{P}$. *In other words,* $c(e)$ *is the* communication delay *between sending the first data item until receiving the last.*

The finish time of a node is thus the node's start time plus its execution time (i.e., its cost).

Definition 4.5 (Node Finish Time) *Let* S *be a schedule for task graph* $G = (\mathbf{V}, \mathbf{E}, w, c)$ *on system* $\mathbf{P}$. *The finish time of node* n *is*

$$t_f(n) = t_s(n) + w(n). \tag{4.2}$$

Again, $t_f(n, P)$ is written to denote $t_f(n)$ and $proc(n) = P$.

Because of Property 2 of the system model, it must be ensured that no two nodes occupy one processor at the same time.

Condition 4.1 (Exclusive Processor Allocation) *Let* S *be a schedule for task graph* $G = (\mathbf{V}, \mathbf{E}, w, c)$ *on system* $\mathbf{P}$. *For any two nodes* $n_i, n_j \in \mathbf{V}$:

$$proc(n_i) = proc(n_j) \Rightarrow \begin{cases} t_s(n_i) < t_f(n_i) \leq t_s(n_j) < t_f(n_j) \\ or \quad t_s(n_j) < t_f(n_j) \leq t_s(n_i) < t_f(n_i) \end{cases}. \tag{4.3}$$

In contrast to the execution time of a task in the target system, which is simply the cost of the respective node, the communication time of an edge e_{ij} depends on the processors of the origin and destination node. As Property 3 states, local communication, that is, communication between tasks on the same processor, is cost free; hence, it takes no time at all—there is no delay. The time of remote communication, that is, between tasks on different processors, is given by the weight of the representing edge. The time at which a communication arrives at the destination processor is defined as the edge finish time.

Definition 4.6 (Edge Finish Time)   *Let $G = (V, E, w, c)$ be a task graph and P a parallel system. The finish time of $e_{ij} \in E$, $n_i, n_j \in V$, communicated from processor P_{src} to P_{dst}, $P_{src}, P_{dst} \in P$, is*

$$t_f(e_{ij}, P_{src}, P_{dst}) = t_f(n_i, P_{src}) + \begin{cases} 0 & \text{if } P_{src} = P_{dst} \\ c(e_{ij}) & \text{otherwise} \end{cases}. \qquad (4.4)$$

So, the arrival time of e_{ij} at processor P_{dst} is given by the finish time of node n_i, that is, the time when the data for e_{ij} is made available by n_i (node strictness, Definition 3.8), plus the communication time of e_{ij}, either local (0) or remote ($c(e_{ij})$). Following Properties 4–6 of the system model, a communication e_{ij} is always initiated as soon as its origin node n_i finishes: there is no delay due to any form of contention.

Finally, a condition can be formulated that warrants the compliance of the schedule S with the precedence constraints of the task graph G.

Condition 4.2 (Precedence Constraint)   *Let S be a schedule for task graph $G = (V, E, w, c)$ on system P. For $n_i, n_j \in V$, $e_{ij} \in E$, $P \in P$,*

$$t_s(n_j, P) \geq t_f(e_{ij}, proc(n_i), P). \qquad (4.5)$$

Condition 4.2 formulates the consequences of a precedence constraint imposed by an edge $e_{ij} \in E$ on the start time of the destination node $n_j \in V$ in schedule S. The node strictness obliges n_j to delay its start until the communication of e_{ij} has completely arrived at processor P.

Condition 4.2 guarantees the execution of the nodes in precedence order.

Lemma 4.1 (Feasible Start Order Is Precedence Order)   *Let S be a schedule for task graph $G = (V, E, w, c)$ on system P. If Condition 4.2 is fulfilled for all edges $e \in E$ then any ordering of the nodes $n \in V$ in ascending start time complies with the precedence constraints of G.*

Proof. Condition 4.2 guarantees that for any edge $e_{ij} \in E$, $n_i, n_j \in V$, $t_s(n_j) \geq t_f(n_i)$, since $t_f(e_{ij}, proc(n_i), proc(n_j)) - t_f(n_i) \geq 0$ (Definitions 4.6 and 4.4). With Definition 4.5 of the node finish time, $t_s(n_j) \geq t_f(n_i) = t_s(n_i) + w(n_i) > t_s(n_i)$, since $w(n_i) > 0$ (Definition 4.4). Hence, in an ascending start time order of the nodes, the origin node n_i of e_{ij} appears before the destination node n_j. Per definition this is a topological order (Definition 3.6), thus a precedence order. $\square$

The two Conditions 4.1 and 4.2 are the only constraints put on a schedule. A very noteworthy consequence is the possible overlap of computation and communication.

While the leaving communications of a node are sent over the network to their destinations, the node's processor can proceed with computation.

A schedule, whose nodes comply with these two conditions, respects the precedence constraints of the task graph and the properties of the system model and such a schedule is called *feasible*.

Definition 4.7 (Feasible Schedule) *Let S be a schedule for task graph $G = (V, E, w, c)$ on system P. S is feasible if and only if all nodes $n \in V$ and edges $e \in E$ comply with Conditions 4.1 and 4.2.*

The feasibility of a schedule can easily be verified, as demonstrated by Algorithm 5. The complexity of verifying Condition 4.1 is $O(V \log V)$ (e.g., with Mergesort, Cormen et al. [42]) and the complexity of verifying Condition 4.2 is $O(V + E)$, as every edge is only considered once, resulting in the total complexity of $O(V \log V + E)$.

Algorithm 5   *Verify Feasibility of Schedule S for $G = (V, E, w, c)$ on P*

  ▷ *Verify Condition 4.1*
  for each $P \in \mathbf{P}$ **do**
    Sort $\{n \in V: proc(n) = P\}$ according to $t_s(n)$. Let sorted nodes be $n_{P,0}, n_{P,1}, \ldots, n_{P,l}$.
    Verify that $t_s(n_{P,i}) < t_f(n_{P,i}) \leq t_s(n_{P,i+1}) < t_f(n_{P,i+1})$ for $i = 0, \ldots, l - 1$.
  end for
  ▷ *Verify Condition 4.2*
  for each $n_j \in V$ **do**
    Verify that $t_s(n_j) \geq t_f(e_{ij}, proc(n_i), proc(n_j)) \; \forall \, e_{ij} \in E, n_i \in \mathbf{pred}(n_j)$.
  end for

From now on, all considered schedules are feasible, unless stated otherwise, and the attribute "feasible" is therefore omitted.

Condition 4.2 requires node n_j to wait until all entering communications $e_{ij} \in E, n_i \in \mathbf{pred}(n_j)$ have arrived at its executing processor. The earliest time when node n_j can start execution is called the *data ready time*.

Definition 4.8 (Data Ready Time (DRT)) *Let S be a schedule for task graph $G = (V, E, w, c)$ on system P. The data ready time of a node $n_j \in V$ on processor $P \in P$ is*

$$t_{dr}(n_j, P) = \max_{n_i \in \mathbf{pred}(n_j)} \{t_f(e_{ij}, proc(n_i), P)\}. \tag{4.6}$$

If $\mathbf{pred}(n_j) = \emptyset$, that is, n_j is a source node, $t_{dr}(n_j) = t_{dr}(n_j, P) = 0$, for all $P \in P$.

Note that the DRT can be determined for any processor P of P, independently of the processor to which n_j is allocated. The constraints on the start time $t_s(n)$ of a node n can now be formulated using the DRT.

Condition 4.3 (DRT Constraint) *Let S be a schedule for task graph $G = (\mathbf{V}, \mathbf{E}, w, c)$ on system $\mathbf{P}$. For $n \in \mathbf{V}, P \in \mathbf{P}$,*

$$t_s(n, P) \geq t_{dr}(n, P). \tag{4.7}$$

Thus, Condition 4.3 merges the constraints imposed on the start time of a node by all entering edges according to Eq. (4.5).

The finish time of a processor P is the time when the last node scheduled on P terminates.

Definition 4.9 (Processor Finish Time) *Let S be a schedule for task graph $G = (\mathbf{V}, \mathbf{E}, w, c)$ on system $\mathbf{P}$. The finish time of processor $P \in \mathbf{P}$ is*

$$t_f(P) = \max_{n \in \mathbf{V}: proc(n) = P} \{t_f(n)\}. \tag{4.8}$$

A schedule terminates when the last of the nodes of the task graph G finishes.

Definition 4.10 (Schedule Length) *Let S be a schedule for task graph $G = (\mathbf{V}, \mathbf{E}, w, c)$ on system $\mathbf{P}$. The schedule length of S is*

$$sl(S) = \max_{n \in \mathbf{V}} \{t_f(n)\} - \min_{n \in \mathbf{V}} \{t_s(n)\}. \tag{4.9}$$

If $\min_{n \in \mathbf{V}} \{t_s(n)\} = 0$, this expression reduces to

$$sl(S) = \max_{n \in \mathbf{V}} \{t_f(n)\}. \tag{4.10}$$

All schedules considered in this text start at time unit 0; thus, expression (4.10) suffices as the definition of the schedule length. An alternative designation for schedule length, which is quite common in the literature, is makespan.

The set of processors used in a given schedule S is defined as the set of all processors on which at least one node is executed.

Definition 4.11 (Used Processors) *Let S be a schedule for task graph $G = (\mathbf{V}, \mathbf{E}, w, c)$ on system $\mathbf{P}$. The set of used processors is*

$$\mathbf{Q} = \bigcup_{n \in \mathbf{V}} proc(n). \tag{4.11}$$

Obviously, for any schedule S

$$|\mathbf{Q}| \leq |\mathbf{P}|. \tag{4.12}$$

To conclude the basic definitions, the sequential time of a task graph is defined.

Definition 4.12 (Sequential Time) *Let $G = (\mathbf{V}, \mathbf{E}, w, c)$ be a task graph. G's sequential time is*

$$seq(G) = \sum_{n \in \mathbf{V}} w(n),\tag{4.13}$$

which corresponds to G's execution time on one processor only (remember, local communication is cost free).

4.2.1 Schedule Example

The above definitions and conditions are illustrated by examining a sample schedule. Figure 4.1(*a*) depicts a Gantt chart of a schedule for the sample task graph of Figure 3.15 on three processors. A Gantt chart is an intuitive and common graphical representation of a schedule, in which each scheduled object (i.e., node) is drawn as a rectangle. The node's position (i.e., the position of its rectangle) in the coordinate system spanned by the time and space axis (i.e., processor axis) is determined by the node's allocated processor and its start time, while the size of the rectangle reflects the node's execution time.

Some examples from this figure illustrate the foregoing definitions and conditions. The start time of node *a* executed on processor P_1 is $t_s(a) = 0$ (i.e., $t_s(a, P_1) = 0$) and, with a computation cost of $w(a) = 2$, its finish time is $t_f(a) = t_f(a, P_1) = 2$.

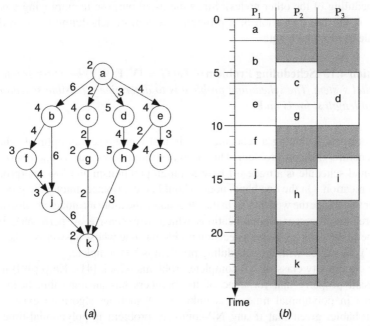

(a)   (b)

Figure 4.1. The Gantt chart (*b*) of a schedule for the sample task graph (*a*) of Figure 3.15 on three processors.

Node d begins at $t_s(d) = 5$ and finishes at $t_f(d) = t_s(d) + w(d) = 5 + 5 = 10$. Its start time $t_s(d) = 5$ is the earliest possible, because it has to wait for the data from node a, $t_f(e_{ad}, P_1, P_3) = t_f(a) + c(e_{ad}) = 2 + 3 = 5$. So in this case, the communication is remote and therefore causes a delay of 3 time units corresponding to the weight of edge e_{ad}. Node b, on the other hand, also receives data from node a, but as the communication is local—both a and b are on the same processor—it takes no time: $t_s(b) = t_f(e_{ab}, P_1, P_1) = t_f(a) = 2$. A good example for the influence of various precedence constraints on the start time of a node is node h. The two entering communications e_{dh} and e_{eh} result in a data ready time of $t_{dr}(h, P_2) = 14$. Responsible is the communication from node d, since d finishes at $t_f(d) = 10$ and the communication of e_{dh} lasts 4 time units, while e_{eh} arrives at P_2 at $t_f(e_{eh}, P_1, P_2) = 10 + 2 = 12$. So node h starts at its DRT. In contrast, node e starts at $t_s(e) = 6$, which is later than its DRT $(t_{dr}(e, P_1) = t_f(e_{ae}, P_1, P_1) = t_f(a) = 2)$, because processor P_1 is busy with node b until that time. The length of the schedule is $sl = t_f(k) = 24$, that is, the finish time of the last node k, as the first node a starts at time unit 0.

4.2.2 Scheduling Complexity

The process of creating a schedule S for a task graph G on a set of processors $\mathbf{P}$ is called *scheduling*. It should be obvious from the example in Figure 4.1 that there is generally more than one possible schedule for a given graph and set of processors (e.g., a could be executed on processor P_2, which of course would have consequences for the scheduling of the other nodes). Since the usual purpose in employing a parallel system is the fast execution of a program, the aim of scheduling is to produce a schedule of minimal length.

Definition 4.13 (Scheduling Problem) *Let $G = (\mathbf{V}, \mathbf{E}, w, c)$ be a task graph and $\mathbf{P}$ a parallel system. The scheduling problem is to determine a feasible schedule S of minimal length sl for G on $\mathbf{P}$.*

Unfortunately, finding a schedule of minimal length (i.e., an optimal schedule) is in general a difficult problem. This becomes intuitively clear as one realizes that an optimal schedule is a trade-off between high parallelism and low interprocessor communication. On the one hand, nodes should be distributed among the processors in order to balance the workload. On the other hand, the more the nodes are distributed, the more interprocessor communications, which are expensive, are performed. In fact, the general decision problem (a decision problem is one whose answer is either "yes" or "no") associated with the scheduling problem is NP-complete.

The complexity class of NP-complete problems (Cook [41], Karp [99]) has the unpleasant property that for none of its members has an algorithm been found that runs in polynomial time. It is unknown if such an algorithm exists, but it is improbable, given that if any NP-complete problem is polynomial-time solvable then all NP-complete problems are polynomial-time solvable (i.e., $P = NP$). Introductions to NP-completeness, its proof techniques, and the discussion of many

NP-complete problems can be found in Coffman [37], Cormen et al. [42], and Garey and Johnson [73].

Theorem 4.1 (NP-Completeness) *Let* $G = (\mathbf{V}, \mathbf{E}, w, c)$ *be a task graph and* $\mathbf{P}$ *a parallel system. The decision problem* SCHED$(G, \mathbf{P})$ *associated with the scheduling problem is as follows. Is there a schedule* $\mathcal{S}$ *for* G *on* $\mathbf{P}$ *with length* $sl(\mathcal{S}) \leq T, T \in \mathbb{Q}^+$? SCHED$(G, \mathbf{P})$ *is NP-complete in the strong sense.*

Proof. First, it is argued that SCHED belongs to NP, then it is shown that SCHED is NP-hard by reducing the well-known NP-complete problem 3-PARTITION (Garey and Johnson [73]) in polynomial time to SCHED. 3-PARTITION is NP-complete in the strong sense. The reduction is inspired by a proof presented in Sarkar [167].

The 3-PARTITION problem is stated as follows. Given a set $\mathbf{A}$ of $3m$ positive integer numbers a_i and a positive integer bound B such that $\sum_{i=1}^{3m} a_i = mB$ with $B/4 < a_i < B/2$ for $i = 1, \ldots, 3m$, can $\mathbf{A}$ be partitioned into m disjoint sets $\mathbf{A}_1, \ldots, \mathbf{A}_m$ (triplets) such that each $\mathbf{A}_i$, $i = 1, \ldots, m$, contains exactly 3 elements of $\mathbf{A}$, whose sum is B?

Clearly, for any given solution $\mathcal{S}$ of SCHED$(G, \mathbf{P})$ it can be verified in polynomial time that $\mathcal{S}$ is feasible (Algorithm 5) and $sl(\mathcal{S}) \leq T$; hence, SCHED$(G, \mathbf{P}) \in$ NP.

From an arbitrary instance of 3-PARTITION $\mathbf{A} = \{a_1, a_2, \ldots, a_{3m}\}$, an instance of SCHED$(G, \mathbf{P})$ is constructed in the following way.

A so-called fork, or send, graph $G = (\mathbf{V}, \mathbf{E}, w, c)$ is constructed as shown in Figure 4.2. It consists of $|\mathbf{V}| = 3m + 1$ nodes $(n_0, n_1, \ldots, n_{3m})$, where node n_0 is the entry node with $3m$ successors. Hence, one edge goes from n_0 to each of the other nodes of G and there are no other edges apart from these. The edge weight of the $3m$ edges is $c(e) = \frac{1}{2} \forall e \in \mathbf{E}$, node n_0 has weight $w(n_0) = 1$, and the $3m$ nodes have weights that correspond to the integer numbers of $\mathbf{A}$, $w(n_i) = a_i, i = 1, \ldots, 3m$. The number of processors of the target system is $|\mathbf{P}| = m$ and the time bound is set to $T = B + 1.5$.

Clearly, the construction of the instance of SCHED is polynomial in the size of the instance of 3-PARTITION.

It is now shown how a schedule $\mathcal{S}$ is derived for SCHED$(G, \mathbf{P})$ from an arbitrary instance of 3-PARTITION $\mathbf{A} = \{a_1, a_2, \ldots, a_{3m}\}$, that admits a solution to 3-PARTITION: let $\mathbf{A}_1, \ldots, \mathbf{A}_m$ (triplets) be m disjoint sets such that each $\mathbf{A}_i$, $i = 1, \ldots, m$, contains exactly 3 elements of $\mathbf{A}$, whose sum is B.

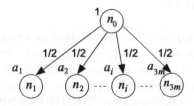

Figure 4.2. The constructed task graph.

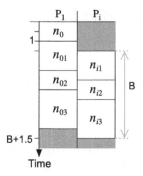

Figure 4.3. Extract of the constructed schedule for P_1 and P_i.

Node n_0 is allocated to a processor, which shall be called P_1. The remaining nodes $n_1, \ldots, n_{3m}$ are allocated in triplets to the processors $P_1, \ldots, P_m$. Let a_{i1}, a_{i2}, a_{i3} be the elements of $\mathbf{A}_i$. The nodes n_{i1}, n_{i2}, n_{i3}, corresponding to the elements of triplet $\mathbf{A}_i$, are allocated to processor P_i. The resulting schedule is illustrated for P_1 and P_i in Figure 4.3.

What is the length of this schedule? The time to execute n_{i1}, n_{i2}, n_{i3} on each processor is B, n_0 is executed in 1 time unit, and the communication from n_0 to its successor nodes takes $\frac{1}{2}$ time unit. Hence, the finish time of all processors P_i, $i = 2, \ldots, m$, is $t_f(P_i) = B + 1.5$. Since the communication on P_1 is local, the finish time on P_1 is $t_f(P_1) = B + 1$. The resulting length of the constructed schedule $\mathcal{S}$ is $sl(\mathcal{S}) = B + 1.5 \leq T$ and therefore the constructed schedule $\mathcal{S}$ is a solution to SCHED(G, $\mathbf{P}$).

Conversely, assume that an instance of SCHED admits a solution, given by the feasible schedule $\mathcal{S}$ with $sl(\mathcal{S}) \leq T$. It will now be shown that $\mathcal{S}$ is necessarily of the same kind as the schedule constructed above.

In this schedule $\mathcal{S}$, each processor can at most spend B time units in executing nodes from $\{n_1, \ldots, n_{3m}\}$. Otherwise the finish time of a processor, say, P_i, spending more that B time units is $t_f(P_i) > w(n_0) + B = 1 + B$. Due to the fact that all a_i's (i.e., node weights) are positive integers, this means that $t_f(P_i) \geq 2 + B$. However, this is larger than the bound T.

Furthermore, with $\sum_{i=1}^{3m} w(n_i) = mB$ and $|\mathbf{P}| = m$, it follows that each processor spends exactly B time units in executing nodes from $\{n_1, \ldots, n_{3m}\}$. Finally, due to $w(n_i) = a_i$, $B/4 < a_i < B/2$, $i = 1, \ldots, 3m$, only three nodes can have the exact execution time of B. Hence, the distribution of the nodes on the m processors corresponds to a solution of 3-PARTITION. $\qquad\square$

This proof demonstrated the NP-completeness of the scheduling problem SCHED in the strong sense (Garey and Johnson [73]). Alternatively, the NP-completeness can also be shown with a proof based on a reduction from PARTITION (Garey and Johnson [73]). Such a proof (Chrétienne [32]) is slightly less involved, but only demonstrates NP-completeness in the weak sense, since PARTITION is NP-complete in the weak sense. In Exercise 4.3 you are asked to devise a proof based on PARTITION.

The NP-completeness of scheduling has been investigated extensively in the literature (e.g., Brucker [27], Chrétienne et al. [34], Coffman [37], Garey and Johnson [73], Ullman [192]). Several further scheduling problems arise by restricting the task graph, for example, by having unit computation or communication costs or by limiting the number of processors. These problems can be treated as special cases, but only a few of them are known to be solvable in polynomial time. A comprehensive overview of scheduling problems and their complexity is given in Section 6.4.

It will be demonstrated in Theorem 5.1 of Section 5.1 that the scheduling problem is equivalent to finding a node order and a processor allocation. Given an optimal order and allocation, the optimal schedule can be constructed in polynomial time. Of course, finding the order and the allocation is still NP-hard.

Task scheduling remains NP-complete even when an unlimited number of processors is available (Chrétienne [32]). A system with equal or more processors than nodes

$$|\mathbf{P}| \geq |\mathbf{V}| \tag{4.14}$$

can be considered as having an unlimited number of processors, because the task execution is nonpreemptive (Property 2 of Definition 4.3 of the target system model) and only entire tasks are scheduled. Hence, if there are more processors than nodes, the surplus processors will have no tasks to execute and can be removed from the system without influencing the schedule.

Theorem 4.2 (NP-Completeness—Unlimited Number of Processors) *Let $G = (\mathbf{V}, \mathbf{E}, w, c)$ be a task graph and $\mathbf{P}_\infty$ a parallel system, with $|\mathbf{P}_\infty| \geq |\mathbf{V}|$. The decision problem* SCHED$(G, \mathbf{P}_\infty)$ *associated with the scheduling problem is as follows. Is there a schedule S for G on $\mathbf{P}_\infty$ with length $sl(S) \leq T, T \in \mathbb{Q}^+$?* SCHED$(G, \mathbf{P}_\infty)$ *is NP-complete.*

The proof for this theorem can be based on a reduction from PARTITION, as mentioned earlier, and is left as an exercise for the reader (Exercise 4.3).

As a logical consequence of the NP-completeness of scheduling, the scientific community has been eager to investigate efficient scheduling algorithms based on heuristics or approximation techniques that produce near optimal solutions. In practice, task scheduling must rely on these algorithms due to the intractability of the problem. In the following chapters, many task scheduling algorithms are analyzed, whereby the focus is on common techniques encountered in many algorithms.

This section finishes with a simple lemma regarding the optimal schedule length and the number of processors. In general, the more processors are *available* for the task graph to be scheduled on, the smaller (or equal) the optimal schedule length (Darte et al. [52]).

Lemma 4.2 (Relation Available Processors—Schedule Length) *Let $G = (\mathbf{V}, \mathbf{E}, w, c)$ be a task graph and $\mathbf{P}, \mathbf{P}_{+1}$ two parallel systems with the only difference* *that* $|\mathbf{P}_{+1}| = |\mathbf{P}| + 1$. *For the optimal schedule of G on the two systems, it holds that*

$$sl(\mathcal{S}_{\text{opt}}(\mathbf{P}_{+1})) \leq sl(\mathcal{S}_{\text{opt}}(\mathbf{P})). \tag{4.15}$$

Proof. Assume the optimal schedule $\mathcal{S}_{\text{opt}}(\mathbf{P})$ for the smaller system is given. By adding one processor to the schedule without using it (i.e., the processor is idle), we are not changing the original schedule. Yet, this schedule is a feasible schedule $\mathcal{S}(\mathbf{P}_{+1})$ for the larger system. Hence, $sl(\mathcal{S}(\mathbf{P}_{+1})) = sl(\mathcal{S}_{\text{opt}}(\mathbf{P}))$ and for the optimal schedule on $\mathbf{P}_{+1}$ it must hold that $sl(\mathcal{S}_{\text{opt}}(\mathbf{P}_{+1})) \leq sl(\mathcal{S}_{\text{opt}}(\mathbf{P}))$. $\qquad\square$

In Section 4.3.2 it will be seen that the situation is less trivial, when the number of *used* processors (Definition 4.11), that is, nonidle, is considered.

4.3 WITHOUT COMMUNICATION COSTS

As was said at the beginning of this chapter, scheduling without communication costs is a special case of the general scheduling problem that was discussed in the previous section. Nevertheless, it is worthwhile to have a closer look at this "special case," since it has some interesting properties. As already mentioned, historically the discussion of scheduling problems started without consideration of communication costs. As a consequence, the literature of the 1970s regarding task scheduling for parallel systems focused on this scheduling problem that will be defined in the following (Coffman [37], Ullman [192]). Only in the late 1980s and the 1990s were communication delays integrated into the scheduling problem (Chrétienne and Picovleav [35], El-Rewini and Ali [61], Rayward-Smith [158]).

Not considering communication costs corresponds to $c(e) = 0 \, \forall e \in \mathbf{E}$ in the task graph model. Thus, a task graph without communication costs can be defined as $G = (\mathbf{V}, \mathbf{E}, w)$ and its edges represent the dependence relations of the nodes, but do not reflect the cost of communication. As a consequence, it can be meaningful to reflect all types of data dependence (flow, output, and antidependence (Section 2.5.1)), not only flow dependence as was done by the general task graph. This was discussed in Section 3.5, when the task graph model was introduced.

Only two small modifications have to be made to the definitions of the foregoing section to completely define the scheduling problem without communication costs. First, the target system model can be significantly simplified and second, the definition of the edge finish time (Definition 4.6) must be modified.

Definition 4.14 (Target Parallel System—Cost-Free Communication) *A target parallel system* $\mathbf{P}_{c0}$ *consists of a set of identical processors connected by a cost-free communication network. This system has the following properties:*

1. Dedicated System. *The parallel system is dedicated to the execution of the scheduled task graph. No other program or task is executed on the system while the scheduled task graph is executed.*

2. Dedicated Processor. *A processor $P \in \mathbf{P}_{c0}$ can execute only one task at a time and the execution is not preemptive.*

3. Cost-Free Communication. *The cost of communication between tasks is negligible and therefore considered zero. Hence, no further considerations regarding communication subsystem, concurrency of communication, and the network structure are necessary for this model.*

The first two properties of this model are identical to Properties 1 and 2 of the target system of the classic model that considers communication costs (Definition 4.3). The third property is new, which supersedes Properties 3–6 of Definition 4.3. There are no communication costs, thus nothing has to be specified or defined regarding the communication subsystem.

Based on this model, it is obvious how to redefine the edge finish time (Definition 4.6), because the communication time is now always zero.

Definition 4.15 (Edge Finish Time—Cost-Free Communication) *Let $G = (\mathbf{V}, \mathbf{E}, w)$ be a task graph and $\mathbf{P}_{c0}$ a parallel system. The finish time of $e_{ij} \in \mathbf{E}$, $n_i, n_j \in \mathbf{V}$, communicated from processor P_{src} to P_{dst}, $P_{\text{src}}, P_{\text{dst}} \in \mathbf{P}$, is*

$$t_f(e_{ij}, P_{\text{src}}, P_{\text{dst}}) = t_f(n_i). \tag{4.16}$$

All other definitions, conditions, and so on of the previous section were carefully made in such a way that no further modifications are necessary to treat the scheduling problem without communication costs.

Setting $t_f(e_{ij}, P_{\text{src}}, P_{\text{dst}})$ to $t_f(n_i)$ has the interesting consequence that the DRT (Definition 4.8) is no longer a function of the processors to which the involved nodes are scheduled. Since this has important consequences for scheduling under this model, as will be seen later, the DRT for scheduling without communication costs is here explicitly redefined.

Definition 4.16 (Data Ready Time (DRT)—Cost-Free Communication) *Let S be a schedule for task graph $G = (\mathbf{V}, \mathbf{E}, w)$ on system $\mathbf{P}_{c0}$. The data ready time of a node $n_j \in \mathbf{V}$ is*

$$t_{\text{dr}}(n_j) = \max_{n_i \in \mathbf{pred}(n_j)} \{t_f(n_i)\}. \tag{4.17}$$

If $\mathbf{pred}(n_j) = \emptyset$, that is, n_j is a source node, $t_{\text{dr}}(n_j) = 0$.

4.3.1 Schedule Example

To illustrate the difference between scheduling with and without communication costs, a schedule is examined where communication costs are zero. Essentially, the same example as in Figure 4.1 is used. Again a schedule for the sample task graph of Figure 3.15 on three processors is depicted in Figure 4.4. The obvious difference is that

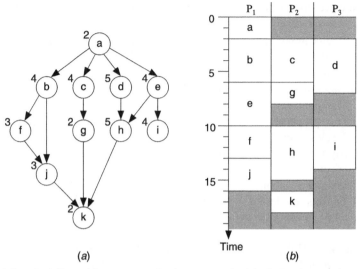

(a)   Time   (b)

Figure 4.4. Scheduling without communication costs: schedule(*b*) for the sample task graph (*a*) (without edge weights) of Figure 3.15 on three processors. Compare to Figure 4.1.

the edges of the task graph have no weights (which is equivalent to $c(e) = 0 \, \forall e \in \mathbf{E}$) as shown in Figure 4.4(*a*). The nodes are scheduled on the same processors and in the same order as in the schedule of Figure 4.1, however with some significant differences in their start times, due to the lack of communication time consideration.

Some examples from Figure 4.4 will illustrate that. The start time of node a on processor P_1 is $t_s(a) = 0$; there is no change in comparison to Figure 4.1, since a is an entry node. In contrast, nodes c and d can now start earlier than in Figure 4.1 and they do so at $t_s(c) = t_s(d) = 2$, immediately after a finishes, $t_f(a) = 2$. This is in compliance with the precedence constraints, $t_s(c) \geq t_f(a)$ and $t_s(d) \geq t_f(a)$, since both node c and node d depend on a through the edges e_{ac} and e_{ad} and the communication time is zero. There is no difference for node b in the schedules of Figure 4.1 and Figure 4.4, since b only depends on a. Hence, the communication is local and thereby no communication costs are inflicted in either of the models; it starts immediately after node a finishes, $t_s(b) = t_f(a) = 2$. Nodes h and i can only start at $t_s(h) = t_s(i) = 10$ (which is, however, much earlier than in Figure 4.1), because both depend on node e, which finishes at $f(e) = 10$. In other words, their DRT is $t_{dr}(h) = t_{dr}(i) = 10$. Note that their DRT does not depend on the processor on which they are scheduled. The same is true for node k, which starts at $t_s(k) = 16$. It depends on the three nodes g, h, j, and node j is the last one to finish; hence, k's DRT is $t_{dr}(k) = \max\{t_f(g), t_f(h), t_f(j)\} = t_f(j) = 16$. The length of the schedule is $sl = t_f(k) = 18$.

4.3.2 Scheduling Complexity

Without communication costs, scheduling remains an NP-complete problem. This is a little bit surprising, given that there is no conflict between high parallelism and low communication costs, as there is in the general problem. On the other hand, problems that are similar to the scheduling problem and seem to be easy at first sight, such as the KNAPSACK or the BINPACKING problems (Cormen et al. [42], Garey and Johnson [73]), are NP-complete, too.

Theorem 4.3 (NP-Completeness—Cost-Free Communication) *Let $G = (\mathbf{V}, \mathbf{E}, w)$ be a task graph and $\mathbf{P}_{c0}$ a parallel system. The decision problem SCHED-C0(G, $\mathbf{P}_{c0}$) associated with the scheduling problem is as follows. Is there a schedule S for G on $\mathbf{P}_{c0}$ with length $sl(S) \leq T, T \in \mathbb{Q}^+$? SCHED-C0($G$, $\mathbf{P}_{c0}$) is NP-complete.*

Proof. The same proof as for Theorem 4.1 is used with one small modification for the construction of an instance of SCHED-C0(G, $\mathbf{P}_{c0}$):

- There are no edge weights, that is, $c(e) = 0 \forall e \in \mathbf{E}$ (instead of $c(e) = \frac{1}{2} \forall e \in \mathbf{E}$ in SCHED(G, $\mathbf{P}$)).

The argumentation is then identical to that in the proof of Theorem 4.1; even the time bound T can remain unchanged. $\square$

Polynomial for Unlimited Processors   While in general the scheduling problem without communication costs is NP-complete, it is solvable in polynomial time for an unlimited number of processors. Note that this is in contrast to the general scheduling problem; see Theorem 4.2. A simple algorithm to find an optimal schedule is based on two ideas:

1. Each node is assigned to a distinct processor.
2. Each node starts execution as soon as possible.

Since there is only one task on each processor, the earliest start time is only determined by the precedence constraints of the task graph; processor constraints cannot occur (Condition 4.1). A node can start its execution as soon as its data is ready. From Eq. (4.17) of Definition 4.16, it is clear that the DRT is not a function of the processors on which the involved nodes are executed. This is the essential difference from the general case, where communication costs must be considered. Hence, the earliest possible start time of a node is its DRT, and each node should start at that time:

$$t_s(n) = t_{dr}(n) \ \forall n \in \mathbf{V}. \tag{4.18}$$

The job of the algorithm is to calculate the DRTs and to assign the start times in the correct order. For the calculation of a node's DRT, the finish time of all its predecessors must be known. This is easily accomplished by handling the nodes in a topological order (Definition 3.6), which guarantees that the finish times of all predecessors $\mathbf{pred}(n_i)$ of a node n_i have already been defined when n_i's DRT is calculated. Algorithm 6 shows this procedure in algorithmic form. It is assumed that the processors and nodes are consecutively indexed, that is, $\mathbf{V} = \{n_1, n_2, \ldots, n_{|\mathbf{V}|}\}$ and $\mathbf{P}_{c0,\infty} = \{P_1, P_2, \ldots, P_{|\mathbf{P}_{c0,\infty}|}\}$.

Algorithm 6   Optimal Scheduling $G = (\mathbf{V}, \mathbf{E}, w)$ on $\mathbf{P}_{c0,\infty}$, with $|\mathbf{P}_{c0,\infty}| \geq |\mathbf{V}|$

> Insert $\{n \in \mathbf{V}\}$ in topological order into sequential list L.
> **for** each n_i in L **do**
> $DRT \leftarrow 0$
>   **for** each $n_j \in \mathbf{pred}(n_i)$ **do**
> $DRT \leftarrow \max\{DRT, t_f(n_j)\}$
>   **end for**
> $t_s(n_i) \leftarrow DRT$; $t_f(n_i) \leftarrow t_s(n_i) + w(n_i)$
> $proc(n_i) \leftarrow P_i, P_i \in \mathbf{P}_{c0,\infty}$
> **end for**

Theorem 4.4 (Optimal Scheduling on Unlimited Processors)   *Let $G = (\mathbf{V}, \mathbf{E}, w)$ be a task graph and $\mathbf{P}_{c0,\infty}$ a parallel system, with $|\mathbf{P}_{c0,\infty}| \geq |\mathbf{V}|$. Algorithm 6 produces an optimal schedule S_{opt} of G on $\mathbf{P}_{c0,\infty}$.*

Proof. The produced schedule S_{opt} is feasible: (1) each node is scheduled on a distinct processor, hence the processor constraint (Condition 4.1) is always adhered to; and (2) the nodes comply with the precedence constraints (Condition 4.2), since the start time of each node is not earlier than its DRT. The correct calculation of the DRTs is guaranteed by the topological order in which the nodes are processed.

The produced schedule S_{opt} is optimal, because each node starts execution at the earliest possible time, the time when its data is ready. The DRT cannot be reduced, since it is not a function of the processors (Definition 4.16); that is, scheduling the involved nodes on other processors does not change the DRT.   □

Algorithm 6 has a time complexity of $O(\mathbf{V} + \mathbf{E})$: calculating a topological order is $O(\mathbf{V} + \mathbf{E})$ (Algorithm 4). To calculate the DRT of each node ($O(\mathbf{V})$), every predecessor is considered, which amortizes to $O(\mathbf{E})$ predecessors for the entire graph. Hence, this part is also $O(\mathbf{V} + \mathbf{E})$.

While Algorithm 6 is simple, it does not utilize the processors in an efficient way. Usually, there will be a schedule S that uses less than $|\mathbf{V}|$ processors, but whose schedule length is also optimal, that is, $sl(S) = sl(S_{\mathrm{opt}})$. Yet again, finding the minimum number of processors for achieving the optimal schedule length is an NP-hard problem (Darte et al. [52]). A simple heuristic to reduce the number of processors, while maintaining the optimal schedule length, is linear clustering, which is presented in Section 5.3.2.

Used Processors   In Lemma 4.2 of Section 4.2.2 it was stated that the more processors are *available* for the task graph to be scheduled on, the smaller (or equal) the optimal schedule length. The situation becomes less trivial and more interesting when one considers the *used* processors (Definition 4.11) in a schedule.

For scheduling without communication costs, it still holds that the more *used* processors, the smaller (or equal) the optimal schedule length.

Theorem 4.5 (Relation Used Processors—Schedule Length) *Let $G = (\mathbf{V}, \mathbf{E}, w)$ be a task graph and $\mathbf{P}_{c0}$ a parallel system. Let $\mathcal{S}^q$ and $\mathcal{S}^{q+1}$ be the classes of all schedules of G on $\mathbf{P}_{c0}$ that use q, $q < |\mathbf{P}_{c0}|$, and $q+1$, $q+1 \leq |\mathbf{P}_{c0}|$, processors, respectively. For an optimal schedule of each of the two classes it holds that*

$$sl(\mathcal{S}_{opt}^{q+1}) \leq sl(\mathcal{S}_{opt}^q). \tag{4.19}$$

Proof. Assume an optimal schedule $\mathcal{S}_{opt}^q$ is given and $\mathbf{Q}$ is the set of processors used by this schedule (Definition 4.11). Let $n \in \mathbf{V}$ be a node scheduled on processor $Q = proc(n), Q \in \mathbf{Q}$ with start time $t_s(n, Q)$. When n is rescheduled from Q onto an idle processor $P \in \mathbf{P}_{c0}, \notin \mathbf{Q}$ then it can start there at its DRT, $t_s(n, P) = t_{dr}(n)$, since it is the only node on P. This start time cannot be later than its start time on Q, since for the start time $t_s(n)$ it must hold on any processor that $t_s(n) \geq t_{dr}(n)$ (Condition 4.2); hence, $t_s(n, P) \leq t_s(n, Q)$. Remember, the DRT remains unaltered, since it is not a function of the processors (Definition 4.16). As n does not start later on P than on Q, it also does not finish later on P than on Q. In turn, the DRTs of all successor nodes of n do not increase, which means the entire schedule remains feasible. This new schedule $\mathcal{S}_{new}$ has $q + 1$ used processors and thus belongs to the class $\mathcal{S}_{opt}^{q+1}$. Its schedule length is the same as that of $\mathcal{S}_{opt}^q$, $sl(\mathcal{S}_{new}) = sl(\mathcal{S}_{opt}^q)$. Thus, for the optimal schedule in $\mathcal{S}_{opt}^{q+1}$ it must hold that $sl(\mathcal{S}_{opt}^{q+1}) \leq sl(\mathcal{S}_{opt}^q)$. □

It should be stressed that this theorem is only valid for scheduling without communication costs. As can be seen, the proof is based on the fact that the DRT is not a function of the involved processors. This does not hold for the general scheduling problem with communication costs (Definition 4.8). The following example illustrates this. Consider a task graph with a chain structure, as illustrated in Figure 4.5(*a*).

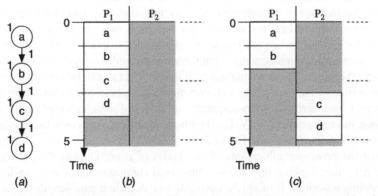

Figure 4.5. Counterexample to Theorem 4.5 when communication costs are considered: chain task graph (*a*) scheduled on one (*b*) and two processors (*c*).

Due to the precedence constraints, the graph must be executed in sequential order, namely, a, b, c, d. Hence, the optimal schedule of this task graph uses only one processor (Figure 4.5(b)). If more than one processor is used (e.g., two as illustrated in Figure 4.5(c)), at least one communication will be remote, which inevitably increases the schedule length by at least one time unit.

4.4 TASK GRAPH PROPERTIES

This section returns to the task graph model as discussed in Section 3.5 of the previous chapter. The concepts presented in the following are employed in the scheduling techniques of the remaining text, in particular, for the assignment of priorities to nodes. In many scheduling algorithms the order in which nodes are considered has a significant influence on the resulting schedule and its length. Gauging the importance of the nodes with a priority scheme is therefore a fundamental part of scheduling.

The discussion begins with the definition of the length of a path in the task graph, based on the computation and communication costs defined in Section 4.1.

Definition 4.17 (Path Length) *Let $G = (V, E, w, c)$ be a task graph. The length of a path p in G is the sum of the weights of its nodes and edges:*

$$len(p) = \sum_{n \in p, V} w(n) + \sum_{e \in p, E} c(e). \tag{4.20}$$

The computation length of a path p in G is the sum of the weights of its nodes:

$$len_w(p) = \sum_{n \in p, V} w(n). \tag{4.21}$$

Note that the definition of the path length in a task graph differs from the general Definition 3.2 in Section 3.1, where the length of a path equals the number of edges. The path length $len(p)$ can be interpreted as the time the path p takes for its execution if all communications between its nodes are interprocessor communications, which happens, for instance, when each node of p is allocated to a different processor. Due to the sequential order inherent in the path, none of the nodes is executed concurrently with any other node. The computation path length $len_w(p)$ can be interpreted as the execution time of the path p when all communications between its nodes are local, that is, they have zero costs. Consequently, all nodes of p are executed on the same processor. In a task graph $G = (V, E, w)$ without communication costs, the path length is necessarily defined as it is in Eq. (4.21).

When the processor allocations of the nodes of p are known, the path length can be determined taking into account that local communications are cost free. In a scheduling algorithm it might be desirable to calculate a path length based on a partial schedule: that is, some processor allocations are given and others are not. The path length determined for a given (partial) processor allocation $\mathcal{A}$ (Definition 4.1) is denoted by $len(p, \mathcal{A})$, the *allocated path length*. Communications between nodes whose processor allocations are unknown are assumed to be remote. The allocated path length $len(p, \mathcal{A})$ is thus in between the path length $len(p)$ and the computation path length $len_w(p)$:

$$len(p) \geq len(p, \mathcal{A}) \geq len_w(p). \qquad (4.22)$$

The above scheme for the distinction between the different path lengths is used throughout this text. All definitions based on path lengths will only be formulated for $len(p)$ but are implicitly valid for $len(p, \mathcal{A})$ and $len_w(p)$, too. The corresponding definitions use the same scheme for distinction: that is, the allocated path length is parameterized with $\mathcal{A}$ and the subscript w is used for the computation length.

4.4.1 Critical Path

An important concept for scheduling is the critical path—the longest path in the task graph.

Definition 4.18 (Critical Path (CP)) *A critical path cp of a task graph $G = (\mathbf{V}, \mathbf{E}, w, c)$ is a longest path in G*

$$len(cp) = \max_{p \in G}\{len(p)\}. \qquad (4.23)$$

The computation critical path cp_w and the allocated critical path $cp(\mathcal{A})$ for a processor allocation $\mathcal{A}$ are defined correspondingly.

The nodes of a critical path cp, consisting of l nodes, are denoted by $n_{cp,1}, n_{cp,2}, \ldots, n_{cp,l}$. Clearly, there might be more than one critical path as several paths can have the same maximum length.

Note that in general

$$cp \neq cp_w \neq cp(\mathcal{A}), \qquad (4.24)$$

from which follows for their path lengths

$$len(cp_w) \leq len(cp) \quad \text{and} \quad len(cp(\mathcal{A})) \leq len(cp). \qquad (4.25)$$

Equivalent inequalities hold for the computation path length and the allocated path length and the corresponding critical paths.

Lemma 4.3 (Critical Path: From Source to Sink) *Let $G = (\mathbf{V}, \mathbf{E}, w, c)$ be a task graph. A critical path cp of G always starts in a source node and finishes in a sink node of G.*

Proof. By contradiction: Suppose cp does not start in a source node. Then, per definition, the first node of cp, here denoted by n_1, has at least one predecessor $n_0 \in \mathbf{pred}(n_1)$; hence, there is the edge $e_{01} \in \mathbf{E}$. A new path q can be concatenated from n_0, e_{01}, and cp, whose length is $len(q) = w(n_0) + c(e_{01}) + len(cp)$. As $w(n_0) > 0$, it follows that $len(q) > len(cp)$—a contradiction. Likewise for the sink node. □

The critical path gains its importance for scheduling from the fact that its length is a lower bound for the schedule length.

Lemma 4.4 (Critical Path Bound on Schedule Length) *Let $G = (\mathbf{V}, \mathbf{E}, w, c)$ be a task graph and cp_w a computation critical path of G. For any schedule S of G on any system $\mathbf{P}$,*

$$sl \geq len_w(cp_w). \tag{4.26}$$

Proof. Due to their precedence constraints (Condition 4.2), the nodes of cp_w can only be executed in sequential order, which takes $len_w(cp_w)$ time, independently of the schedule or the number of processors. Thus, the duration of G's execution is at least $len_w(cp_w)$. □

In the worst case that all communications among nodes are remote,

$$sl \geq len(cp), \tag{4.27}$$

yet Eq. (4.26) is also fulfilled.

For the special case of scheduling without communication costs on an unlimited number of processors (Section 4.3.2), the lower bound of the schedule length established by Eq. (4.26) is tight. In other words, the length of the optimal schedule is the length of the critical path.

Lemma 4.5 (Optimal sl on Unlimited Processors—Cost-Free Communication) *Let $G = (\mathbf{V}, \mathbf{E}, w)$ be a task graph, cp_w a computation critical path of G, and $\mathbf{P}_{c0,\infty}$ a parallel system, with $|\mathbf{P}_{c0,\infty}| \geq |\mathbf{V}|$. For an optimal length schedule S_{opt} of G on system $\mathbf{P}_{c0,\infty}$,*

$$sl(S_{opt}) = len_w(cp_w). \tag{4.28}$$

Proof. Theorem 4.4 establishes that Algorithm 6 produces an optimal schedule S_{opt} of G on $\mathbf{P}_{c0,\infty}$. In this algorithm, the start time $t_s(n)$ of each node n is set to its DRT

$$t_s(n) = t_{dr}(n) \;\forall n \in \mathbf{V}. \tag{4.29}$$

By Definition 4.16, node n's DRT is the maximum of the finish times of its predecessor nodes (Eq. (4.17)). Together with $t_f(n) = t_s(n) + w(n)$ (Eq. (4.2) of Definition 4.5), it holds that

$$t_{dr}(n) = \max_{n_i \in \mathbf{pred}(n)} \{t_{dr}(n_i\} + w(n_i)\}. \tag{4.30}$$

By induction it is clear that node n's DRT is a sum of ancestor node weights. As per Definition 4.17, this is the computation length len_w of a path composed of ancestor nodes of n. In particular, for the node with the maximum finish time n_{last}, that is, $sl(S_{opt}) = t_f(n_{last})$ (Definition 4.10), it holds that

$$t_f(n_{last}) = t_{dr}(n_{last}) + w(n_{last}).$$

Thus, $t_f(n_{last})$ is the length of a path ending with node n_{last}. Per definition of the computation critical path, this path cannot be longer than the critical path, $t_f(n_{last}) \leq len(cp_w)$. With Eq. (4.26), it follows that $sl(S_{opt}) = len(cp_w)$.   □

The critical path of the sample task graph (e.g., in Figure 4.1) is $cp = \langle a, b, f, j, k \rangle$ with length $len(cp) = 34$, while there are two computation critical paths cp_w, $\langle a, b, f, j, k \rangle$ and $\langle a, d, h, k \rangle$, with computation length $len_w(cp_w) = 14$.

4.4.2   Node Levels

The critical path identifies the nodes of a task graph, whose constrained sequential execution takes at least the execution time of any other path in the task graph. This makes the nodes of the critical path important, as a late execution start of any of these nodes directly results in an extended schedule length. The scheduling of a less important node, that is, a node not belonging to the critical path, is not that important, as long as it does not delay the execution of any critical path node. But how can one differentiate between nodes not belonging to the critical path? In general, some nodes are more important than others.

This problem is tackled by the notion of node levels, which is a natural extention of the critical path concept on a node-specific basis. Given a node $n \in \mathbf{V}$, in general, paths exist in G that end in n, that is, whose last node is n, and that start with n, that is, whose first node is n (see also the notion of ancestors and descendants in Definition 3.4, Section 3.1). Analogous to the critical path, the longest path in each of the two sets of paths can be distinguished. A node level is then the length of the longest path.

Definition 4.19 (Bottom and Top Levels)   *Let $G = (\mathbf{V}, \mathbf{E}, w, c)$ be a task graph and $n \in \mathbf{V}$.*

- Bottom Level. *The bottom level $bl(n)$ of n is the length of the longest path starting with n:*

$$bl(n) = \max_{n_i \in \mathbf{desc}(n) \cap \mathbf{sink}(G)} \{len(p(n \to n_i))\}. \tag{4.31}$$

If $\mathbf{desc}(n) = \emptyset$, then $bl(n) = w(n)$. A path starting with n of length $bl(n)$ is called a bottom path of n and denoted by $p_{bl(n)}$.

- Top Level. *The top level $tl(n)$ of n is the length of the longest path ending in n, excluding $w(n)$:*

$$tl(n) = \max_{n_i \in \mathbf{ance}(n) \cap \mathbf{source}(G)} \{len(p(n_i \to n))\} - w(n). \qquad (4.32)$$

If $\mathbf{ance}(n) = \emptyset$, then $tl(n) = 0$. A path ending in n of length $tl(n) + w(n)$ is called a top path of n and denoted by $p_{tl(n)}$.

Considering the proof of Lemma 4.3, it is evident that the longest path starting with a node n must always terminate in a sink node (hence, $n_i \in \mathbf{desc}(n) \cap \mathbf{sink}(G)$ in Eq. (4.31)) and the longest path terminating in n must always start in a source node (hence, $n_i \in \mathbf{ance}(n) \cap \mathbf{source}(G)$ in Eq. (4.32)). This fact gave name to the top level and the bottom level, as the longest paths start at the "top" of the task graph and end at its "bottom." Figure 4.6 illustrates the top and bottom levels of a node n with the corresponding paths in a task graph with several source and sink nodes.

It should be noted that the paths corresponding to the computation node levels—computation bottom level $bl_w(n)$ and computation top level $tl_w(n)$—are not identical to the paths of the levels defined earlier. In general,

$$p_{bl(n)} \neq p_{bl_w(n)} \quad \text{and} \quad p_{tl(n)} \neq p_{tl_w(n)}. \qquad (4.33)$$

The significance of the node levels becomes apparent by the following considerations. At the time the execution of a node $n \in G$ is initiated, the minimum remaining time until the execution of G is completed is given by n's bottom level $bl(n)$ (assuming the worst case where all communications are interprocessor communications). The

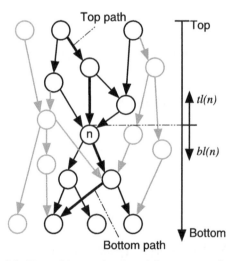

Figure 4.6. Top and bottom levels and the corresponding paths.

precedence constraints of G do not permit any earlier termination, independent of the number of involved processors. The computation bottom level $bl_w(n)$ reflects the best-case minimum time until the termination of G's execution, since the costs of communication are neglected, so they must all be local.

Likewise, a node cannot start earlier than at the time given by its top level $tl(n)$ (once more, assuming that all communications are remote; $tl_w(n)$ corresponds to the case that all communications are local). Recall that the top level does not include the cost $w(n)$ of n, while the bottom level does.

The above observations are formalized in the following lemma.

Lemma 4.6 (Level Bounds on Start Time) *Let S be a schedule for task graph $G = (\mathbf{V}, \mathbf{E}, w, c)$ on system* $\mathbf{P}$. *For* $n \in \mathbf{V}$,

$$sl \geq t_s(n) + bl_w(n) \tag{4.34}$$

$$t_s(n) \geq tl_w(n) \tag{4.35}$$

Proof. Both Eqs. (4.34) and (4.35) are obvious from the definition of $bl_w(n)$ and $tl_w(n)$. For Eq. (4.34), due to precedence constraints, all descendant nodes of n must be executed after n (Condition 4.2). Their execution in precedence order takes at least $bl_w(n)$ time, including the execution time of n. For Eq. (4.35), due to precedence constraints, all ancestors of n must be executed before n (Condition 4.2). Their execution in precedence order takes at least $tl_w(n)$ time. $\square$

Note that Lemma 4.6 relies on the computation levels, to reflect the best possible case. A lemma based on the top and bottom levels needs the additional restriction that all communications among the ancestor and descendent nodes are remote. Lemma 4.6, however, is always true.

The top path $p_{tl(n)}$ and the bottom path $p_{bl(n)}$ of a node $n \in G$ together form a path

$$p_{tb(n)} = p_{tl(n)} \cup p_{bl(n)}, \tag{4.36}$$

which is a longest path of G through n, starting in a source node and ending in a sink node. This path is called a *level path* of n and its length is

$$len(p_{tb(n)}) = tl(n) + bl(n). \tag{4.37}$$

Equation (4.37) holds for every node $n_i \in p_{tb(n)}$, $n_i \in \mathbf{V}$. For the nodes of a critical path, this lemma follows.

Lemma 4.7 (Critical Path Length and Node Levels) *Let $G = (\mathbf{V}, \mathbf{E}, w, c)$ be a task graph. For any node $n_{cp,i}$ of a critical path cp*

$$len(cp) = tl(n_{cp,i}) + bl(n_{cp,i}). \tag{4.38}$$

Proof. By contradiction: Divide cp into two subpaths: cp_t ending in $n_{cp,i}$ and cp_b starting with $n_{cp,i}$. The length of cp_t must be $len(cp_t) = tl(n_{cp,i}) + w(n_{cp,i})$: if it was shorter, cp_t could be substituted with the top path $p_{tl(n_{cp,i})}$ of n and the path concatenated from $p_{tl(n_{cp,i})}$ and cp_b would be longer than cp; if it was longer, $tl(n_{cp,i})$ would not be the top level of $n_{cp,i}$, but instead $len(cp_t) - w(n_{cp,i})$. The length of cp_b must be $len(cp_b) = bl(n_{cp,i})$ with the analogous argumentation. Then, $len(cp) = len(cp_t) - w(n_{cp,i}) + len(cp_b) = tl(n_{cp,i}) + bl(n_{cp,i})$.   □

With the argument of the above proof, it can also be seen that cp is a level path of each $n_{cp,i}$.

For any source node $n_{\text{src}} \in G$,

$$bl(n_{\text{src}}) = len(p_{tb(n_{\text{src}})}), \qquad (4.39)$$

since by definition $tl(n_{\text{src}}) = 0$ if $\mathbf{pred}(n_{\text{src}}) = \emptyset$. Consequently, a source node with the highest bottom level of all nodes is the first node $n_{cp,1}$ of a critical path cp of G:

$$bl(n_{cp,1}) = len(cp) \geq bl(n_i) \, \forall \, n_i \in \mathbf{V}. \qquad (4.40)$$

As-Soon/Late-as-Possible Start Times   As a result of Eq. (4.35), the top level $tl(n)$ of a node n is sometimes called the as-soon-as-possible (ASAP) start time of n,

$$\text{ASAP}(n) = tl(n). \qquad (4.41)$$

This property of the (computation) top level was already used in Theorem 4.4. It establishes that an optimal schedule on an unlimited number of processors for a task graph without communication costs can be constructed by starting each node as-soon-as-possible, Eq. (4.18); hence at its computation top level.

The counterpart, the as-late-as-possible (ALAP) start time of a node n, directly correlates to the bottom level of n. As stated by the inequality Eq. (4.27), the critical path of a task graph is a lower bound for the schedule length. A node n ought to start early enough so that it does not increase the schedule length beyond that bound. Together with Eq. (4.34) the ALAP start time of n is thus

$$\text{ALAP}(n) = len(cp) - bl(n). \qquad (4.42)$$

Consequently, arranging nodes in decreasing bottom level order is equivalent to arranging them in increasing ALAP order.

For consistency, the notation of top and bottom levels will be used as a substitute for ASAP and ALAP, respectively, even when algorithms are described that were proposed with the latter notations.

Table 4.1. Node Levels for All Nodes of the Sample Task Graph[a]

Nodes	tl/ASAP	bl	$tl + bl$	ALAP	tl_w/ASAP$_w$	bl_w	$tl_w + bl_w$	ALAP$_w$
a	0	34	34	0	0	14	14	0
b	8	26	34	8	2	12	14	2
c	4	12	16	22	2	8	10	6
d	5	19	24	15	2	12	14	2
e	6	16	22	18	2	11	13	3
f	16	18	34	16	6	8	14	6
g	10	6	16	28	6	4	10	10
h	14	10	24	24	7	7	14	7
i	13	4	17	30	6	4	10	10
j	23	11	34	23	9	5	14	9
k	32	2	34	32	12	2	14	12

[a]For example, as in Figure 4.1.

Examples Table 4.1 displays the various node levels for all nodes of the sample task graph (e.g., in Figure 4.1). The value of node a's bottom level $bl(a)$ is the length of the cp, $len(cp) = 34$ (see Eq. (4.40)). All nodes n_i, whose sum of $tl(n_i)$ and $bl(n_i)$ is 34, are critical path nodes (Lemma 4.7). In comparison, there are more computation critical path nodes: that is, nodes whose sum of $tl_w(n_i) + bl_w(n_i)$ is the length of the computation critical path $len_w(cp_w) = 14$ (see also Section 4.4.1).

Computing Levels and Critical Path To compute node levels, the following recursive definition of the levels is convenient. For a task graph $G = (\mathbf{V}, \mathbf{E}, w, c)$ and $n_i \in \mathbf{V}$,

$$bl(n_i) = w(n_i) + \max_{n_j \in \mathbf{succ}(n_i)} \{c(e_{ij}) + bl(n_j)\}, \tag{4.43}$$

$$tl(n_i) = \max_{n_j \in \mathbf{pred}(n_i)} \{tl(n_j) + w(n_j) + c(e_{ji})\}. \tag{4.44}$$

It can easily be shown by contradiction that these definitions are equivalent to Definition 4.19. For the computation levels (bl_w and tl_w), $c(e_{ij}) = 0$ in the above equations, and for the allocated levels ($bl(n_i, \mathcal{A})$ and $tl(n_i, \mathcal{A})$), $c(e_{ij}) = 0$ if the communication e_{ij} is local; that is, $proc(n_i) = proc(n_j)$.

To determine the levels with the above equations, a correct order of the nodes must be warranted, so that all levels on the right side of the expressions are already defined when calculating a level. For the bottom level this order is inverse topological and for the top level it is topological. Algorithms for the calculation of the levels can then be formulated as in Algorithms 7 and 8.

The time complexity of both algorithms is $O(\mathbf{V} + \mathbf{E})$: calculating a topological order is $O(\mathbf{V} + \mathbf{E})$ (Algorithm 4). Furthermore, the level of each node ($O(\mathbf{V})$) is calculated by considering all of its predecessors or successors, which amortizes to $O(\mathbf{E})$ predecessors or successors for the entire graph. Hence, this part is also $O(\mathbf{V} + \mathbf{E})$.

Algorithm 7   Compute Bottom Levels of $G = (\mathbf{V}, \mathbf{E}, w, c)$

Insert $\{n \in \mathbf{V}\}$ in inverse topological order into sequential list L.

for each n_i in L **do**

$\quad max \leftarrow 0; n_{bl_{\text{succ}}}(n_i) \leftarrow NULL$

$\quad$ **for** each $n_j \in \mathbf{succ}(n_i)$ **do**

$\quad\quad$ **if** $c(e_{ij}) + bl(n_j) > max$ **then**

$\quad\quad\quad max \leftarrow c(e_{ij}) + bl(n_j); n_{bl_{\text{succ}}}(n_i) \leftarrow n_j$

$\quad\quad$ **end if**

$\quad\quad bl(n_i) \leftarrow w(n_i) + max$

$\quad$ **end for**

end for

Algorithm 8   Compute Top Levels of $G = (\mathbf{V}, \mathbf{E}, w, c)$

Insert $\{n \in \mathbf{V}\}$ in topological order into sequential list L.

for each n_i in L **do**

$\quad max \leftarrow 0; n_{tl_{\text{pred}}}(n_i) \leftarrow NULL$

$\quad$ **for** each $n_j \in \mathbf{pred}(n_i)$ **do**

$\quad\quad$ **if** $tl(n_j) + w(n_j) + c(e_{ji}) > max$ **then**

$\quad\quad\quad max \leftarrow tl(n_j) + w(n_j) + c(e_{ji}); n_{tl_{\text{pred}}}(n_i) \leftarrow n_j$

$\quad\quad$ **end if**

$\quad\quad tl(n_i) \leftarrow max$

$\quad$ **end for**

end for

Observe that the top and bottom paths are also computed with the presented algorithms. For each node n, the algorithm stores the first successor (predecessor) as $n_{bl_{\text{succ}}}(n)$ ($n_{tl_{\text{pred}}}(n)$) that led to the maximum value. Upon termination of the algorithm, a bottom path of a node n, for instance, is given by the recursive list of successors: $p_{bl(n)} = \langle n, n_{bl_{\text{succ}}}(n), n_{bl_{\text{succ}}}(n_{bl_{\text{succ}}}(n)), \ldots \rangle$.

Moreover, a critical path and its length are also computed by Algorithm 7. It suffices to store a node with the highest bottom level during the run of Algorithm 7. From Eq. (4.40), it is known that this node $n_{cp,1}$ is the first node of a critical path and the path cp is given by the recursive list of $n_{cp,1}$'s successors, $cp = \langle n_{cp,1}, n_{bl_{\text{succ}}}(n_{cp,1}), n_{bl_{\text{succ}}}(n_{bl_{\text{succ}}}(n_{cp,1})), \ldots \rangle$.

As stated before, the paths of node levels and the critical path are in general not unique. For instance, the sample task graph (e.g., in Figure 4.1) possesses two different computation critical paths. To avoid ambiguity, additional criteria must be used to distinguish between the various paths. The two algorithms presented return only one bottom path and one top path for each node, since they store only the first successor (predecessor) with the maximum value they encounter. If the nodes are considered in the same order in each run, the algorithms always compute the same level paths for a given task graph.

4.4.3 Granularity

For a task graph, the notion of granularity (Section 2.4.1) usually describes the relation between the computation and the communication costs. Often, the granularity of a task graph is defined as the relation between the minimal node weight and the maximal edge weight.

Definition 4.20 (Task Graph Granularity) *Let $G = (\mathbf{V}, \mathbf{E}, w, c)$ be a task graph. G's granularity is*

$$g(G) = \frac{\min_{n \in \mathbf{V}} w(n)}{\max_{e \in \mathbf{E}} c(e)}. \tag{4.45}$$

With this definition, a task graph is said to be coarse grained if $g(G) \geq 1$. Coarse granularity is a desirable property of a task graph. One objective of task scheduling is always to minimize the cost of communication. This is achieved by having as much local communication (i.e., communication between tasks on the same processor) as possible. Unfortunately, this objective conflicts with the other objective of scheduling, namely, the distribution of the tasks among the processors. Graphs with coarse granularity can be parallelized (i.e., scheduled) more efficiently, since the inflicted cost of communication through parallelization is reasonable to small relative to the cost of computation.

It is therefore not surprising that certain theoretical results of task scheduling can be established for coarse grained task graphs. For example, bounds on the schedule length in relation to the optimal schedule length can be established for coarse grained graphs (Darte et al. [52], Gerasoulis and Yang [77], Hanen and Munier [86]). See also Theorem 4.6 and Theorem 5.3 of Section 5.3.2.

Some of the algorithms that will be presented later in this text essentially try to improve (i.e., increase) the granularity of a task graph, for example, clustering (Section 5.3) and grain packing (Section 5.3.5). Communication costs are eliminated by grouping nodes together and thereby making communication local, thus cost free.

Weak Granularity A local ratio between the computation and communication costs is the task or node grain (Gerasoulis and Yang [77]). This ratio measures the communication received and sent by a node in relation to the computation of its predecessors and successors, respectively.

Definition 4.21 (Grain) *Let $G = (\mathbf{V}, \mathbf{E}, w, c)$ be a task graph. The grain of node $n_i \in \mathbf{V}$ is*

$$grain(n_i) = \min \left\{ \frac{\min_{n_j \in \mathbf{pred}(n_i)} w(n_j)}{\max_{e_{ji} \in \mathbf{E}, n_j \in \mathbf{pred}(n_i)} c(e_{ji})}, \frac{\min_{n_j \in \mathbf{succ}(n_i)} w(n_j)}{\max_{e_{ij} \in \mathbf{E}, n_j \in \mathbf{succ}(n_i)} c(e_{ij})} \right\}. \tag{4.46}$$

If $\mathbf{pred}(n_i) \cup \mathbf{succ}(n_i) = \emptyset$, that is, the node is independent, the grain is not defined.

As an example for this definition consider Figure 4.7, where node n has

$$grain(n) = \min \left\{ \frac{\min\{2, 3, 4\}}{\max\{1, 1, 1\}}, \frac{\min\{5, 6\}}{\max\{2, 2\}} \right\} = \min \left\{ \frac{2}{1}, \frac{5}{2} \right\} = 2.$$

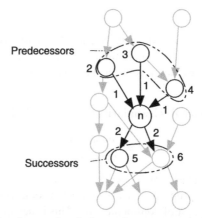

Predecessors

Successors

Figure 4.7. A task graph, where node n is featured as an example for Definition 4.21 of the grain; node n has $grain(n) = \min\{\min\{2,3,4\}/\max\{1,1,1\}, \min\{5,6\}/\max\{2,2\}\} = 2$.

So the grain is a local definition of granularity in the same way the node levels are local definitions of the critical path concept (Section 4.4.2). Of course, other definitions of the node grain are possible; for example, the incoming and outgoing communications could be set into relation with the weight $w(n)$ of the node n itself. However, the above definition of the grain leads to the following, weaker definition of global granularity, which is defined as the minimum of all node grains (Gerasoulis and Yang [77]).

Definition 4.22 (Task Graph Weak Granularity) *Let $G = (\mathbf{V}, \mathbf{E}, w, c)$ be a task graph. G's weak granularity is*

$$g_{\text{weak}}(G) = \min_{n \in \mathbf{V}, \textbf{pred}(n) \cup \textbf{succ}(n) \neq \emptyset} grain(n). \tag{4.47}$$

This definition of granularity is called weak granularity because

$$g(G) \leq g_{\text{weak}}(G), \tag{4.48}$$

as can be seen directly from the definitions. Hence, the desirable property of coarse granularity is easier to achieve with this latter definition. For some theoretical results of task scheduling, coarse granularity in this weak sense is enough (Gerasoulis and Yang [77]). This is the main motivation for defining the node grain as done in Definition 4.21.

While it might seem that the two definitions of granularity do not differ much, their values might be hugely different in practice. To comprehend this, consider the task graph G given in Figure 4.8. Its granularity is

$$g(G) = \frac{\min\{1, 100\}}{\max\{1, 100\}} = \frac{1}{100},$$

whereas its weak granularity is $g_{\text{weak}}(G) = \min\{grain(a), grain(b), grain(c)\} = \min\{1, 1, 1\} = 1$. Hence, the granularities differ by a factor of 100 in this example!

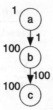

Figure 4.8. Example task graph G to illustrate difference between $g(G)$ and $g_{\text{weak}}(G)$.

Note that the task graph of Figure 4.8 is not unrealistic: in real programs there is often (but not always) some form of relation between the computation time and the communicated data. This is the case in the example graph: node a has a short computation time $w(a) = 1$ and consequently a short outgoing communication time $c(e_{ab}) = 1$ and the same principle holds for nodes b and c. In conclusion, the concept of weak granularity seems to reflect this quite well.

Granularity and Critical Paths   Using the granularity, a relation between the critical path cp and the computation critical path cp_w of a task graph G can be established.

First, the ratio between the weight of an edge and weight of its origin node is studied. Let $G = (\mathbf{V}, \mathbf{E}, w, c)$ be a task graph. For any edge $e_{ij} \in \mathbf{E}$, $n_i, n_j \in \mathbf{V}$, it follows directly from Definitions 4.21 and 4.22 that

$$g_{\text{weak}}(G) \le grain(n_j) \le \frac{w(n_i)}{c(e_{ij})}. \tag{4.49}$$

With this consideration, the following theorem can be established.

Theorem 4.6 (Relation Between Critical Path and Computation Critical Path)
Let $G = (\mathbf{V}, \mathbf{E}, w, c)$ be a task graph, cp its critical path, and cp_w its computation critical path. The nodes of cp are denoted by $\mathbf{V}_{cp} \subseteq \mathbf{V}$, where $n_{\text{last}} \in \mathbf{V}_{cp}$ is the last node of cp, and its edges by $\mathbf{E}_c \subseteq \mathbf{E}$. It holds that

$$len(cp) \le \left(1 + \frac{1}{g_{\text{weak}}(G)}\right) len_w(cp_w). \tag{4.50}$$

Proof. Inequality (4.50) follows from Definition 4.17 of the path length, by substituting the edge weights using Eq. (4.49).

$$len(cp) = \sum_{n \in \mathbf{V}_{cp}} w(n) + \sum_{e_{ij} \in \mathbf{E}_{cp}} c(e_{ij})$$

$$\le \sum_{n \in \mathbf{V}_{cp}} w(n) + \sum_{e_{ij} \in \mathbf{E}_{cp}} \frac{w(n_i)}{g_{\text{weak}}(G)} \quad \text{with Eq. (4.49)}$$

$$= \sum_{n \in \mathbf{V}_{cp}} w(n) + \sum_{n \in \mathbf{V}_{cp} - n_{\text{last}}} \frac{w(n)}{g_{\text{weak}}(G)}$$

$$\leq \sum_{n \in \mathbf{V}_{cp}} w(n) + \frac{1}{g_{\text{weak}}(G)} \sum_{n \in \mathbf{V}_{cp}} w(n) \quad \text{adding } \frac{w(n_{\text{last}})}{g_{\text{weak}}(G)} \qquad (4.51)$$

$$= \left(1 + \frac{1}{g_{\text{weak}}(G)}\right) \sum_{n \in \mathbf{V}_{cp}} w(n)$$

$$= \left(1 + \frac{1}{g_{\text{weak}}(G)}\right) len_w(cp)$$

$$\leq \left(1 + \frac{1}{g_{\text{weak}}(G)}\right) len_w(cp_w) \qquad \text{with Eq. (4.25).} \qquad \square$$

A corresponding inequality is valid for any path p in task graph G:

$$len(p) \leq \left(1 + \frac{1}{g_{\text{weak}}(G)}\right) len_w(p), \qquad (4.52)$$

which can be shown with the above proof, leaving out the very last step of Eq. (4.51). Obviously, Theorem 4.6 is also valid when the weak granularity is substituted by the granularity $g(G)$, but using weak granularity results in a tighter inequality, as should be clear from the example of Figure 4.8. Actually, an even weaker granularity could be defined, where the node grain reflects only the relation between the node weight and its *outgoing* edge weights. The foregoing proof would also hold for such a definition of granularity. For the sake of comparability with other results, this is not done here.

Theorem 4.6 states that the larger the weak granularity, the smaller is the possible difference between the critical path and the computation critical path:

$$g_{\text{weak}}(G) \to \infty \Rightarrow len(cp) \to len_w(cp_w). \qquad (4.53)$$

The theorem is valuable for establishing bounds on the length of certain schedules (see Theorem 5.3 of Section 5.3.2).

Communication to Computation Ratio   The measure of granularity considers extreme values (minimums and maximums) and consequently guarantees certain properties of a task graph. For example, selecting arbitrarily one node $n \in \mathbf{V}$ and one edge $e \in \mathbf{E}$ of a task graph $G = (\mathbf{V}, \mathbf{E}, w, c)$ with granularity $g(G)$, it always holds that

$$\frac{w(n)}{c(e)} \geq g(G). \qquad (4.54)$$

However, the general scheduling behavior of a task graph is not necessarily related to the granularity of the graph. After all, two almost identical graphs can have very different granularities due to the fact that one node or edge weight differs significantly between them. Average or total measures are usually better suited to reflect the scheduling behavior. An often employed measure, especially in experimental comparisons, is the ratio of the total costs.

Definition 4.23 (Communication to Computation Ratio (CCR)) *Let* $G = (\mathbf{V}, \mathbf{E}, w, c)$ *be a task graph. G's communication to computation ratio is*

$$CCR(G) = \frac{\sum_{e \in \mathbf{E}} c(e)}{\sum_{n \in \mathbf{V}} w(n)}. \tag{4.55}$$

Usually, a task graph is said to have high, medium, and low communication for CCRs of about 10, 1, and 0.1, respectively. With the help of the CCR, one can judge the importance of communication in a task graph, which strongly determines the scheduling behavior. In many experiments, the CCR is used to characterize the task graphs of the workload (Khan et al. [103], Macey and Zomaya [132], McCreary et al. [136], Sinnen and Sousa [179]). In the literature the CCR is sometimes the ratio of the average edge to the average node weight (Kwok and Ahmad [111]). This definition is insensible to the number of edges and thus does not reflect the total communication volume.

4.5 CONCLUDING REMARKS

This chapter was devoted to the formal definition of the task scheduling problem and its theoretical analysis. The chapter's aim is to provide the reader with a consistent framework of task scheduling. Using this framework, the following chapters will study heuristics and techniques of task scheduling.

Focus has been placed on the general scheduling problem, namely, the one including communication costs, without any restriction on the task graph and for a target parallel system with a finite number of processors. Later in Section 6.4 it will be seen that there are many variations of this problem, which often result from some form of restriction. Apart from these "special cases," there are many other scheduling problems, for instance, scheduling with deadlines or with preemption. They sometimes differ quite strongly from the ones discussed in this text and are often not related to parallel systems. The reader interested in such scheduling problems should refer to Brucker [27], Chrétienne et al. [34], Leung [121], Pinedo [151], and Veltman et al. [196].

4.6 EXERCISES

4.1 Figure 4.9(*a*) depicts a small task graph *G*, according to Definition 3.13. It is scheduled on a target system as defined in Definition 4.3, consisting of three processors. Three different schedules are shown in Figures 4.9(*b*) to 4.9(*d*).

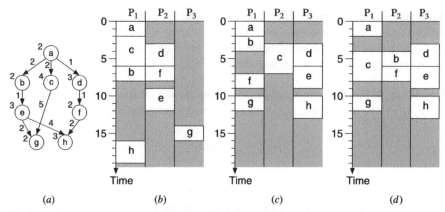

Figure 4.9. Task graph (*a*) and three different, possibly infeasible, schedules on three processors (*b*)–(*d*).

(a) For each schedule check if it is feasible.

(b) In case a schedule is infeasible, make it feasible with as little modifications as possible. What is the resulting schedule length?

(c) Create a feasible schedule for task graph *G* that is of equal length or shorter than the ones of Figure 4.9.

4.2 In Exercise 4.1 you are asked to evaluate the feasibility of the schedules depicted in Figure 4.9. The feasibility is to be checked for the general scheduling model that considers communication costs. Consider now scheduling without communication costs (Section 4.3), which is based on the target system model formulated in Definition 4.14.

(a) For each schedule of Figure 4.9 check if it is feasible when communication costs are ignored.

(b) In case a schedule is infeasible, make it feasible. Furthermore, try to reduce the schedule length for each schedule without changing either processor allocation or the node order on each processor. What is the resulting schedule length?

4.3 The proof of Theorem 4.1 demonstrated NP-completeness of the scheduling problem SCHED with a reduction from 3-PARTITION. A less involved proof, which shows NP-completeness in the weak sense, can be based on a reduction from the well known NP-complete problem PARTITION (Garey and Johnson [73]):

> Given *n* positive integer numbers $\{a_1, a_2, \ldots, a_n\}$, is there a subset **I** of indices such that $\sum_{i \in \mathbf{I}} a_i = \sum_{i \notin \mathbf{I}} a_i$?

Devise such a proof. (*Hint*: Use a fork-join graph.) Why does this proof also show NP-completeness for an unlimited number of processors?

4.4 Consider again the task graph $G = (\mathbf{V}, \mathbf{E}, w, c)$ of Figure 4.9(a). Determine for each node $n \in \mathbf{V}$:

(a) Top level $tl(n)$ and computation top level $tl_w(n)$.

(b) Bottom level $bl(n)$ and computation bottom level $bl_w(n)$.

What is the graph's critical path and what is its computation critical path?

4.5 In Exercise 4.4, you are asked to calculate the node levels for the task graph $G = (\mathbf{V}, \mathbf{E}, w, c)$ of Figure 4.9(a). When the nodes are allocated to processors, these levels can change because local communication has zero costs. Figures 4.9(b) to 4.9(d) show three schedules for the task graph of Figure 4.9(a). Using the processor allocation given by these three schedules, calculate the following for each schedule S:

(a) The allocated top level $tl(n, S)$ for each $n \in \mathbf{V}$.

(b) The allocated bottom level $bl(n, S)$ for each $n \in \mathbf{V}$.

What is the graph's allocated critical path for each schedule?

4.6 Given is this task graph:

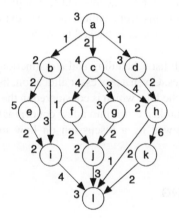

Order its nodes $n \in \mathbf{V}$ in the following orders, while adhering to the precedence constraints:

(a) In descending order of the node weights $w(n)$.

(b) In descending bottom level $bl(n)$ order.

(c) In ascending top level $tl(n)$ order.

(d) In descending top level $tl(n)$ order.

(e) In descending order of $tl(n) + bl(n)$.

4.7 For the task graphs of Exercises 4.1 and 4.6, determine the following:

(a) Granularity $g(G)$.

(b) Weak granularity $g_{\text{weak}}(G)$.

(c) Communication to computation ratio $CCR(G)$.

CHAPTER 5

Fundamental Heuristics

Equipped with the definitions and conditions established in the previous chapter, this chapter studies the two fundamental heuristics of task scheduling: *list scheduling* and *clustering*. These two heuristics are classes or categories rather than simple algorithms. Most of the algorithms that have been proposed for task scheduling fall into one of these two classes.

Both classes are discussed in general terms, following the expressed intention of this book to focus on common concepts and techniques. These are encountered in many scheduling algorithms from which they are extracted and treated separately. Nevertheless, the discussion is backed up with references to many proposed algorithms.

This chapter starts with list scheduling and a distinction is made between static and dynamic node priorities. Given a processor allocation, list scheduling can also be employed to construct a schedule, which is considered in the subsequent section. The area of clustering can be broken down into a few conceptually different approaches. Those are analyzed, followed by a discussion on how to go from clustering to scheduling.

5.1 LIST SCHEDULING

This section is devoted to the dominant heuristic technique encountered in scheduling algorithms, the so-called list scheduling. In its simplest form, the first part of list scheduling sorts the nodes of the task graph to be scheduled according to a priority scheme, while respecting the precedence constraints of the nodes—that is, the resulting node list is in topological order. In the second part, each node of the list is successively scheduled to a processor chosen for the node. Usually, the chosen processor is the one that allows the earliest start time of the node. Algorithm 9 outlines this simplest form of list scheduling.

List scheduling can be considered a heuristic skeleton. An algorithm applying the list scheduling technique has the freedom to define the two, so far unspecified, criteria: the priority scheme for the nodes and the choice criterion for the processor.

Task Scheduling for Parallel Systems, by Oliver Sinnen
Copyright © 2007 John Wiley & Sons, Inc.

Algorithm 9   *Simple List Scheduling—Static Priorities (G = (V, E, w, c), P)*
1: ▷ *1. Part:*
2: Sort nodes $n \in V$ into list L, according to priority scheme and precedence constraints.
3: ▷ *2. Part:*
4: **for** each $n \in L$ **do**
5:    Choose a processor $P \in \mathbf{P}$ for n.
6:    Schedule n on P.
7: **end for**

Section 5.1.3 studies various schemes for the attribution of priorities to nodes, which are mainly based on the task graph related concepts discussed in Section 4.4. Now the concepts behind list scheduling are discussed and a more generic list scheduling heuristic is developed.

Partial Schedules   List scheduling creates the final schedule by successively scheduling each node of the list onto a processor of the target system. Thus, each node n is added to the current partial schedule, denoted by $\mathcal{S}_{\mathrm{cur}}$, which consists of the nodes scheduled before n. As each node is only scheduled once, that is, the start time and the allocated processor are never changed in a later step of the algorithm, the partial schedules must be feasible in order to achieve a feasible final schedule.

Free Nodes   Scheduling a node means to allocate it to a processor and to set its start time. As elaborated in Section 4.1, in order to produce a feasible schedule the start time on the allocated processor must obey Condition 4.1 (Exclusive Processor Allocation) and Condition 4.2 (Precedence Constraint). However, Condition 4.2 can only be verified for a node n if all predecessors—and in consequence all ancestors—of n have already been scheduled: that is, they are part of the current partial schedule. A node for which this is true is called a *free node*.

Definition 5.1 (Free Node)   *Let $G = (V, E, w, c)$ be a task graph, $\mathbf{P}$ a parallel system, and $\mathcal{S}_{\mathrm{cur}}$ a partial feasible schedule for a subset of nodes $V_{\mathrm{cur}} \subset V$ on $\mathbf{P}$. A node $n \in V$ is said to be free if $n \notin V_{\mathrm{cur}}$ and $\mathbf{ance}(n) \subset V_{\mathrm{cur}}$.*

An essential characteristic of list scheduling is that it guarantees the feasibility of all partial schedules and the final schedule by scheduling only free nodes and choosing an appropriate start time for each node. Every node to be scheduled is free, because the nodes are processed in precedence order (i.e., in topological order). Hence, by definition, at the time a node is scheduled all ancestor nodes have already been processed.

End Technique   With the requisite of node n being free, the partial schedule remains feasible if the scheduling of n complies with the two feasibility conditions. The most common way to determine an appropriate start time of n is defined next.

Definition 5.2 (End Technique) *Let $G = (\mathbf{V}, \mathbf{E}, w, c)$ be a task graph, $\mathbf{P}$ a parallel system, and S_{cur} a partial feasible schedule for a subset of nodes $\mathbf{V}_{\text{cur}} \subset \mathbf{V}$ on $\mathbf{P}$. The start time of the free node $n \in \mathbf{V}$, on a given processor P, is determined by*

$$t_s(n, P) = \max\{t_{\text{dr}}(n, P), t_f(P)\}. \tag{5.1}$$

This determination of the start time is here called "end technique," as node n is scheduled at the end of all other nodes scheduled on processor P. In Section 6.1, another technique is presented—the insertion technique—which can compute earlier start times, yet with a higher complexity. Since $t_s(n, P) \geq t_{\text{dr}}(n, P)$, $t_s(n, P)$ complies with Condition 4.3—that is, it complies with Condition 4.2 for all predecessors of node n. Furthermore, as $t_s(n, P) \geq t_f(P)$ (here $t_f(P)$ is the finish time of P in the partial schedule S_{cur}), $t_s(n, P)$ cannot violate Condition 4.1 either.

An interesting property of the end technique is that it produces a schedule of optimal length if the nodes are scheduled in the "right" order on the "right" processor.

Theorem 5.1 (End Technique Is Optimal) *Let S be a feasible schedule for task graph $G = (\mathbf{V}, \mathbf{E}, w, c)$ on system $\mathbf{P}$. Using simple list scheduling (Algorithm 9), the schedule S_{end} is created employing the end technique (Definition 5.2), whereby the nodes are scheduled in nondecreasing order of their start times in S and allocated to the same processors as in S. Then*

$$sl(S_{\text{end}}) \leq sl(S). \tag{5.2}$$

Proof. First, S_{end} is a feasible schedule, since the node order is a topological order according to Lemma 4.1. Without loss of generality, suppose that both schedules start at time unit 0. It then suffices to show that $t_{f,S_{\text{end}}}(n) \leq t_{f,S}(n) \, \forall n \in \mathbf{V}$ (see Definition 4.10 of schedule length).

Since the processor allocations are identical for both schedules, communications that are remote in one schedule are also remote in the other schedule and likewise for the local communications. Thus, the DRT $t_{\text{dr}}(n)$ of a node n can only differ between the schedules through different finish times of the predecessors (i.e., through different start times) as the execution time is identical in both schedules.

By induction, it is shown now that $t_{s,S_{\text{end}}}(n) \leq t_{s,S}(n) \, \forall n \in \mathbf{V}$. Evidently, this is true for the first node to be scheduled, as it starts in both schedules at time unit 0. Now let $S_{\text{end,cur}}$ be a partial schedule of the nodes of $\mathbf{V}_{\text{cur}}$ of G, for which $t_{s,S_{\text{end,cur}}}(n) \leq t_{s,S}(n) \, \forall n \in \mathbf{V}_{\text{cur}}$ is true. For the node $n_i \in \mathbf{V}, \notin \mathbf{V}_{\text{cur}}$ to be scheduled next on processor $proc_S(n_i) = P$, it holds that $t_{\text{dr},S_{\text{end,cur}}}(n_i, P) \leq t_{\text{dr},S}(n_i, P)$, with the above argumentation. With the end technique, n_i is scheduled at the end of the last node already scheduled on P. But this cannot be later than in S, because the nodes on P in $S_{\text{end,cur}}$ are the same nodes that are executed before n_i on P in S, due to the schedule order of the nodes, and, according to the assumption, no node starts later than in S. Thus, $t_{s,S_{\text{end,cur}}}(n_i, P) \leq t_{s,S}(n_i, P)$ for node n_i. By induction this is true for all nodes of the schedule, in particular, the last node, which proves the theorem. $\quad\square$

This theorem is valid for any given schedule $\mathcal{S}$, in particular, for a schedule of optimal length $\mathcal{S}_{opt}$. In turn, this means that an optimal schedule for a given task graph and target system is defined by the processor allocation and the nodes' execution order. Given these two inputs, a schedule of optimal length can be constructed with list scheduling and the end technique. In other words, the scheduling problem reduces to allocating the nodes to the processors and to ordering the nodes. This is a very important result as it "simplifies" the scheduling problem. But of course, finding the optimal node order and allocation is still NP-hard.

5.1.1 Start Time Minimization

As mentioned at the beginning of this section, the most common processor choice for a node to be scheduled is the processor allowing the earliest start time of the node. This processor is found by simply computing the start time $t_s(n, P)$ for every $P \in \mathbf{P}$ according to Eq. (5.1) and selecting the processor for which the start time is minimal:

$$P_{\min} \in \mathbf{P} : t_s(n, P_{\min}) = \min_{P \in \mathbf{P}}\{\max\{t_{dr}(n, P), t_f(P)\}\}. \tag{5.3}$$

Algorithm 10 displays the procedure for this processor choice and the subsequent scheduling of the free node n on $P_{\min}$. This procedure is an implementation of lines 5 and 6 of the simple list scheduling (Algorithm 9).

Algorithm 10 *Start Time Minimization: Schedule Free Node n on Earliest-Start-Time Processor*
Require: n is a free node
    $t_{\min} \leftarrow \infty; P_{\min} \leftarrow NULL$
    for each $P \in \mathbf{P}$ **do**
        if $t_{\min} > \max\{t_{dr}(n, P), t_f(P)\}$ **then**
            $t_{\min} \leftarrow \max\{t_{dr}(n, P), t_f(P)\}; P_{\min} \leftarrow P$
        end if
    end for
    $t_s(n) \leftarrow t_{\min}; proc(n) \leftarrow P_{\min}$

In the literature, list scheduling usually implies the above start time minimization method. Many algorithms have been proposed for this kind of list scheduling (Adam et al. [2], Coffman and Graham [38], Graham [80], Hu [93], Kasahara and Nartia [102], Lee et al. [117], Liu et al. [128], Wu and Gajski [207], Yang and Gerasoulis [210]). Some of these publications, especially the earlier ones, are based on restricted scheduling problems as discussed in Section 6.4.

An Example To illustrate, simple list scheduling with start time minimization is applied to the sample task graph (e.g., in Figure 4.1) and three processors. Suppose the first part of list scheduling established the node order $a, b, c, d, e, f, g, i, h, j, k$. Figure 5.1 visualizes the scheduling process described in the following.

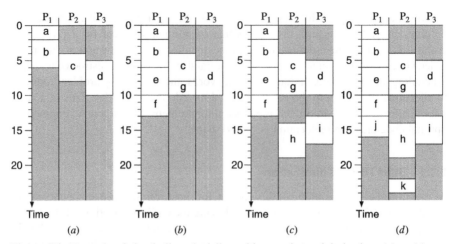

Figure 5.1. Example of simple list scheduling with start time minimization: (*a*) to (*c*) are snapshots of partial schedules; (*d*) shows the final schedule. The node order of the sample task graph (e.g., in Figure 4.1) is $a, b, c, d, e, f, g, i, h, j, k$.

It is irrelevant to which processor node a is scheduled and P_1 is chosen. The next node, b, is also scheduled on P_1, as local communication between a and b permits node b's earliest DRT on P_1 and thus its earliest start time. Nodes c and d are best scheduled on processors P_2 and P_3, despite the remote communication with a on these processors; on P_1 they had to wait until b finishes, which is later than their start times on P_2 and P_3. The partial schedule at this point is shown in Figure 5.1(*a*). Node e is now best scheduled on P_1, due to P_1's early finish time. Node f's start time is identical on all processors and P_1 is chosen. Next, node g is scheduled on P_2, since its predecessor node c is scheduled on this processor (Figure 5.1(*b*)). For node i, all processors allow the same earliest start time and the node is scheduled on P_3. Node h depends on nodes d and e, whereby node d sends the larger message. This determines the earliest start time on the two processors P_1 and P_2, of which P_2 is chosen (Figure 5.1(*c*)). The next node j is best scheduled on processor P_1 and node k's start must be delayed on every processor in the wait for communications from nodes h and j. The final schedule, with k on P_2, is depicted in Figure 5.1(*d*).

Complexity   The complexity of the simple list scheduling can be broken down into the complexity of the first and the second part (Algorithm 9). As for the first part, its complexity is analyzed in Section 5.1.3, since it depends on the employed priority scheme. The complexity of the second part depends on the way in which a processor is chosen for a node. With start time minimization (Algorithm 10) the complexity is as follows. To calculate the start time of a node n according to Eq. (5.1), the data ready time $t_{dr}(n)$ is computed, which involves one calculation for every predecessor of n. For all nodes, calculating the DRT amortizes to $O(\mathbf{E})$, as the sum of the predecessors of all nodes is $|\mathbf{E}|$. Since this is done for each processor, the total complexity of the DRT calculation is $O(\mathbf{PE})$. The start time of every node is computed on every processor, that is, $O(\mathbf{PV})$ times; hence, the total complexity is $O(\mathbf{P}(\mathbf{V} + \mathbf{E}))$. It is possible to achieve a slightly lower complexity for list scheduling with static priorities, as analyzed by Radulescu and Gemund [157].

List scheduling with start time minimization belongs to the class of greedy algorithms (Cormen et al. [42]). At each step, the heuristic tries to create a new partial schedule of short length, with the conjecture that this will eventually result in a short final schedule. A mistake regarding communication made in an early step cannot be remedied later. Graham [80] shows that, for task graphs without communication costs, the worst-case length of a schedule produced by list scheduling with start time minimization is twice the optimal length. For task graphs with communication costs, however, no such guarantee on the schedule length exists.

Theorem 5.2 (No List Schedule Bound) *For any list scheduling algorithm with start time minimization, no constant $C \in \mathbb{Q}^+$ exists such that*

$$sl(\mathcal{S}) \leq C \times sl(\mathcal{S}_{\text{opt}}) \qquad (5.4)$$

is true for every task graph $G = (\mathbf{V}, \mathbf{E}, w, c)$ on every parallel system $\mathbf{P}$. $\mathcal{S}$ is the schedule produced with the list scheduling and $\mathcal{S}_{\text{opt}}$ is a schedule of optimal length.

Proof. The theorem is proved by showing that a task graph can be constructed for any constant C, for which the list schedule length is $sl(\mathcal{S}) > C \times sl(\mathcal{S}_{\text{opt}})$. Consider the task graph in Figure 5.2(a) with $k \in \mathbb{Q}^+$. For $k > 1$, an optimal schedule on two processors, of length 3, is visualized in Figure 5.2(b); if the nodes are not scheduled all on one processor, at least one communication is sent across the network resulting in $sl(\mathcal{S}) \geq 3k$, for example, as illustrated in Figure 5.2(c). For list scheduling there are two possible node orders: A, B, C or B, A, C. However, both orders result in a schedule where A and B are scheduled on different processors, due to the start time minimization. Independent of which processor node C is scheduled on, the schedule

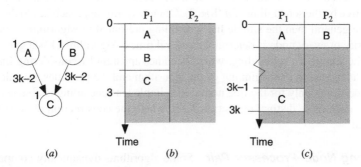

Figure 5.2. (a) A task graph; (b) its optimal schedule ($k > 1$); and (c) a list schedule.

length is $3k$ (Figure 5.2(c)). For any C a k with $k > 1$, $k > C$ can be chosen so that $sl(S) = 3k > 3C = C \times sl(S_{\text{opt}})$. $\qquad\qquad\qquad\qquad\qquad\qquad\qquad\qquad\qquad\qquad$ $\square$

In particular, list scheduling with start time minimization does not even guarantee a schedule length shorter than the sequential time of a task graph. Still, numerous experiments have demonstrated that list scheduling produces good schedules for task graphs with low to medium communication (e.g., Khan et al. [103], Kwok and Ahmad [111], Sinnen and Sousa [177]).

5.1.2 With Dynamic Priorities

The simple list scheduling outlined in Algorithm 9 establishes the schedule order of the nodes before the actual scheduling process. During the node scheduling in the second part, this order remains unaltered, so the node priorities are *static*. To achieve better schedules, it might be beneficial to consider the state of the partial schedule in the schedule order of the nodes. Changing the order of the nodes is equivalent to changing their priorities, hence the node priorities are *dynamic*. It remains that the node order must be compatible with the precedence constraints of the task graph, which is obeyed if only free nodes are scheduled. A list scheduling algorithm for dynamic node priorities is given in Algorithm 11.

Algorithm 11 *List Scheduling—Dynamic Priorities* $(G = (\mathbf{V}, \mathbf{E}, w, c), \mathbf{P})$
1: Put source nodes $\{n_i \in \mathbf{V} : \mathbf{pred}(n_i) = \emptyset\}$ into set S.
2: **while** $S \neq \emptyset$ **do** $\qquad\qquad\qquad\qquad\qquad\qquad\qquad$ $\triangleright$ S *only contains free nodes*
3: $\quad$ Calculate priorities for nodes $n_i \in S$.
4: $\quad$ Choose a node n from S; $S \leftarrow S - n$.
5: $\quad$ Choose a processor P for n.
6: $\quad$ Schedule n on P.
7: $\quad$ $S \leftarrow S \cup \{n_i \in \mathbf{succ}(n) : \mathbf{pred}(n_i) \subseteq S_{\text{cur}}\}$
8: **end while**

Instead of iterating over the node list, Algorithm 11 calculates in each step the priority of all free nodes and only then selects a node and a processor for scheduling. After the scheduling of node n, the set S of free nodes is updated with the nodes that became free by the scheduling of n (line 7). Obviously, this approach has higher complexity in general than the simple list scheduling presented in Algorithm 9, because the priority of every node is determined several times. Algorithm 11 is a generalization of the simple list scheduling, which becomes apparent by considering line 3. If this line is moved to the beginning of the algorithm and the priorities are calculated for all nodes, not only for the free nodes, the simple list scheduling is obtained. This becomes even more distinct in Section 5.1.3, where the construction of node lists is explained.

Choosing Node–Processor Pair Some algorithms dynamically compute the priorities of the free nodes by evaluating the start time (see Eq. (5.1)) for every free node on every processor. That node of all free nodes that allows the earliest start time is selected together with the processor on which this time is achieved. The choice of the node and the processor is made simultaneously, based on the state of the partial schedule. Two examples for dynamic list scheduling algorithms employing this technique are ETF (earliest time first) by Hwang et al. [94] and DLS (dynamic level scheduling) by Sih and Lee [169].

In comparison to the simple list scheduling, the complexity increases by a factor of $|\mathbf{V}|$. Evaluating the start time of every free node on every processor is in the worst case $O(\mathbf{P}(\mathbf{V} + \mathbf{E}))$, which is already the complexity of the simple list scheduling. As this step is done $|\mathbf{V}|$ times, the total complexity is $O(\mathbf{P}(\mathbf{V}^2 + \mathbf{VE}))$.

5.1.3 Node Priorities

Discussion of the list scheduling technique distinguished between static and dynamic node priorities. How these priority schemes are realized is discussed in the next paragraphs.

Static Static priorities are based on the characteristics of the task graph. In the simplest case, the priority metric itself establishes a precedence order among the nodes. It is then sufficient to order the nodes according to their priorities with any sort algorithm, for example, with Mergesort (Cormen et al. [42]), which has a complexity of $O(\mathbf{V} \log \mathbf{V})$. Prominent examples for such priority metrics are the top and the bottom level of a node (Section 4.4.2).

Lemma 5.1 (Level Orders Are Precedence Orders) *Let $G = (\mathbf{V}, \mathbf{E}, w, c)$ be a task graph. The nonincreasing bottom level (bl) order and the nondecreasing top level (tl) order of the nodes $n \in \mathbf{V}$ are precedence orders.*

Proof. The lemma follows immediately from the recursive definition of the bottom level and top level in Eqs. (4.43) and (4.44), respectively. A node always has a smaller bottom level than any of its predecessors and a higher top level than any of its successors.   □

If the priority metric is not compatible with the precedence constraints of the task graph, a correct order can be established by employing the free node concept. Algorithm 12 shows how to create a priority ordered node list that complies with the precedence constraints. As mentioned before, the general structure is similar to Algorithm 11, but a priority queue (Q) is used instead of a set (S), because the priorities assigned to the nodes are static.

The simple list scheduling outlined in Algorithm 9 could be rewritten by using the below algorithm and substituting line 5 with lines 5 and 6 from Algorithm 9. However, then the conceptually nice separation of the first and the second part of simple list scheduling is lost.

Suppose the priority queue in Algorithm 12 is a heap (Cormen et al. [42]); the algorithm's complexity is $O(\mathbf{V} \log \mathbf{V} + \mathbf{E})$: every node is inserted once in the heap,

Algorithm 12   Create Node List
1: Assign a priority to each $n \in \mathbf{V}$.
2: Put source nodes $n \in \mathbf{V} : \mathbf{pred}(n) = \emptyset$ into priority queue Q.
3: **while** $Q \neq \emptyset$ **do**   ▷ *Q only contains free nodes*
4:      Let n be the node of Q with highest priority; remove n from Q.
5:      Append n to list L.
6:      Put $\{n_i \in \mathbf{succ}(n) : \mathbf{pred}(n_i) \subseteq L\}$ into Q.
7: **end while**

which costs $O(\log \mathbf{V})$ and the determination of the free nodes amortizes to $O(\mathbf{E})$. Removing the node with the highest priority from the heap is $O(1)$.

An example for a metric that is not compatible with the precedence order is $tl + bl$—a metric that gives preference to the nodes of the critical path. Table 5.1 displays several node orders of the sample task graph obtained with node level based priority metrics. The values of the node levels are given in Table 4.1. As analyzed in Section 4.4, calculating these levels has a complexity of $O(\mathbf{V} + \mathbf{E})$. For all these metrics the first part of the simple list scheduling is thus $O(\mathbf{V} \log \mathbf{V} + \mathbf{E})$ (calculating levels: $O(\mathbf{V} + \mathbf{E})$ + sorting: $O(\mathbf{V} \log \mathbf{V})$ or $O(\mathbf{V} \log \mathbf{V} + \mathbf{E})$).

Already the few examples in Table 5.1 illustrate the large variety of possible node orders. In experiments, the relevance of the node order for the schedule length was demonstrated (e.g., Adam et al. [2], Sinnen and Sousa [177]). Unfortunately, most comparisons of scheduling algorithms (e.g., Ahmad et al. [6], Gerasoulis and Yang [76], Khan et al. [103], Kwok and Ahmad [111], McCreary et al. [136]) analyze entire algorithms: that is, the algorithms differ not only in one but many aspects. This generally impedes one from drawing conclusions on discrete aspects of the algorithms (e.g., on the node order). Still, most algorithms apply orders based on node levels and the bottom level order exhibits good average performance (e.g., Adam et al. [2], Sinnen and Sousa [177]).

Breaking Ties   Table 5.1 also shows examples where the node order is not fully defined, indicated in the table by parentheses around the respective nodes. Incidentally,

Table 5.1. Common Priority Metrics and the Corresponding Precedence Compatible Orders of the Nodes of the Sample Task Graph[a]

Priority Metric	Node Order
tl (ASAP)	$a, c, d, e, b, g, i, h, f, j, k$
bl (ALAP)	$a, b, d, f, e, c, j, h, g, i, k$
$tl + bl$	$a, b, f, j, d, e, h, i, c, g, k$
tl_w (ASAP$_w$)	$a, (b, c, d, e), (f, g, i), h, j, k$
bl_w (ALAP$_w$)	$a, (b, d), e, (c, f), h, j, (g, i), k$
$tl_w + bl_w$	$a, ([b, f, j], d), e, h, ([c, g], i), k$

[a]For example, in Figure 4.1. Node levels are given in Table 4.1; order of nodes within parentheses is not defined.

only computation levels are affected here, but this is a general problem that can occur for most level based priority metrics, especially in regular task graphs or for integer task graph weights. Scheduling algorithms therefore use additional criteria to decide the node order. For example, in a bottom level based order, such a tie breaker can be the top level of the nodes. But often ties are broken randomly.

Example Priority Schemes Given the enormous number of proposed scheduling algorithms, the following list of priority schemes represents only a small, though important part, of the existing ones. In the HLF (highest level first) (Hu [93]) and HLFET (highest level first with estimated times) (Adem et al. [2]) algorithms, the nodes are ordered according to the computation bottom level bl_w. The CP/MISF (critical path/most immediate successors first) algorithm by Kasahara and Nartia [102] is also bl_w based but breaks the ties by giving precedence to the node with the higher number of successors. Wu and Gajski [207] propose the MCP (modified critical path) heuristic, where the nodes are ordered by their bl and for nodes with equal bl, the bl values of the successors are considered and so on. A critical path based order is proposed by Ahmad and Kwok [4, 114], where non-CP nodes are scheduled according to their bl and ties are broken with tl. Sinnen and Sousa analyze [177] various priority schemes of two different types: (1) node orders according to the (computation) bottom level augmented with metrics based on the entering communication of a node; and (2) critical path based orders as proposed by Ahmad and Kwok [4, 115].

Dynamic Level or critical path based priority schemes turn into dynamic priorities when they are recalculated in each scheduling step based on the current partial schedule. Section 4.4 explained how the allocated path length is calculated given the (partial) processor allocation of the nodes. For instance, the allocated critical path might be recalculated for each partial schedule.

It should be noted that the allocated bottom level $bl(n, \mathcal{S}_{cur})$ of a node n remains unaltered in list scheduling until n is scheduled; that is, $bl(n, \mathcal{S}_{cur}) = bl(n)$—the bottom level of n depends only on its descendants (see Eq. (4.43)).

In general, recalculating the levels or the CP in each step of scheduling multiplies the costs for the level or CP determination by a factor of $|\mathbf{V}|$. However, an efficient implementation might only update the levels for those nodes that are affected by the scheduling of a node.

Start Times Apart from node orders based on task graph characteristics, priority schemes often use the potential start time of free nodes as a preference criterion. The node priority thereby becomes a function of the processor. Choosing the node–processor pair as discussed in Section 5.1.2 is such a priority scheme. Another technique selects the free node in each step that has the earliest DRT, assuming all communications are remote, sometimes called *ready node* (Yang and Gerasoulis [210]).

Dynamic priorities are often a mixture of task graph based metrics and the earliest start time, whereby the task graph characteristics are only quantified at the beginning of the algorithm and remain unmodified. For instance, the DLS algorithm (Sih and Lee [169]) defines its dynamic level using bl_w and the node's earliest start time on all processors. The ETF algorithm by Hwang et al. [94] also selects the node–processor pair with the earliest start time, breaking ties with bl_w.

5.2 SCHEDULING WITH GIVEN PROCESSOR ALLOCATION

At the definition of a schedule (Definition 4.2) in Section 4.2, it was stated that scheduling is the spatial and temporal assignment of the tasks to the processors. List scheduling does this assignment in an integrated process, that is, at the same time a node is assigned to a processor and attributed a start time. In fact, when referring to scheduling, normally both the spatial and temporal assignments are meant. Yet, it is a natural idea to split this process into its two fundamental aspects: processor allocation and attribution of start times.

When doing so, there are two alternatives: (1) first processor allocation, then attribution of start times; and (2) first attribution of start times, then processor allocation. For a limited number of processors, the second alternative is extremely difficult, because nodes with overlapping execution times have to be executed on different processors. Furthermore, without knowledge of the processor allocation, it is not known which communications will be local and which remote. Hence, adhering to the processor and precedence constraints (Conditions 4.1 and 4.2) would be very difficult. As a consequence, the two-phase scheduling starts with the allocation (or mapping) of the nodes to the processors. Based on these considerations, the generic structure of a two-phase heuristic is shown in Algorithm 13.

Algorithm 13   Generic Two-Phase Scheduling Algorithm $(G = (V, E, w, c), P)$
▷ *Processor allocation/mapping*
(1) Assign each node $n \in V$ to a processor $P \in P$
▷ *Scheduling/ordering nodes*
(2) Attribute a start time $t_s(n)$ to each node $n \in V$, adhering to Condition 4.1 (processor constraint) and Condition 4.2 (precedence constraint)

Next it is analyzed how list scheduling can be utilized to handle the second phase of this algorithm. The processor allocation (i.e., the first phase) can be determined in many different ways. One possibility is to extract it from another schedule, as is done, for example, in the proof of Theorem 5.1. This is especially interesting when a more sophisticated scheduling model is used in the second phase, like those discussed in Chapters 7 and 8. With such an approach, the scheduling in the first phase can be considered an estimate, which is refined in the second phase. A genetic algorithm based heuristic for finding a processor allocation is referenced in Section 8.4.2. More on processor allocation heuristics can be found in El-Rewini et al. [65] and Rayward-Smith et al. [159].

5.2.1 Phase Two

The second phase of this generic two-phase scheduling algorithm can easily be performed by a list scheduling heuristic. In each iteration of (the second part of) list scheduling (i.e., for each node $n \in \mathbf{V}$), the heuristic first chooses a processor P on which to schedule n. This is done in line 5 of both static list scheduling (Algorithm 9) and dynamic list scheduling (Algorithm 11). When the processor allocation $\mathcal{A}$ is already given, this line can simply be omitted, as $proc(n)$ is already determined for all $n \in \mathbf{V}$.

It suffices to schedule each node at the earliest possible start time,

$$t_s(n, proc(n)) = \max\{t_{\mathrm{dr}}(n, proc(n)), t_f(proc(n))\}, \tag{5.5}$$

hence using the end technique. Thus, Eq. (5.5) is an implementation of lines 5 and 6 of the simple list scheduling (Algorithm 9) or the dynamic list scheduling (Algorithm 11). As the selection of the processor is not performed anymore in these algorithms, the complexity of the second part of list scheduling reduces by the factor $|\mathbf{P}|$. The second part of simple list scheduling is then $O(\mathbf{V} + \mathbf{E})$.

From Theorem 5.1 it is known that the end technique is optimal for a given node order and processor allocation. As a result, scheduling with a given processor allocation reduces to finding the best node order.

Both static and dynamic priorities can be employed to order the nodes for their scheduling. However, since the processor allocations are already determined, task graph characteristics, like node levels and the critical path, can be computed using the allocated path length (Section 4.4), that is, the path length based on the known processor allocations. These characteristics do not change during the entire scheduling; hence, dynamic priorities are only sensible when considering the state of the partial schedules, for example, choosing the node among the free nodes that can start earliest, that is, the ready node (Section 5.1.3).

One might wonder whether this scheduling problem with a given preallocation is still NP-hard. After all, it is "only" about finding the best node order. Unfortunately it is still NP-hard, even for task graphs without communication costs, unit execution time, and very simple graph structures, such as forest (Goyal [79]) or chains (Rayward-Smith et al. [159]); see also Hoogeveen et al. [91].

5.3 CLUSTERING

As mentioned before in Section 4.2.2, task scheduling under the classic model is a trade-off between minimizing interprocessor communication costs and maximizing the concurrency of the task execution. A natural idea is therefore to determine first— before the actual scheduling—which nodes should always be executed on the same processor. Obvious candidates for grouping are nodes that depend on each other, especially nodes of the critical path.

Clustering is a technique that follows this idea. It is therefore only suitable for scheduling with communication costs. In its core it is a scheduling technique for an unlimited number of processors. Nevertheless, it is often proposed as an initial step in scheduling for a limited number of processors. To distinguish between the limited number of (physical) processors and the unlimited number of (virtual) processors assumed in the clustering step, the latter are called clusters, hence the term "clustering." In clustering, the nodes of the task graph are mapped *and* scheduled into these clusters.

Definition 5.3 (Clustering)   *Let $G = (V, E, w, c)$ be a task graph. A clustering C is a schedule of G on an implicit parallel system* **C** *(Definition 4.3) with an "unlimited" number of processors; that is, $|C| = |V|$. The processors $C \in C$ are called clusters.*

Clustering based scheduling algorithms for a limited number of processors consist of several steps, similar to the two-phase scheduling outlined by Algorithm 13 in Section 5.2. Algorithm 13 comprises two phases, where the first phase is the mapping of the nodes and the second is their scheduling. In clustering based algorithms, three steps are necessary: (1) clustering, (2) mapping of the clusters to the (physical) processors, and (3) scheduling of the nodes. Algorithm 14 outlines a generic three-step clustering based scheduling heuristic for a limited number of processors.

Algorithm 14   Generic Three-Step Clustering Based Scheduling Algorithm ($G = (V, E, w, c)$, P)
  ▷ *Clustering nodes*
  (1) Find a clustering C of G
  ▷ *Mapping clusters to processors*
  (2) Assign clusters of **C** to (physical) processors of **P**
  ▷ *Scheduling/ordering nodes*
  (3) Attribute start time $t_s(n)$ to each node $n \in V$, adhering to Condition 4.1 (processor constraint) and Condition 4.2 (precedence constraint)

In contrast to the two-phase Algorithm 13, where the first step is the pure processor allocation of the nodes, clustering also includes the scheduling of the nodes in the clusters. As will be seen later, this is done for an accurate estimation of the execution time of the task graph. It also makes clustering a complete scheduling algorithm for an unlimited number of processors.

While the third step is theoretically identical to the second phase of Algorithm 13, its actual implementation may differ, since the partial node orders as established by the clustering C might be considered in determining the final node order.

In terms of parallel programming terminology, clustering correlates to the step of parallelization designated by orchestration (Culler and Singh [48]) or agglomeration (Foster [69]) as described in Section 2.3.

The often cited motivation for clustering was given by Sarkar [167]: if tasks are best executed in the same processor (cluster) of an ideal system, that is, a system that possesses more processors (clusters) than tasks, they should also be executed on the same processor in any real system. Due to the NP-hardness of scheduling, it cannot be expected that the foregoing conjecture is always true, yet it is an intuitive reasoning for a heuristic.

In the following, the first step of Algorithm 14, the actual clustering, is studied. Steps 2 and 3 are treated in Section 5.4.

5.3.1 Clustering Algorithms

Strictly speaking, the first step of Algorithm 14 can be performed by any scheduling algorithm suitable for an unlimited number of processors. However, in the literature the term clustering designates a certain kind of algorithm, with several characteristic aspects. The following discussion is based on Darte et al. [52], El-Rewini et al. [65], and Gerasoulis and Yang [76].

Principle of Clustering Algorithms   Clustering algorithms start with an *initial clustering* C_0 of the task graph G. Usually each node $n \in V$ is allocated to a distinct cluster $C \in C$.

The clustering algorithm then performs *incremental steps of refinement* going from clustering C_{i-1} to clustering C_i, in which clusters are *merged*. This means that the nodes of these clusters are merged into one single cluster. If communicating nodes are executed in the same cluster, their communication becomes local and hence its cost is zero according to the target system model (Definition 4.3). This can be beneficial for the total execution time of the graph, as it eliminates communication costs. Normally, a merging of clusters is performed only if the schedule length of the new clustering C_i decreases or at least remains the same compared to the schedule length of the current clustering C_{i-1}. The steps of refinement are performed until all candidates for merging have been considered. Algorithm 15 summarizes this principle of clustering algorithms.

Algorithm 15   ***Principle of Clustering Algorithms*** ($G = (V, E, w, c)$)

> Create initial clustering C_0: allocate each node $n \in V$ to a distinct cluster $C \in C$, $|C| = |V|$
> $i \leftarrow 0$
> **repeat**
>> $i \leftarrow i + 1$
>> Select candidate clusters for merging
>> Create new clustering C_i: merge candidate clusters into one cluster
>> **if** $sl(C_i) > sl(C_{i-1})$ **then**   ▷ *merging increases current schedule length*
>>> $C_i \leftarrow C_{i-1}$ ▷ *reject new clustering C_i*
>> **end if**
> **until** all candidates for merging have been considered

Figure 5.3 shows some example clusterings of a simple task graph (Figure 5.3(*a*)), which has often been used to illustrate clustering (Gerasoulis and Yang [76]). An initial clustering is depicted in Figure 5.3(*b*). Each node is allocated to a distinct cluster,

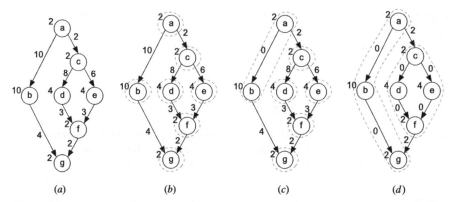

Figure 5.3. Clusterings of a simple task graph: (*a*) simple task graph for clustering algorithms (Gerasoulis and Yang [76]); (*b*) initial clustering; (*c*) clustering after clusters of nodes *a* and *b* have been merged; (*d*) clustering with only two clusters.

symbolized by the gray dashed line enclosing each node. Figure 5.3(*c*) depicts the clustering, where the clusters of nodes *a* and *b* of the initial clustering have been merged into one cluster. This might be the clustering produced by an algorithm after the first refinement step. Lastly, Figure 5.3(*d*) illustrates a clustering with only two clusters.

Implicit Schedule   An essential characteristic of clustering algorithms is that the clustering obtained at each step is a feasible schedule. Otherwise it would not be possible to determine the current schedule length, on which clustering decisions are based.

In clustering algorithms, the scheduling of the nodes is often given implicitly by the allocation of the nodes to the clusters. The first example for this is the initial clustering, where each node starts as-soon-as-possible, that is, at its top level (Section 4.4.2; see also Section 4.3.2):

$$t_s(n) = tl(n) \ \forall n \in \mathbf{V}. \tag{5.6}$$

The resulting schedule is shown in Figure 5.4(*a*), which is the implicit schedule of the clustering in Figure 5.3(*b*).

In general, the starting times of the nodes (i.e., their scheduling) are implicitly given whenever the node order within each cluster is well defined. This is the case when there are no independent nodes in the same cluster. If this criterion is met, the start time of each node in clustering C_i is its *allocated* top level (Section 4.4):

$$t_s(n) = tl(n, C_i) \ \forall n \in \mathbf{V}. \tag{5.7}$$

Figure 5.4(*b*) displays the implicit schedule for the clustering of Figure 5.3(*c*). It is not clear at this point what the implicit schedule of the clustering shown in

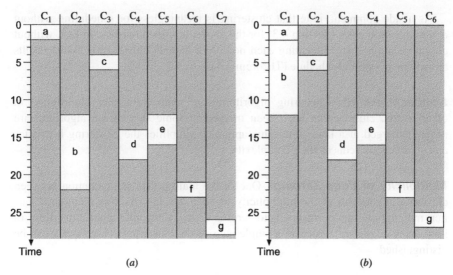

Figure 5.4. Implicit schedules of (a) the initial clustering as shown in Figure 5.3(b) and (b) the clustering in Figure 5.3(c).

Figure 5.3(d) is, since nodes d and e are independent and in the same cluster. A clustering algorithm must establish a rule on how to order (i.e., schedule) such nodes.

In order to keep the complexity of clustering algorithms as low as possible, these algorithms try to be smart about the scheduling of the nodes in each incremental step. Actually, it is not necessary to generate a schedule at each step; it is merely necessary to determine the current schedule length. For this reason, the term *estimated parallel time* (EPT) was introduced for clustering algorithms (Gerasoulis and Yang [76]), which simply is the length of the (implicit) schedule of the present clustering. The aim of the algorithms is to obtain the EPT with as little effort as possible. Different clustering algorithms pursue different strategies and techniques for the ordering (i.e., the scheduling) of the nodes, as will be seen later.

Edge Zeroing   It is important to realize that merging two clusters C_x and C_y can only be beneficial if there is at least one edge e_{ij} going from a node n_i in cluster C_x to a node n_j in cluster C_y (or vice versa). Such a merging will turn the remote communication associated with e_{ij} into a local communication. For this reason, the merging of the clusters is also called *edge zeroing*, as the communication costs of the corresponding edges become zero. Merging clusters that do not result in edge zeroing cannot improve the schedule length (i.e., the EPT) of the clustering.

The attentive reader might have observed that edge zeroing is already reflected in Figure 5.3. The edge weights are 0, whenever the incident nodes are in the same cluster. For example, in Figure 5.3(c) the edge weight $c(e_{ab})$ is zero, since both a and b are in the same cluster, while it is $c(e_{ab}) = 10$ in the initial clustering of Figure 5.3(b). In Figure 5.3(d) most edges are zero, because there are only two clusters resulting in mainly local communication.

As a side comment, note that clustering is meaningless for scheduling without communication costs (Section 4.3) for the above described reasons. In fact, without communication costs, allocating each node to a distinct cluster (processor) results already in an optimal schedule (Theorem 4.4).

Nonbacktracking   Clustering algorithms are nonbacktracking algorithms. In other words, clusters that have been merged in some step of the algorithm are never unmerged at a later time. This approach simplifies the clustering heuristics and significantly reduces their complexity.

Multiplicity of Edge Zeroing   One of the main points of distinction between clustering algorithms is the multiplicity of edge zeroing. As was mentioned earlier, merging clusters corresponds to zeroing edges. Clustering algorithms differ in how many edges are zeroed in a single step of the clustering. Three types can be distinguished:

1. *Path.* All edges of a path in the task graph are zeroed when beneficial.
2. *Edge.* One single edge is zeroed when beneficial.
3. *Node.* One or more incident edges of a single node are zeroed when beneficial.

The above distinction only applies to the edges that are the primary candidates for edge zeroing. In a heuristic, the zeroing of an edge can trigger the implicit zeroing of other edges that are incident on the same node. Heuristics based on each of these three types of multiplicity will be studied next.

5.3.2 Linear Clustering

Linear clustering is a special class of clustering. In linear clustering only dependent nodes are grouped into one cluster. If two nodes in one cluster are independent, this cluster, and with it the entire clustering, is nonlinear.

Definition 5.4 (Linear Cluster and Clustering)   *Let $C \in \mathbf{C}$ be a cluster of a clustering C of task graph $G = (\mathbf{V}, \mathbf{E}, w, c)$. Let $n_1, n_2, \ldots, n_m$ be all the nodes allocated to C, that is, $\{n_1, n_2, \ldots, n_m\} = \{n \in \mathbf{V} : proc(n) = C\}$, arranged in topological order. The cluster is linear if and only if there is a path $p(n_1 \rightarrow n_m)$ from n_1 to n_m to which all nodes allocated to C belong, that is, $\{n_1, n_2, \ldots, n_m\} \subseteq \{n \in \mathbf{V} : n \in p(n_1 \rightarrow n_m)\}$. A clustering C is said to be linear if and only if all of its clusters $\mathbf{C}$ are linear; otherwise it is nonlinear.*

For example, in Figure 5.3 the clusterings (*b*) and (*c*) are linear, while the clustering of (*d*) is nonlinear (*d* and *e* are independent and in the same cluster). In general, initial clusterings with one node per cluster are always linear.

From the considerations of implicit schedules, it is clear that linear clusterings have well-defined implicit schedules. In any linear clustering C_{linear} of a task graph

$G = (\mathbf{V}, \mathbf{E}, w, c)$, each node starts execution at its allocated top level, that is, at the earliest possible time, according to Eq. (5.7). As a result, the schedule length of $\mathcal{C}_{\text{linear}}$ is the *allocated* length of the *allocated* critical path $cp(\mathcal{C}_{\text{linear}})$,

$$
\begin{aligned}
sl(\mathcal{C}_{\text{linear}}) &= len(cp(\mathcal{C}_{\text{linear}}), \mathcal{C}_{\text{linear}}) \\
&= tl(n_{cp(\mathcal{C}_{\text{linear}})}, \mathcal{C}_{\text{linear}}) + bl(n_{cp(\mathcal{C}_{\text{linear}})}, \mathcal{C}_{\text{linear}}),
\end{aligned}
\tag{5.8}
$$

with $n_{cp(\mathcal{C}_{\text{linear}})} \in cp(\mathcal{C}_{\text{linear}})$ being a node of the allocated critical path. A direct consequence of this is formulated in Lemma 5.2.

Lemma 5.2 (Linear Clustering: Schedule Length and Zeroed Edges) *Let $G = (\mathbf{V}, \mathbf{E}, c)$ be a task graph, and $\mathcal{C}_{\text{linear}}$ and $\mathcal{C}_{\text{linear},+}$ two linear clusterings of G. Let $\mathbf{E}_{\text{zero}}$ be the set of zeroed edges of $\mathcal{C}_{\text{linear}}$, that is, $\mathbf{E}_{\text{zero}} = \{e_{ij} \in \mathbf{E} : proc(n_i) = proc(n_j)\}$, and $\mathbf{E}_{\text{zero},+}$ the set of zeroed edges of $\mathcal{C}_{\text{linear},+}$, with $|\mathbf{E}_{\text{zero},+}| > |\mathbf{E}_{\text{zero}}|$ and $\mathbf{E}_{\text{zero}} \subset \mathbf{E}_{\text{zero},+}$. For the schedule lengths of the clusterings it holds that*

$$
sl(\mathcal{C}_{\text{linear},+}) \leq sl(\mathcal{C}_{\text{linear}}).
\tag{5.9}
$$

Proof. Since in $\mathcal{C}_{\text{linear},+}$ at least one more edge is zeroed than in $\mathcal{C}_{\text{linear}}$, it holds for the allocated top and bottom levels that

$$
\begin{aligned}
tl(n, \mathcal{C}_{\text{linear},+}) &\leq tl(n, \mathcal{C}_{\text{linear}}) \\
bl(n, \mathcal{C}_{\text{linear},+}) &\leq bl(n, \mathcal{C}_{\text{linear}})
\end{aligned}
\quad \forall n \in \mathbf{V}.
\tag{5.10}
$$

Since the schedule length of a clustering is the sum of the allocated top and bottom levels of an allocated critical path node, Eq. (5.8), this proves the lemma. $\square$

In other words, given a linear clustering, zeroing more edges, while maintaining the linear property, does not increase the schedule length. Based on these observations, it is easy to establish a bound on the schedule length of a linear clustering relative to the granularity of the task graph.

Theorem 5.3 (Bound on Linear Clustering) *Let $G = (\mathbf{V}, \mathbf{E}, w, c)$ be a task graph, $\mathcal{C}_{\text{linear}}$ a linear clustering of G, and $\mathcal{C}_{\text{opt}}$ a clustering of G with optimal schedule length. It holds that*

$$
sl(\mathcal{C}_{\text{linear}}) \leq \left(1 + \frac{1}{g_{\text{weak}}(G)}\right) sl(\mathcal{C}_{\text{opt}}).
\tag{5.11}
$$

Proof. Let $\mathcal{C}_0$ be an initial clustering, where each task $n \in \mathbf{V}$ is allocated to a different cluster $C \in \mathbf{C}$. Consequently, the allocated length of the allocated critical path $cp(\mathcal{C}_0)$, that is, $\mathcal{C}_0$'s schedule length (see Eq. (5.8)), is identical to the length of the critical path cp of G

$$
sl(\mathcal{C}_0) = len(cp(\mathcal{C}_0), \mathcal{C}_0) = len(cp).
\tag{5.12}
$$

From Lemma 5.2 it is clear that

$$sl(\mathcal{C}_{\text{linear}}) \leq sl(\mathcal{C}_0) \tag{5.13}$$

for any linear clustering $\mathcal{C}_{\text{linear}}$. Theorem 4.6 states that

$$len(cp) \leq \left(1 + \frac{1}{g_{\text{weak}}(G)}\right) len_w(cp_w),$$

from which it follows that

$$sl(\mathcal{C}_{\text{linear}}) \leq sl(\mathcal{C}_0) = len(cp) \leq \left(1 + \frac{1}{g_{\text{weak}}(G)}\right) len_w(cp_w). \tag{5.14}$$

Finally, with $len_w(cp_w) \leq sl(\mathcal{S})$ for any schedule $\mathcal{S}$ (Lemma 4.4), in particular, $len_w(cp_w) \leq sl(\mathcal{S}_{\text{opt}})$, one gets

$$sl(\mathcal{C}_{\text{linear}}) \leq \left(1 + \frac{1}{g_{\text{weak}}(G)}\right) len_w(cp_w) \leq \left(1 + \frac{1}{g_{\text{weak}}(G)}\right) sl(\mathcal{C}_{\text{opt}}). \tag{5.15}$$

$\square$

From Theorem 5.3 it follows that the larger the weak granularity, the smaller is the possible difference between the schedule length of the linear clustering and the optimal schedule length:

$$g_{\text{weak}}(G) \to \infty \Rightarrow sl(\mathcal{C}_{\text{linear}}) \to sl(\mathcal{C}_{\text{opt}}). \tag{5.16}$$

This observation is no surprise, given that Theorem 4.4 establishes that an optimal schedule can be created in polynomial time for scheduling without communication costs on an unlimited number of processors. Only the minimization of the number of used processors is NP-hard. While clustering in general is not a technique useful for scheduling without communication costs, linear clustering can be employed as a heuristic for the reduction of the processor number. Starting with an initial clustering, which is optimal for scheduling without communication costs (Theorem 4.4), Lemma 5.2 guarantees that any linear clustering is still optimal for this scheduling problem.

For coarse grained task graphs G, that is, $g_{\text{weak}}(G) \geq 1$, Theorem 5.3 establishes a 2-optimal bound

$$sl(\mathcal{C}_{\text{linear}}) \leq 2sl(\mathcal{C}_{\text{opt}}). \tag{5.17}$$

For coarse grained task graphs, Gerasoulis and Yang [77] even demonstrated that there is at least one linear clustering that is optimal. Unfortunately, the algorithm to obtain such a clustering in polynomial time is unknown.

Algorithm A linear clustering can be achieved with a simple and intuitive algorithm proposed by Kim and Browne [104]. Consider as a basis the generic clustering Algorithm 15: in each refinement step a new longest path p, consisting of previously unexamined edges, is selected in task graph G. The nodes of p are merged into one cluster and the edges incident on these nodes are marked examined. This is repeated until all edges of G have been examined. Clearly, such an algorithm belongs to the path type in terms of the multiplicity of edge zeroing.

Algorithm 16 outlines the above described procedure. $E_{unex} \subseteq E$ is the set of unexamined edges and $G_{unex} = (V_{unex}, E_{unex})$ is the task graph spawned by these edges and the nodes $V_{unex} \subseteq V$, which are incident on these edges. Hence, each G_{unex} is a subgraph of G. In each step the critical path cp of the current unexamined graph G_{unex} is determined and its edges are merged into one cluster. Observe that all edges incident on the nodes of cp, $\{e_{ij} \in E_{unex} : n_i \in V_{cp} \vee n_j \in V_{cp}\}$, are marked as examined at the end of each step, and not only those edges $e_{ij} \in E_{cp}$ that are part of cp. This is important, because otherwise, in a later step, one of the edges $e_{ij} \notin E_{cp}$ not being part of cp could be zeroed. This would mean that the edge's two incident nodes n_i and n_j, of which at least one is part of a cluster with more than one node, are merged into the same cluster, potentially resulting in a nonlinear cluster.

Algorithm 16   *Linear Clustering Algorithm* $(G = (V, E, w, c))$
    Create initial clustering C_0: allocate each node $n \in V$ to a distinct cluster $C \in C$, $|C| = |V|$
    $E_{unex} \leftarrow E$ ▷ E_{unex}: *set of unexamined edges,* $E_{unex} \subseteq E$
    while $E_{unex} \neq \emptyset$ **do**   ▷ *there is at least one unexamined edge*
        Find a critical path cp of graph $G_{unex} = (V_{unex}, E_{unex})$ ▷ V_{unex}: *set of nodes*
        $n \in V$ *on which edges* E_{unex} *are incident*
        Merge all nodes V_{cp} of cp into one cluster
        $E_{unex} \leftarrow E_{unex} - \{e_{ij} \in E_{unex} : n_i \in V_{cp} \vee n_j \in V_{cp}\}$ ▷ *Mark edges that are incident on nodes of* V_{cp} *as examined*
    end while

In comparison to the general structure of a clustering algorithm, as outlined in Algorithm 15, there is no check of whether the schedule length of the new clustering is not worse than the current one. The check is unnecessary, as Lemma 5.2 guarantees that this is the case.

Finding the critical path of the task graph G_{unex} in each step is essentially the determination of the nodes' top and bottom levels. This can be performed using the algorithms presented in Section 4.4.2, with a complexity of $O(V + E)$. Merging the nodes and marking the corresponding nodes and edges as examined has lower complexity than $O(V + E)$. As there are at most $O(V)$ steps, Algorithm 16's total complexity is thus $O(V(V + E))$.

As a last comment on linear clustering, note the similarity between linear projections of dependence graphs as analyzed in Section 3.5.1 and linear clustering. The linear projection of DGs for example, is used in VLSI array processor design (Kung [109]).

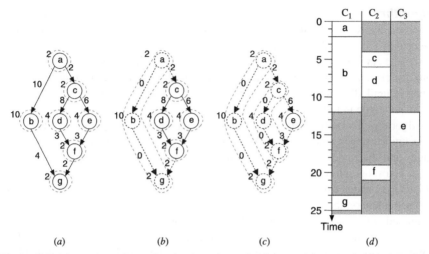

(a)   *(b)*   *(c)*   *(d)*

Figure 5.5. Linear clusterings of a simple task graph (Figure 5.3(*a*)): (*a*) initial clustering; (*b*) clustering after first step; (*c*) final clustering; (*d*) schedule of final clustering. (Examined elements are dotted.)

Example As an example, the linear clustering algorithm is applied to the task graph of Figure 5.3(*a*). The usual initial clustering is depicted in Figure 5.5(*a*).

- All edges are unexamined at this stage and the critical path is $cp = \langle a, b, g \rangle$ with a length of $len(cp) = 28$. These nodes are merged into one cluster and they are marked examined as are their incident edges, $e_{ab}, e_{bg}, e_{ac}, e_{fg}$. The resulting clustering is illustrated in Figure 5.5(*b*), where the examined nodes and edges are drawn with dotted lines.
- The current unexamined graph G_{unex} consists of the nodes $V_{unex} = \{c, d, e, f\}$ and the edges $E_{unex} = \{e_{cd}, e_{ce}, e_{df}, e_{ef}\}$. Its critical path is $cp = \langle c, d, f \rangle$ with a length of $len(cp) = 19$. These nodes are merged into one cluster and they are marked examined as are their incident edges, $e_{cd}, e_{ce}, e_{df}, e_{ef}$. The resulting clustering is illustrated in Figure 5.5(*c*).
- There are no unexamined edges left, $E_{unex} = \emptyset$, and the algorithm terminates. Therefore, the final clustering $\mathcal{C}_{final}$ is the one given in Figure 5.5(*c*), with a schedule length of

$$sl(\mathcal{C}_{final}) = len(cp(\mathcal{C}_{final}), \mathcal{C}_{final}) = len(\langle a, c, e, f, g \rangle, \mathcal{C}_{final}) = 25.$$

The resulting implicit schedule is displayed in Figure 5.5(*d*).

5.3.3 Single Edge Clustering

The second approach to clustering discussed here considers one single edge at a time for zeroing. A simple algorithm based on this approach was proposed by Sarkar

[167]. At the beginning of the algorithm the edges of the task graph are assigned a priority, which determines the order in which they are considered. The main part of the algorithm iterates over all edges **E** and checks for each edge if its zeroing results in a schedule length that is smaller than or equal to the schedule length of the present clustering. If yes, the zeroing is accepted: that is, the clusters of the incident nodes of the edge are merged. This procedure is repeated until all edges have been considered. Algorithm 17 outlines a generic version of the above described algorithm.

Algorithm 17   Single Edge Clustering Algorithm $(G = (V, E, w, c))$

    Create initial clustering C_0: allocate each node $n \in V$ to a distinct cluster $C \in \mathbf{C}$,
    $|\mathbf{C}| = |V|$
    $i \leftarrow 0$
    $\mathbf{E}_{unex} \leftarrow \mathbf{E}$ ▷ $\mathbf{E}_{unex}$: *set of unexamined edges*, $\mathbf{E}_{unex} \subseteq \mathbf{E}$
    while $\mathbf{E}_{unex} \neq \emptyset$ **do**   ▷ *there is at least one unexamined edge*
        $i \leftarrow i + 1$
        Let e_{ij} be the edge of $\mathbf{E}_{unex}$ with highest priority
        if $proc(n_i) = proc(n_j)$ **then**   ▷ *incident nodes are in same cluster*
            $C_i \leftarrow C_{i-1}$ ▷ *nothing to do*
        else
            Create new clustering C_i: merge clusters $proc(n_i)$ and $proc(n_j)$ into one cluster
            Schedule C_i using heuristic
            if $sl(C_i) > sl(C_{i-1})$ **then**   ▷ *merging increases current schedule length*
                $C_i \leftarrow C_{i-1}$ ▷ *reject new clustering C_i*
            end if
        end if
        $\mathbf{E}_{unex} \leftarrow \mathbf{E}_{unex} - \{e_{ij}\}$ ▷ *Mark edge e_{ij} as examined*
    end while

Observe the `if` statement that checks whether the incident nodes n_i and n_j of the currently considered edge are already in the same cluster $(proc(n_i) = proc(n_j))$. Even though all nodes are initially allocated to a distinct cluster, this situation can in fact occur due to the zeroing of previous edges in the list. To illustrate this, regard Figure 5.6, where a three-node task graph is clustered by zeroing one edge at a time. The edges are considered in the order e_{ab}, e_{ac}, e_{bc}. Figure 5.6(a) shows the initial clustering, Figure 5.6(b) the clustering after the zeroing of e_{ab}, and Figure 5.6(c) after the zeroing of e_{ac}. The zeroing of e_{ac} merges all three nodes a, b, c into one cluster, which implicitly zeros edge e_{bc} as its communication is now also local. Therefore, the correct clustering after the zeroing of e_{ac} is the one shown in Figure 5.6(d).

A key characteristic of this approach to clustering is that there is no clear implicit schedule. This is always a problem, irrespective of the order in which the edges are considered. Independent nodes might be allocated to the same cluster. Consequently, Algorithm 17 contains a statement for the determination of the schedule of each new clustering C_i. In essence, this is scheduling with a given processor allocation as discussed in Section 5.2, where the processor allocation is given by the clustering C_i of the nodes. As a consequence, the determination of the schedule reduces to the

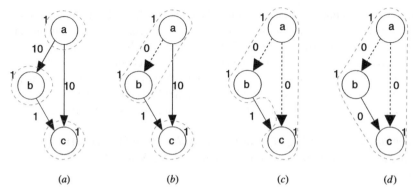

Figure 5.6. Implicit zeroing of edges: (*a*) initial clustering; (*b*) clustering after zeroing e_{ab}; (c) clustering after zeroing e_{ab} and e_{ac}; (*d*) e_{bc} was implicitly zeroed in (*c*).

ordering of the nodes. Any priority scheme for ordering the nodes, as discussed in Section 5.1.3, can be employed. Not surprisingly, Gerasoulis and Yang [76] suggest employing the allocated bottom level (Definition 4.19) of the current clustering C_{i-1}.

The last issue to address in this clustering approach is the order in which edges are considered. There is no constraint on the possible orders; however, an argument for a good heuristic is to zero "important" edges first. This is the same argument as in list scheduling, where "important" nodes should be considered first. Another analogy to list scheduling is that the priority of the edges can be static or dynamic. Algorithm 17 is formulated in a generic way that allows one to handle both static and dynamic priorities.

Static Edge Order In case static priorities are used, the edges can be sorted into a list before the main part of the algorithm starts, just like in static list scheduling (Algorithm 9). An idea that comes to mind is to sort the edges in decreasing order of their weights (as suggested by Sarkar [167]); that is, edges with high costs are considered earlier than those with low costs. Such an ordering will substantially reduce the total communication cost of the resulting clustering, with the implicit assumption that this also reduces the schedule length. As with the ordering of the nodes in, for example, list scheduling (Section 5.1.3), many other orderings are conceivable. Exercise 5.7 asks you to suggest priority schemes to order edges.

The complexity of single edge clustering with a static edge order is as follows. In each step one edge is considered for zeroing; that is, there are $|\mathbf{E}|$ steps. When an edge is zeroed, the node order must be determined, which usually is $O(\mathbf{V} + \mathbf{E})$ (e.g., using the allocated bottom level as the node priority; see above). Then the schedule length is determined, which implies in the worst case a construction of a schedule, which is also $O(\mathbf{V} + \mathbf{E})$ (see Section 5.2). It results in a complexity of $O(\mathbf{V} + \mathbf{E})$ for each step and a total complexity of $O(\mathbf{E}(\mathbf{V} + \mathbf{E}))$. This complexity result implies that the ordering of the edges has lower or equal complexity. The total complexity might increase for the case where higher complexity scheduling algorithms are used in each step.

Example To illustrate Algorithm 17 with a static edge order, it is applied to the simple graph of Figure 5.3(a). The edges are sorted in decreasing order of their weights, resulting in the order $e_{ab}, e_{cd}, e_{ce}, e_{bg}, e_{df}, e_{ef}, e_{ac}, e_{fg}$. Note that edges e_{df}, e_{ef} and e_{ac}, e_{fg} have the same weight value, and therefore their order was chosen at random. Independent nodes are scheduled in decreasing order of their allocated bottom level of the current clustering C_{i-1}. The usual initial clustering is depicted in Figure 5.7(a).

- e_{ab}, e_{cd}. The zeroing of the first two edges e_{ab} and e_{cd} is straightforward as only dependent nodes are merged. The resulting clusterings are linear and the schedule length decreases. Figure 5.7(b) depicts the clustering C_2 after these two steps, with an implicit schedule length of $sl(C_2) = 25$.

- e_{ce}. Zeroing the edge e_{ce} results in a nonlinear cluster as it merges the nodes c, d, e into a single cluster, and d and e are independent. The allocated bottom levels of both are $bl(d, C_2) = bl(e, C_2) = 13$, and priority is given here at random to d. So the nodes are scheduled in order c, d, e in their cluster, with the start times $t_s(c) = 4$, $t_s(d) = 6$, $t_s(e) = 10$. This results in a schedule length of $sl(C_3) = 23$, which is less than the current schedule length $sl(C_2) = 25$. Hence, the new clustering C_3, shown in Figure 5.7(c), is accepted.

- e_{bg}. By zeroing edge e_{bg}, the nodes a, b, g are merged into one cluster, which is linear. This does not decrease the schedule length, but it also does not increase it; hence, the clustering C_4 is accepted (Figure 5.7(d)).

- e_{df}. The zeroing of edge e_{df} merges the nodes c, d, e, f into one cluster. As node f depends on d and e, it is scheduled after them, resulting in a schedule length of $sl(C_5) = 20$ for this clustering C_5 (Figure 5.7(e)).

- e_{ef}. Edge e_{ef} has been zeroed implicitly (its weight is 0 in Figure 5.7(e)), when edge e_{df} was zeroed, so nothing is done in this step.

- e_{ac}, e_{fg}. The zeroing of either e_{ac} or e_{fg}, or both, would merge all nodes into one single cluster, increasing the schedule length to 26. Hence, they are not accepted, which makes the clustering of Figure 5.7(e) the final clustering, with the schedule length $sl(C_{\text{final}}) = 20$.

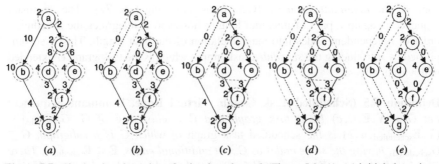

(a) (b) (c) (d) (e)

Figure 5.7. Single edge clustering of a simple task graph (Figure 5.3(a)): (a) initial clustering; (b) clustering after zeroing e_{ab}, e_{cd}; (c) after zeroing e_{ce}; (d) after zeroing e_{bg}; (e) after zeroing e_{df}, also final clustering. (Examined edges are dotted.)

Dynamic Edge Order   In dynamic list scheduling, the priority of all free nodes is recalculated in each step of the algorithm (Section 5.1.2). The same principle applies here for dynamic edge orders. In each step of the clustering, the priorities of the unexamined edges are recalculated and the edge with the highest priority is considered for zeroing.

In terms of priority metrics, a natural idea is again to attribute higher priority to the edges of an allocated critical path of the current clustering C_{i-1}, which naturally changes in each step and therefore requires a recalculation. While this seems similar to the approach of the linear clustering algorithm presented earlier in this section (Algorithm 16), it differs in two substantial points. First, only one edge of an allocated CP is zeroed at each step, and a heuristic has to be employed to determine which (as discussed later). Second, by only zeroing one edge of the current allocated CP, the allocated CP of the next clustering C_i might already be different. As a consequence, the clusterings produced by this algorithm are generally not linear.

Since computing an allocated CP (and the allocated node levels for that matter) has a complexity of $O(V + E)$, the runtime complexity of the clustering algorithm is usually not increased by using dynamic priorities in comparison to using static.

Zeroing an edge of the allocated CP, even if it is unique, does not necessarily mean that the schedule length of the resulting clustering C_i can be decreased. The reason lies in the fact that the schedule length of C_{i-1} might be longer than the allocated length of an allocated CP of the task graph,

$$sl(C_{i-1}) \geq len(cp(C_{i-1}), C_{i-1}). \tag{5.18}$$

Only in linear clusterings does the scheduling length equal the allocated length of the allocated critical path, Eq. (5.8). The ordering of independent nodes allocated to the same cluster has an impact on the schedule length of C_{i-1}, without having any influence on the allocated CP. To support this reasoning, regard the scheduling with a given processor allocation discussed in Section 5.2. The allocated CP is the same for all possible schedules, while finding a schedule with optimal length is NP-hard.

Dominant Sequence   The foregoing thoughts lead to the introduction of a new concept, the *dominant sequence* (Gerasoulis and Yang [76, 77]). The dominant sequence is again a path of maximal length; however, it integrates the scheduling order of independent nodes into the calculation of the path length. This is accomplished by introducing the *scheduled* task graph, which is the task graph G extended with virtual edges.

Definition 5.5 (Scheduled Task Graph, Virtual Edge, Dominant Sequence)
Let $G = (\mathbf{V}, \mathbf{E}, w, c)$ be a task graph and C a clustering of G. $G_{\text{scheduled}} = (\mathbf{V}, \mathbf{E}_{\text{scheduled}}, w, c)$ is the scheduled task graph of which G is a subgraph, $G \subseteq G_{\text{scheduled}}$, having the same nodes as G, but additional edges, $\mathbf{E} \subseteq \mathbf{E}_{\text{scheduled}}$. There is a virtual edge $e_{ij} \in \mathbf{E}_{\text{scheduled}}, \notin \mathbf{E}$ for each independent node pair $n_i, n_j \in \mathbf{V}$ (i.e., there is no path $p(n_i \rightarrow n_j)$ or $p(n_j \rightarrow n_i)$ in G) that is allocated to the same cluster $C = proc(n_i) = proc(n_j)$, and where n_j is scheduled directly after node n_i on

C (i.e., there is no node scheduled between them). The weight of e_{ij} is $c(e_{ij}) = 0$. Given $G_{\text{scheduled}}$, the dominant sequence *(DS) of G and clustering C is defined as the allocated critical path of $G_{\text{scheduled}}$ and clustering C:*

$$ds(\mathcal{C}) = cp(G_{\text{scheduled}}, \mathcal{C}). \tag{5.19}$$

To illustrate the concept of the scheduled task graph and its virtual edges, return to the clustering example of Figure 5.7. In Figure 5.7(e), two independent nodes, *d* and *e*, are allocated to the same cluster. During the application of the single edge clustering algorithm, it was established that in the execution order *d* precedes *e*. Using the concept of virtual edges, the edge e_{de} with weight $c(e_{de}) = 0$ is therefore inserted into the scheduled task graph, as depicted in Figure 5.8. The allocated critical path of this scheduled task graph is $\langle a, c, d, e, f, g \rangle$ with length 20, which is the dominant sequence $ds(\mathcal{C}_{\text{final}})$ of this clustering $\mathcal{C}_{\text{final}}$.

So the length of the dominant sequence $ds(\mathcal{C}_{\text{final}})$ of $\mathcal{C}_{\text{final}}$ is equal to its schedule length $sl(\mathcal{C}_{\text{final}}) = 20$. This reminds one of Eq. (5.8), which establishes for linear clusterings that the schedule length is identical to the allocated CP. This is no coincidence, as Lemma 5.3 demonstrates.

Lemma 5.3 (Clustering Is Linear for Scheduled Task Graph) *Let $G = (\mathbf{V}, \mathbf{E}, w, c)$ be a task graph, $\mathcal{C}$ a clustering of G, and $G_{\text{scheduled}} = (\mathbf{V}, \mathbf{E}_{\text{scheduled}}, w, c)$ the corresponding scheduled task graph. Then $\mathcal{C}$ is a linear clustering of $G_{\text{scheduled}}$.*

Proof. The proof follows from the definition of the scheduled task graph. Let $n_1, n_2, \ldots, n_m$ be all the nodes allocated to any cluster C, that is, $\{n_1, n_2, \ldots, n_m\} = \{n \in \mathbf{V} : proc(n) = C\}$, arranged in topological order of $G_{\text{scheduled}}$. For any node pair n_i, n_{i+1}, $i = 1, m - 1$, there is a path in $G_{\text{scheduled}}$ from n_i to n_{i+1}. Either this path already exists in G or, if it does not, there is a virtual edge $e_{i,i+1}$ representing this path,

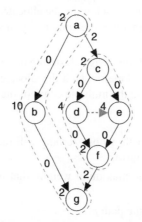

Figure 5.8. Scheduled task graph of clustering $\mathcal{C}_{\text{final}}$ from Figure 5.7(e): virtual edge e_{de} (gray and dashes) indicates scheduling order between nodes *d* and *e*.

as per definition of $G_{scheduled}$. By induction, there is a path $p(n_1 \rightarrow n_m)$ from n_1 to n_m in $G_{scheduled}$ to which all nodes allocated to C belong, that is, $\{n_1, n_2, \ldots, n_m\} \subseteq \{n \in \mathbf{V} : n \in p(n_1 \rightarrow n_m)$ in $G_{scheduled}\}$. Per Definition 5.4, this makes C a linear cluster and the clustering C linear, because it holds for any of its clusters. $\square$

A direct consequence of this lemma is that the relation between the schedule length and the length of the DS, as observed earlier, is generally true:

$$sl(C) = len(ds(C), C). \tag{5.20}$$

So what the allocated CP is to a linear clustering the DS is to a nonlinear one. Note that this is valid, because virtual edges have zero weight and thus add nothing to the length of a path.

As the dominant sequence determines the schedule length of a clustering C_{i-1}, zeroing one of its edges $e \in ds(C_{i-1})$ can reduce the schedule length $sl(C_{i-1})$. Or considered the other way around, if an edge is not part of a DS, zeroing it cannot reduce the schedule length.[1] In particular, if an edge belongs to an allocated CP, but not to a DS, its zeroing cannot reduce the schedule length of the current clustering C_{i-1}.

Unfortunately, there is no guarantee that the schedule length $sl(C_{i-1})$ of the current clustering C_{i-1} decreases or at least remains the same if an edge $e \in ds(C_{i-1})$ of the DS is zeroed. Zeroing an edge of the DS can result in the insertion of one or more new virtual edges into the scheduled task graph $G_{scheduled}$. These edges might become part of the DS of the new scheduled task graph, potentially increasing its length. It also means that the schedule length of a clustering C_{i-1} is reduced, if no new virtual edges are inserted into the scheduled task graph $G_{scheduled}$ of C_{i-1} and the DS was unique.

Given the above reasoning, it seems very appropriate for a dynamic single edge clustering heuristic to aim at zeroing an edge of the current dominant sequence. Gerasoulis and Yang [76] showed analytically and experimentally that doing so often leads to better clusterings when compared to other heuristics. The complexity of determining the DS is identical to computing the allocated CP, namely, $O(\mathbf{V} + \mathbf{E})$ for each step. Hence, it does not increase the complexity of single edge clustering.

Selecting Edge in Path   Single edge clustering heuristics that dynamically order the edges based on a path-based metric, be it the allocated CP, the DS, or something else, still have to choose among the multiple edges of the path p. Darte et al. [52] see three alternatives:

1. Select the edge of the path, whose zeroing will reduce the schedule length of the resulting clustering most.
2. Select the edge with the maximum weight, that is, $e_{max} \in p$ with $c(e_{max}) = \max_{e \in p} c(e)$.
3. Select the first edge of the path p.

---

[1] Unless an edge of the dominant sequence is implicitly zeroed as a consequence.

Obviously, only nonzero edges are considered in all cases. The first alternative is of course very costly, as the schedule length must be determined for the zeroing of each edge of the path, which generally is $O(V + E)$. In the worst case, there are $O(E)$ edges in the path; thus, the total complexity of single edge clustering increases by the same factor, resulting in $O(E^2(V + E))$. The second alternative is similar to statically ordering the edges by their weights, as proposed by Sarkar [167] (see earlier example). Interestingly, the total complexity is the same in both cases, dynamic and static, namely, $O(E(V + E))$. The third alternative is simple and cheap, but it does not have a lower total complexity.

5.3.4 List Scheduling as Clustering

Regarding the multiplicity of edge zeroing, two of the three types (Section 5.3.1) have been discussed so far. Linear clustering belongs to the path type, whereas single edge clustering considers only one edge at a time. In this section, a heuristic approach with node multiplicity is analyzed, where in each step of the heuristic the edges incident on a given node are considered for zeroing.

Essentially, the approach analyzed in the following is an adapted list scheduling. This might come as a surprise, given that this chapter makes a clear distinction between both approaches. However, it will be seen that list scheduling can in fact be interpreted as a clustering heuristic if certain conditions are met.

Number of Processors Clustering algorithms have no limit on the number of processors, while list scheduling generally has. However, by making the target system consist of $|V|$ processors in a list scheduling algorithm, there is virtually an unlimited number of processors (see Section 4.2.2).

Edge Zeroing A distinct characteristic of clustering is that nodes are only merged into the same cluster if an edge is zeroed as a consequence (Section 5.3.1). In contrast, list scheduling might in general allocate a node n to a processor P, even if none of n's predecessors is on P. So n is merged with the nodes already on P, $\{n_i \in V : proc(n_i) = P\}$, yet no edge can be zeroed, as there are no edges between n and these nodes. Using start time minimization (Section 5.1.1), the starting time of n on such a processor P must be (Condition 4.2)

$$t_s(n, P) \geq t_{dr}(n, P) = t_{dr}(n), \tag{5.21}$$

where $t_{dr}(n)$ is the data ready time of n, assuming all of its entering communications are remote. If instead node n is scheduled on an empty processor P_{empty}, $\{n_i \in V : proc(n_i) = P_{empty}\} = \emptyset$, its start time is not later

$$t_s(n, P_{empty}) = \max\{t_{dr}(n, P_{empty}), t_f(P)\} = t_{dr}(n). \tag{5.22}$$

Consequently, scheduling node n on an empty processor P_{empty} instead of on processor P does not change the heuristic approach of start time minimization. Furthermore, when $|\mathbf{V}|$ processors are used in list scheduling (as proposed earlier), there will always be at least one empty processor.

So list scheduling with start time minimization can be employed as a clustering heuristic by merely adding a tie breaker. If the start time of $t_s(n) = t_{dr}(n)$ or less cannot be achieved on any of the processors to which n's predecessors are scheduled (which would at least zero one edge), schedule n onto an empty processor. There might be other processors that could achieve the same start time of $t_s(n) = t_{dr}(n)$, but the tie is broken by always using an empty processor.

Algorithm 18 shows the modified start time minimization of list scheduling. Compared to the original Algorithm 10, only two points differ: (1) P_{empty} is made the default processor and (2) the `for` loop only iterates over the processors of the predecessors of n, $\bigcup_{n_i \in \mathbf{pred}(n)} proc(n_i)$, and not over all processors.

Algorithm 18 *List Scheduling's Start Time Minimization for Clustering (See Algorithm 10)*

Require: n is a free node
  $t_{min} \leftarrow t_{dr}(n); P_{min} \leftarrow P_{empty}$
  for each $P \in \bigcup_{n_i \in \mathbf{pred}(n)} proc(n_i)$ **do**
    if $t_{min} \geq \max\{t_{dr}(n, P), t_f(P)\}$ **then**
      $t_{min} \leftarrow \max\{t_{dr}(n, P), t_f(P)\}; P_{min} \leftarrow P$
    end if
  end for
  $t_s(n) \leftarrow t_{min}; proc(n) \leftarrow P_{min}$

In each step of list scheduling one node n_j is scheduled on a processor. From a clustering point of view, this means that the incoming edges of n_j, $\{e_{ij} \in \mathbf{E} : n_i \in \mathbf{pred}(n_j)\}$, are considered for zeroing. If n_j is scheduled on an empty processor, none of these edges is zeroed.

Implicit Schedules   List scheduling was presented in Section 5.1 as an algorithm that incrementally builds a schedule by extending the current partial schedule with one node at a time. Only the final schedule is a complete and feasible schedule. In contrast, clustering heuristics start with an initial clustering and construct intermediate clusterings that are all feasible. This conflict can be solved by making a similar assumption for list scheduling: simply assume that each node is initially allocated to a distinct processor, with the corresponding implicit schedule. In each step of list scheduling, one node is considered for merging with its predecessors. If it is scheduled on an empty processor, that can be interpreted as rejecting the merge and leaving it on the processor it was (implicitly) scheduled on.

Complexity   List scheduling has a significantly lower complexity than the clustering algorithms presented so far. As stated in Section 5.1, the complexity of the second part of list scheduling with start time minimization (Algorithm 10) is usually $O(\mathbf{P}(\mathbf{V} + \mathbf{E}))$. With the proposed start time minimization (Algorithm 18), this changes slightly. The start time of a node n is not evaluated for all processors as in Algorithm 10, but only for the processors holding at least one of n predecessors, that is, $O(\textbf{pred}(n))$ times. Since determining n's start time involves the calculation of the DRT on each processor, which is $O(\textbf{pred}(n))$, this step has a complexity of $O(\textbf{pred}(n)\,\textbf{pred}(n))$. However, an efficient implementation can bring this down to $O(\textbf{pred}(n)\log\textbf{pred}(n))$, using a priority queue (e.g., heap, Cormen et al. [42]) for the arrival times of the incoming communications. The data arrival time only changes between processors if it becomes local; otherwise it remains the same for all processors. For all nodes, this amortizes to $O(\textbf{E}\log\textbf{V})$ (with $O(\log\textbf{E}) = O(\log\textbf{V})$). Since there are $|\textbf{V}|$ steps, one for each node, the total complexity of the second part of this list scheduling is $O(\textbf{V} + \textbf{E}\log\textbf{V})$. Assuming a level-based node order calculated in the first part of list scheduling, which is $O(\textbf{V}\log\textbf{V} + \textbf{E})$ (Section 5.1.3), the resulting list scheduling complexity is $O((\textbf{V} + \textbf{E})\log\textbf{V})$. In comparison, linear clustering is $O(\textbf{V}(\textbf{V} + \textbf{E}))$, Algorithm 16, and single edge clustering is at least $O(\textbf{E}(\textbf{V} + \textbf{E}))$, Algorithm 17.

One of the reasons for the higher complexity of many clustering algorithms is the fact that nodes might need to be reordered and rescheduled in each step after merging. This is not the case in list scheduling (with the end technique), because a node is added to the end of the nodes already allocated to a processor. Another contributing factor can also be the recalculation of the node levels in each step.

Dynamic Sequence Clustering Gerasoulis and Yang [76] classified the MCP (modified critical path) (Wu and Gajski [207]) list scheduling heuristic as a clustering heuristic (its node priority scheme is mentioned in Section 5.1.3). Another heuristic analyzed in their comparison of clustering algorithms is the DSC (dynamic sequence clustering) (Yang and Gerasoulis [211]). Like MCP, it can be interpreted as a list scheduling heuristic.

A very special feature of DSC is that it zeros the edges of the dynamic sequence, as its name suggests. In experiments, it was shown to outperform the other compared algorithms. Moreover, it produces optimal results for task graphs with fork or join structure. Despite the good performance, it only has a complexity of $O((\textbf{V} + \textbf{E})\log\textbf{V})$. This performance and complexity are achieved through several key points:

- The algorithm concentrates on the next node of the dynamic sequence n_{DS}, even if it is only partially free. The free nodes that are scheduled until n_{DS} becomes free are not allowed to have a negative effect on the scheduling of n_{DS}. Hence, DSC implements a list scheduling with a lookahead technique.
- The node priority is the sum of the allocated bottom level and allocated top level of the *scheduled* task graph, $tl(n, C_{i-1}) + bl(n, C_{i-1})$ in $G_{\text{scheduled}}$. Note that the bottom level of a node does not change during list scheduling (Section 5.1.3), that is, $(bl(n, C_{i-1})$ in $G_{\text{scheduled}}) = (bl(n)$ in $G)$.
- To reduce complexity, the top level is estimated in a form that does not change the behavior of the algorithm, but does not require a complete traversal of the task graph.

For a detailed description of DSC and its theoretical properties please refer to Yang and Gerasoulis [211].

5.3.5 Other Algorithms

As mentioned at the beginning of this clustering section, virtually any scheduling algorithm that is suitable for an unlimited number of processors can be employed for the first phase of the three-step clustering based scheduling Algorithm 14.

For example, Hanen and Munier [86] proposed an integer linear program (ILP) formulation of scheduling a task graph on an unlimited number of processors. For coarse grained task graphs, that is, $g(G) \geq 1$, their relaxed ILP heuristic produces clusterings with a length that is within a factor of $\frac{4}{3}$ of the optimal solution, $sl(C) \leq \frac{4}{3} sl(C_{opt})$.

Apart from the above mentioned MCP algorithm, Wu and Gajski [207] proposed another algorithm called MD (mobility directed), which is a list scheduling algorithm for an unlimited number of processors with dynamic node priorities, using a node insertion technique (Section 6.1).

A clustering algorithm for heterogeneous processors was proposed by Cirou and Jeannot [36], using the concept of node triplets.

In a survey paper, Kwok and Ahmad [113] go as far as to classify scheduling algorithms that are designed for an unlimited number of processors into one class.

Grain Packing Another problem related to task graphs, the so-called grain packing problem (El-Rewini et al. [65]), is essentially the same as the clustering problem. Grain packing addresses the problem of how to partition a program into subtasks (grains) in order to obtain the shortest possible execution time. Generally, this is the subtask decomposition step of parallel programming (Section 2.4). However, when subtasks are grouped together (i.e., packed), grain packing rather relates, like clustering, to orchestration (Culler and Singh [48]) or agglomeration (Foster [69]) as described in Section 2.3.

Given a task graph, grain packing algorithms determine how to group the nodes together into larger nodes so that the program can run in minimum time. Well, that is exactly what clustering algorithms are about. Some algorithms that were proposed for the grain packing problem are in fact scheduling algorithms, for example, the ISH (insertion scheduling heuristic) and the DSH (duplication scheduling heuristic) by Kruatrachue and Lewis [105, 106].

A noteworthy approach in this context is a grain packing algorithm proposed by McCreary and Gill [135]. Nodes are grouped into so-called clans, done by a graph partitioning algorithm (i.e., a pure graph theoretic approach). As a consequence, there is no implicit schedule associated with the resulting node packing. Hence, this is a grain packing algorithm, which is indeed different from clustering. In effect, the algorithm changes the granularity of the task graph before it is scheduled.

Clustering with Node Duplication One of the objectives of clustering algorithms is the reduction of communication costs. Another scheduling technique, whose only objective is the elimination of interprocessor communication, is node duplication. This technique is studied in Section 6.2. The combination of clustering and node duplication is very powerful in reducing communication costs and can lead to even better results than pure clustering. Some algorithms integrating both techniques have been proposed (e.g., Liou and Palis [125], Palis et al. [140]).

5.4 FROM CLUSTERING TO SCHEDULING

The previous section discussed algorithms for the first phase (i.e., the clustering phase) of Algorithm 14. For a complete scheduling algorithm for an arbitrary number of processors, it remains to study the second and third phases of such a clustering based scheduling algorithm.

In the second phase, the clusters C of the obtained clustering $\mathcal{C}$ are mapped onto the available physical processors $\mathbf{P}$. If the number of clusters $|C|$ is less than or equal to the number of physical processors $|\mathbf{P}|$, $|C| \leq |\mathbf{P}|$, this is trivial. As per Definition 4.3, all processors are identical and fully connected, which means that it does not matter how the clusters are mapped onto the processors. Moreover, even the scheduling of phase three is unnecessary, as the scheduling is given implicitly by the clustering $\mathcal{C}$.

This is of course different, in the more likely case that there are more clusters than processors, $|C| > |\mathbf{P}|$, which is assumed in the following. Note that all presented clustering algorithms do not try to minimize the number of clusters.

5.4.1 Assigning Clusters to Processors

As was mentioned before, scheduling is a trade-off between minimizing the communication costs and maximizing the concurrency of the task execution. The clustering phase is all about minimizing the communication costs; hence, it is intuitive to maximize the concurrency of the task execution in the cluster to processor mapping. Achieving high concurrency is equivalent to balancing the load across the processors.

This reminds one of a variant of the classic bin packing problem (Cormen et al. [42], Garey and Johnson [73]): given n objects, with s_i the size of object i, and m bins, the problem is to find a distribution of the objects over the m bins so that the maximum fill height of the bins is minimal. While bin packing is also an NP-hard problem (Garey and Johnson [73]), there exist many heuristics for its near optimal solution.

In order to benefit from this wealth of heuristics, it is only necessary to cast the cluster mapping problem as a bin packing problem. This is straightforward:

- The clusters are the objects.
- The size s_i of a cluster C_i is the total computation load of its nodes,

$$w(C_i) = \sum_{n \in \mathbf{V}: proc(n) = C_i} w(n). \tag{5.23}$$

- The processors are the bins.

With this formulation, bin packing heuristics can be employed for the cluster to processor mappings. Consider the two examples given in the following.

Wrap Mapping   First, the clusters are ordered in decreasing order of their total weight $w(C)$. Then, starting with the first, each cluster is assigned to a distinct processor. When all processors have already been assigned one cluster, the algorithm returns to the first processor (i.e., it wraps around) and repeats the process until all clusters have been assigned (Darte et al. [52]). Algorithm 19 outlines this approach. Such a load balancing approach is also referred to as round robin. Its complexity is $O(\mathbf{V} + \mathbf{C} \log \mathbf{C})$, where calculating the total weights for all clusters is $O(\mathbf{V})$, sorting the clusters is $O(\mathbf{C} \log \mathbf{C})$ (e.g., Mergesort, Cormen et al., [42]), and the actual cluster to processor assignment is $O(\mathbf{C})$.

Algorithm 19   Wrap Mapping of Clusters $\mathbf{C}$ onto Processors $\mathbf{P}$
Require: Let $\{P_1, P_2, \ldots, P_{|\mathbf{P}|}\}$ be set of processors $\mathbf{P}$
    Sort clusters $\mathbf{C}$ in decreasing order of their total computation weight $w(C)$; let $C_1, C_2, \ldots, C_{|\mathbf{C}|}$ be the final order
    for $i = 1$ to $|\mathbf{C}|$ **do**
        Map cluster C_i onto processor $P_{i \bmod |\mathbf{P}|}$
    end for

Load Minimization Mapping   Essentially, this algorithm is just a special case of list scheduling with start time minimization (Section 5.1). To apply list scheduling to this problem, it suffices to consider each cluster as an independent node. In the first phase they are ordered in decreasing order of their total weight $w(C)$, identical to the wrap mapping. In the second phase they are assigned to the processor with the least load, that is, earliest finish time in list scheduling terminology. In comparison to wrap mapping, the actual assignment of the clusters to the processors increases to $O(\mathbf{C} \log \mathbf{P})$, using a priority queue for the processors (e.g., heap, Cormen et al. [42]). With the above conjecture that $|\mathbf{C}| > |\mathbf{P}|$, however, the total complexity is not affected and remains $O(\mathbf{V} + \mathbf{C} \log \mathbf{C})$.

Gerasoulis et al. [75], proposed a variant of wrap mapping, where the clusters with above average total weight are separated from the other clusters.

A quite different technique was proposed by Sarkar [167]. Essentially, he suggested to use list scheduling to assign the clusters to the processors. As usual, the nodes are first ordered into a priority list. While iterating over the list, each node is assigned to the processor that results in the smallest increase of the schedule length. The crucial difference from ordinary list scheduling is that as soon as one node of a cluster C is assigned to a processor P, all other nodes belonging to the same cluster C are also assigned to processor P. A major difference from the load balancing heuristics discussed earlier stems from the fact that communication costs are also taken into account, namely, when determining the schedule length increase.

5.4.2 Scheduling on Processors

The third phase of Algorithm 14 is algorithmically identical to the second phase of scheduling with a given processor allocation, Algorithm 13, Section 5.2. At this point each node has already been allocated to a physical processor. It only remains to order the nodes, for which any node priority scheme can be utilized.

Potentially, the node ordering can make use of the already existing partial node orders as established by the clustering C. However, to the author's best knowledge, no such scheme has yet been proposed.

Performance A question that naturally imposes itself is how clustering based scheduling algorithms, as outlined by Algorithm 14, perform in comparison to single-phase algorithms like list scheduling. The striking answer is that it is unknown. To the author's best knowledge, no direct experimental comparison has been performed for such algorithms and a limited number of target system processors.

While there are many comparisons of scheduling algorithms (e.g., Ahmad et al. [5, 6] Gerasoulis and Yang [76], Khan et al. [103], Kwok and Ahmad [111, 112] McCreary et al. [136]), heuristics are compared in classes. Clustering algorithms are only compared against other clustering algorithms or against algorithms for an unlimited number of processors. They are not compared as being the first phase of multiphase scheduling algorithms for a limited number of processors.

As discussed at the beginning of Section 5.3, there are good intuitive arguments that multiphase scheduling algorithms can produce better schedules than single-phase algorithms; yet, to the author's best knowledge, there is no experimental evidence. Hence, the reader is impudently asked for help with Exercise 5.8, which requests such a comparison.

5.5 CONCLUDING REMARKS

In this chapter, the two major classes of scheduling heuristics have been analyzed, namely, list scheduling and clustering. A myriad of algorithms belonging to either one of these two classes (or to both as discussed in Section 5.3.4) has been proposed in the past. In face of this, the discussion concentrated on the essential and common concepts and techniques encountered in these algorithms. In Section 6.3, Chapter 7, and Chapter 8, it will be seen how these fundamental heuristics, especially list scheduling, can be employed under more sophisticated system models.

Chapter 6 looks at more advanced scheduling techniques, which can be used in combination with the fundamental heuristics. Furthermore, it returns to the more theoretic aspects of task scheduling, like integrating heterogeneous processors and studying the complexity of variations of the general scheduling problem.

This chapter concludes with some bibliographic notes. A comprehensive survey of task scheduling algorithms can be found in Kwok and Ahmad [113]. Comparisons of algorithms are often a good starting and orientation point for studying task scheduling; for example, see Ahmad et al. [6], Gerasoulis and Yang [76], Khan et al. [103], Kwok and Ahmad [111], and McCreary et al. [136]. Other comprehensive publications that (at least in part) study task scheduling algorithms and its theory are by Chrétienne et al. [34], Coffman [37], Cosnard and Trystram [45], Darte et al. [52], El-Rewini and Abd-El-Barr [60], El-Rewini and Lewis [64], and El-Rewini, Lewis, and Ali [65].

5.6 EXERCISES

5.1 When nodes are ordered according to a priority scheme (e.g., bottom level), it is possible that some nodes have the same priority. Propose tie breaking metrics for these cases. (Suggestion: Consider the edges that are incident on the nodes in question.)

5.2 Use list scheduling with start time minimization to schedule the following task graph on four processors:

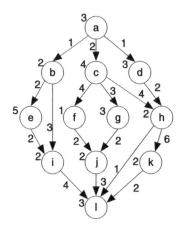

(a) The nodes shall be ordered in alphabetical order. What is the resulting schedule length?

(b) Now order the nodes according to their bottom levels and repeat the scheduling. What is the resulting schedule length?

5.3 Use list scheduling with dynamic priorities to schedule the following task graph on three processors:

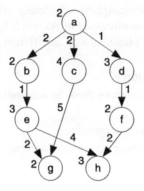

Dynamically compute the priorities of the free nodes by evaluating the start time for every free node on every processor. That node of all free nodes that allows the earliest start time is selected together with the processor on which this time is achieved. What is the resulting schedule length?

5.4 Create a linear clustering for the task graph of Exercise 5.2. Draw the *scheduled* task graph (Definition 5.5) for the final clustering. What is its dominant sequence?

5.5 The following questions are about linear clustering (Section 5.3.2):

(a) What is the general advantage of linear clustering in comparison with other clustering approaches?

(b) What is the practical advantage of merging in each step the critical path of the unexamined graph?

(c) After merging the nodes of a critical path *cp* into one cluster, why is it necessary to zero all edges incident on the nodes, and not only the ones that are part of the path?

5.6 Use single edge clustering to cluster the task graph of Exercise 5.2. Consider the edges in decreasing order of their weights. Break ties between equally weighted edges by giving preference to the edge whose origin node identifier comes earlier in the alphabet. Break a further tie by considering the target nodes in the same way.

Draw the *scheduled* task graph (Definition 5.5) for the final clustering. What is its dominant sequence?

5.7 In single edge clustering, the edges are ordered according to a priority scheme, which can be static or dynamic. As a static scheme, sorting the edges in decreasing order of their weights is intuitive and simple. Suggest other static edge orderings or priority schemes. Justify your suggestions.

You might want to review how nodes are ordered; see Sections 4.4 and 5.1.3.

5.8   Research: Experimentally compare single-phase with multiple-phase scheduling heuristics for a limited number of processors. For single-phase heuristics, considere variants of list scheduling with different priority schemes, static as well as dynamic. For multiphase scheduling, consider the three-phase clustering based algorithm as discussed in Section 5.3.1. Employ different clustering heuristics in the first phase, as well as different cluster-to-processor mappings in the second phase.

   (a) Implement algorithms.

   (b) Experimentally compare them by scheduling a large set of different task graphs.

   (c) Document results in an article.

   (d) Publish article in peer-reviewed conference or journal.

Advanced Task Scheduling

The previous chapter introduced the two fundamental scheduling algorithms: list scheduling and clustering. In this chapter, the focus is on more advanced aspects of task scheduling.

The first sections are dedicated to general techniques, which are not scheduling algorithms as such but are techniques that are used in combination with the fundamental algorithms. Following the framework approach utilized so far, they are studied detached from concrete algorithms in a general manner. The insertion technique (Section 6.1) is a generalization of the end technique (Definition 5.2), where nodes no longer need to be scheduled after already scheduled nodes. With node duplication (Section 6.2), one node might be executed on more than one processor. The goal is to save communication costs, as node duplication can make more communications local.

The classic target system model (Definition 4.3) used so far assumes that all processors are identical regarding their processing speed. In real parallel systems, this is not always given. Section 6.3 extends the scheduling model toward heterogeneous processors. This is done in a simple but powerful way, which leaves almost all scheduling concepts studied so far unaffected.

Section 6.4 returns to the more theoretical side of task scheduling. It studies the NP-hardness of many variations of the scheduling problem introduced in Chapter 4.

A completely different approach to scheduling, based on genetic algorithms, is analyzed in Section 6.5. A genetic algorithm is a stochastic search based on the principles of evolution.

The concluding remarks of this chapter will provide some references to further task scheduling publications.

6.1 INSERTION TECHNIQUE

In Section 5.1, the most common technique for the node's start time determination—the end technique (Definition 5.2)—was presented. On a given processor P, a node is scheduled at the end of all nodes already scheduled on this processor.

Task Scheduling for Parallel Systems, by Oliver Sinnen

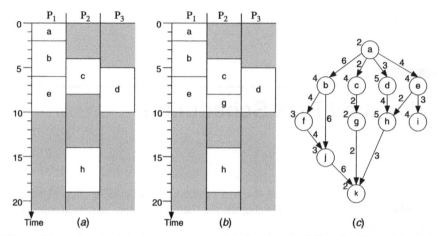

Figure 6.1. Partial schedule of sample task graph before the scheduling of g (a); partial schedule after node g is inserted in the idle period between c and h on P_2 (b); the sample task graph (c).

Under certain circumstances, a node can be scheduled earlier, if there is a time period between two nodes already scheduled on P, where P runs idle. This becomes clear with the following example. Consider the partial schedule of the sample task graph displayed in the Gantt chart of Figure 6.1(a): nodes a, b, c, d, e, h have already been scheduled and the next node is g. With the end technique g's start time on each processor is (see Eq. (5.1))

$$t_s(g, P_1) = \max\{t_{dr}(g, P_1), t_f(P_1)\} = \max\{10, 10\} = 10,$$

$$t_s(g, P_2) = \max\{8, 19\} = 19 \text{ or } t_s(g, P_3) = \max\{10, 10\} = 10.$$

Unfortunately, g's earliest DRT is on P_2, $t_{dr}(g, P_2) = 8$, where the processor only finishes at time unit 19. However, there is a large period between c and h during which P_2 runs idle, since h has to wait for its communications from nodes e and d. Node g fits easily into this period and can start immediately after c, because the communication e_{cg} is local. A corresponding partial schedule, where g is inserted between c and h, is shown in the Gantt chart of Figure 6.1(b). Of course, g cannot be placed into the space before node c on P_2 or d on P_3 since it depends on c.

It is intuitive that this insertion approach to scheduling has the potential of reducing the schedule length, as nodes might be scheduled earlier and time during which a processor runs idle might be eliminated. Experimental evidence for this intuition is given by Kwok and Ahmad [111] and by Sinnen and Sousa [177].

Clearly, Conditions 4.1 and 4.2 for a feasible schedule also apply when inserting a free node between other already scheduled nodes. This is independent of the scheduling algorithm, as long as the insertion is supposed to preserve the feasibility of the partial schedule. The subsequently defined condition unites both Conditions 4.1 and 4.2 for a feasible schedule and it also includes the end technique case. Therefore, it is simply called *scheduling condition*.

Condition 6.1 (Scheduling Condition) *Let $G = (V, E, w, c)$ be a task graph, **P** a parallel system, and S_{cur} a partial feasible schedule for a subset of nodes $V_{cur} \subset V$ on **P**. Let $n_{P,1}, n_{P,2}, \ldots, n_{P,l}$ be the nodes scheduled on processor $P \in \mathbf{P}$, that is, $\{n_{P,1}, n_{P,2}, \ldots, n_{P,l}\} = \{n \in V_{cur} : proc(n) = P\}$, with $t_s(n_{P,i}) < t_s(n_{P,i+1})$ for $i = 1, 2, \ldots, l - 1$.*

Two fictive nodes $n_{P,0}$ and $n_{P,l+1}$ are defined for which $t_f(n_{P,0}) = 0$ and $t_s(n_{P,l+1}) = \infty$. A free node $n \in V$ can be scheduled on P between the nodes $n_{P,i}$ and $n_{P,i+1}$, $0 \le i \le l$, if

$$\max\{t_f(n_{P,i}), t_{dr}(n, P)\} + w(n) \le t_s(n_{P,i+1}). \tag{6.1}$$

Inequality (6.1) warrants that the time period between two consecutive nodes is large enough so that node n can start its execution after its DRT and the finish time of the first node (term $\max\{t_f(n_{P,i}), t_{dr}(n, P)\}$), and complete (term $+w(n)$) before the second node starts execution (term $\le t_s(n_{P,i+1})$). Clearly, Condition 6.1 is also valid for scheduling without communication costs, due to its generic formulation using the DRT. Figure 6.2 illustrates the scheduling condition as it shows the scheduling of a node n between the two consecutive nodes $n_{P,i}$ and $n_{P,i+1}$ on processor P.

The end technique is included in this condition, as the start time of the last, fictive node $n_{P,l+1}$ is $t_s(n_{P,l+1}) = \infty$; in other words, n always fits into the time period starting with the finish time $t_f(P) = t_f(n_{P,l})$ of processor P.

Analogous to the end technique, the start time of a node is set to the earliest time in the first slot that complies with scheduling Condition 6.1.

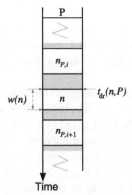

Figure 6.2. Scheduling condition: node n is scheduled in the gap between two consecutive nodes on P.

Definition 6.1 (Insertion Technique)   *Let k be the smallest value of i, $0 \le i \le l$, for which Eq. (6.1) is true. The start time of n on P is then defined as*

$$t_s(n, P) = \max\{t_f(n_{P,k}), t_{dr}(n, P)\}. \tag{6.2}$$

Again, this definition also applies to scheduling without communication costs.

6.1.1   List Scheduling with Node Insertion

Even though employing the insertion technique in list scheduling with start time minimization has the potential to reduce the schedule length in comparison to the end technique, there is in general no guarantee of improvement. This becomes clear when one realizes that the insertion technique might lead to other processor allocations than the end technique. For instance, with the insertion technique node g is scheduled on P_2 in the example of Figure 6.1; with the end technique it would be allocated to either P_1 or P_2. As a consequence, communications might be remote, when they would be local with the end technique, resulting in a longer schedule. However, with the same node order and processor allocation the insertion technique is at least as good as the end technique. This is formulated in the next theorem, which is an extension of Theorem 5.1 of Section 5.1 with an almost identical proof.

Theorem 6.1 (Insertion Better than End Technique)   *Let S be a feasible schedule for task graph $G = (\mathbf{V}, \mathbf{E}, w, c)$ on system $\mathbf{P}$. Using simple list scheduling (Algorithm 9), with the nodes scheduled in nondecreasing order of their start times in S and allocated to the same processors as in S, two schedules are created: S_{end}, employing the end technique (Definition 5.2) and S_{insert} employing the insertion technique (Definition 6.1). Then,*

$$sl(S_{insert}) \le sl(S_{end}) \le sl(S). \tag{6.3}$$

Proof. First, S_{end} and S_{insert} are feasible schedules, since the node order is a topological order according to Lemma 4.1. Following the same argumentation as in the proof of Theorem 5.1, it suffices to shown that $t_{f,S_{insert}}(n) \le t_{f,S_{end}}(n) \le t_{f,S}(n)$ $\forall n \in \mathbf{V}$.

Since the processor allocations are identical for all schedules, communications that are remote in one schedule are also remote in the other schedules and likewise for the local communications. Thus, the DRT $t_{dr}(n)$ of a node n can only differ between the schedules through different finish times of the predecessors, that is, through different start times, as the execution time is identical in all schedules.

By induction, it is shown now that $t_{s,S_{insert}}(n) \le t_{s,S_{end}}(n) \le t_{s,S}(n)$ $\forall n \in \mathbf{V}$. Evidently, this is true for the first node to be scheduled, as it starts in all schedules at time unit 0. Now, let $S_{insert,cur}$ and $S_{end,cur}$ be two partial schedules of the same nodes of $\mathbf{V}_{cur}$ of G, for which $t_{s,S_{insert,cur}}(n) \le t_{s,S_{end,cur}}(n) \le t_{s,S}(n)$ $\forall n \in \mathbf{V}_{cur}$ is true. For the node $n_i \in \mathbf{V}, \notin \mathbf{V}_{\text{cur}}$ to be scheduled next on processor $proc_{\mathcal{S}}(n_i) = P$, it holds that $t_{\text{dr},\mathcal{S}_{\text{insert,cur}}}(n_i, P) \le t_{\text{dr},\mathcal{S}_{\text{end,cur}}}(n_i, P) \le t_{\text{dr},\mathcal{S}}(n_i, P)$, with the above argumentation. With the end technique, n_i is scheduled at the end of the last node already scheduled on P. But this cannot be later than in $\mathcal{S}$, because the nodes on P in $\mathcal{S}_{\text{end,cur}}$ are the same nodes that are executed before n_i on P in $\mathcal{S}$, due to the schedule order of the nodes, and, according to the assumption, no node starts later than in $\mathcal{S}$. With the insertion technique, n_i is scheduled in the worst case (i.e., there is no idle period between two nodes already scheduled on P complying with Eq. (6.1)) also at the end of the last node already scheduled on P. Thus, $t_{s,\mathcal{S}_{\text{insert,cur}}}(n_i, P) \le t_{s,\mathcal{S}_{\text{end,cur}}}(n_i, P) \le t_{s,\mathcal{S}}(n_i, P)$ for node n_i. By induction this is true for all nodes of the schedule, in particular, the last node, which proves the theorem. $\qquad\square$

Theorem 5.1 of Section 5.1 establishes that an optimal schedule is defined by the processor allocation and the nodes' execution order. List scheduling with the end technique can construct an optimal schedule from these inputs.

While this result is included in Theorem 6.1, it also establishes that the insertion technique might improve a given nonoptimal schedule. *Rescheduling* a given schedule with list scheduling and the insertion technique, using the processor allocation and the node order of the original schedule, might improve the schedule length. In particular, schedules produced with the end technique might be improved. What first sounds like a contradiction to Theorem 5.1, after all it states that the end technique is optimal given a processor allocation and the nodes' execution order, becomes clear when one realizes that rescheduling with the insertion technique can only reduce the length of a schedule by reordering the nodes. Inserting a node in an earlier slot changes the node order on the corresponding processor.

Complexity Regarding the complexity of list scheduling with the insertion technique, the second part of the corresponding Algorithm 9 is examined. Determining the data ready time remains $O(\mathbf{PE})$ for all nodes on all processors (see Section 5.1). What changes is the time complexity of the start time calculation. In the worst case, it must be checked for every consecutive node pair on a processor, if the time period between them is large enough to accommodate the node to be scheduled. At the time a node is scheduled, there are at most $O(\mathbf{V})$ nodes scheduled on all processors. So if the start time is determined for every processor—as in the typical case of start time minimization—this amortizes to $O(\max(\mathbf{V}, \mathbf{P}))$, which is of course $O(\mathbf{V})$, because it is meaningless to schedule on more processors than the task graph has nodes. The start time is calculated for every node; thus, the final complexity of the second part of simple list scheduling with start time minimization is $O(\mathbf{V}^2 + \mathbf{PE})$. For comparison, with the end technique it is $O(\mathbf{P}(\mathbf{V} + \mathbf{E}))$ (see Section 5.1).

The insertion technique is employed in various scheduling heuristics, for example, ISH (insertion scheduling heuristic) by Kruatrachue [105], (E)CPFD ((economical) critical path fast duplication) by Ahmad and Kwok [4], BSA (bubble scheduling and allocation) by Kwok and Ahmad [114], and MD (mobility directed) by Wu and Gajski [207].

6.2 NODE DUPLICATION

A major challenge in scheduling is to avoid interprocessor communications. A node often has to wait for its entering communications, while its executing processor runs idle. For instance, in the schedule for the sample task graph visualized by Figure 6.1(a), nodes c and d are delayed due to the communication from node a and both processors P_2 and P_3 run idle during that time. One is tempted to move node a, for example, to P_2 in order to eliminate the cost of communication e_{ac}. However, this move creates communication costs between nodes a and b, as illustrated by the partial schedule in Figure 6.3(a).

A solution that has been exploited to reduce communication costs, while avoiding the above described problem, is node duplication. In this approach, some nodes of a task graph are allocated to more than one processor of the target system. For the previous example, node a is scheduled three times, once on each processor as shown in Figure 6.3(b). The communications from node a are now local on each processor of the target system and nodes b, c, d can start immediately after a finishes. Since nodes c and d start earlier, subsequent nodes can also start earlier, potentially leading to a shorter schedule length. Figure 6.3(c) displays a schedule similar to that of Figure 4.1, yet with the duplication of nodes a and j. The schedule length is reduced from 24 to 22 time units. But the potential of duplication is even higher, as demonstrated by the schedule in Figure 6.3(d), where the duplication of a can reduce the schedule length to 18 time units.

For node duplication, some formal changes must be carried out for the definitions and conditions of Section 4.1. The definition of a schedule (Definition 4.2) changes as follows.

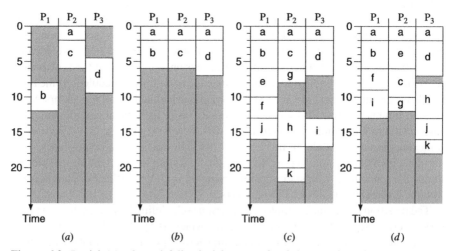

Figure 6.3. Partial (a), (b) and full schedules (c), (d) of the sample task graph (e.g., in Figure 6.1); the schedules of (b), (c), and (d) use node duplication.

- The function $proc(n)$ for the processor allocation of node n becomes a subset of P, denoted by **proc**(n), because with duplication a node can be allocated to more than one processor. Of course, **proc**$(n) \subseteq \mathbf{P}$ and $|\mathbf{proc}(n)| \geq 1$.
- The function $t_s(n)$ for the start time assignment to node n is ambiguous with node duplication. Only with specification of a processor is the start time of a node defined. Thus, it is now obligatory to write $t_s(n, P)$ (see Eq. (4.1)) to denote the start time of n on P; t_s becomes a function of the nodes *and* the processors, $t_s : \mathbf{V} \times \mathbf{P} \rightarrow \mathbb{Q}_0^+$.

Node duplication also has an impact on Condition 4.2 regarding the precedence constraints of the task graph.

Condition 6.2 (Precedence Constraint—Node Duplication) *For a schedule $\mathcal{S}_{\text{dup}}$ with node duplication, Eq. (4.5) of Condition 4.2 becomes*

$$t_s(n_j, P) \geq \min_{P_x \in \mathbf{proc}(n_i)} \{t_f(e_{ij}, P_x, P)\}. \tag{6.4}$$

Given the communication e_{ij}, node n_j cannot start until *at least one* instance of the duplicated nodes of n_i has provided the communication e_{ij}. In the schedule of Figure 6.3(c), for example, the communication e_{jk} is received from the duplicated node j on processor P_2; the same communication from the instance of node j on P_1 does not arrive on time on P_2 for the start of k at $t_s(k, P_2) = 20$ ($t_f(e_{jk}, P_1, P_2) = t_f(j, P_1) + c(e_{jk}) = 16 + 6 = 22$).

Following from this altered precedence constraint condition, the definition of the data ready time (Definition 4.8) must be adapted.

Definition 6.2 (Data Ready Time (DRT)—Node Duplication) *For a schedule $\mathcal{S}_{\text{dup}}$ with node duplication, Eq. (4.6) of Definition 4.8 becomes*

$$t_{\text{dr}}(n_j, P) = \max_{n_i \in \mathbf{pred}(n_j)} \{ \min_{P_x \in \mathbf{proc}(n_i)} \{t_f(e_{ij}, P_x, P)\}\}. \tag{6.5}$$

The rest of the definitions and conditions of Section 4.1 remain unmodified, except for the necessary formal adaptation for the altered definition of a schedule (see above). Even though all forgoing conditions and definitions were again formulated in a generic form and could be applied to scheduling without communication costs, it makes no sense to use node duplication when communication has no costs.

Node Order Anomaly A consequence of the modified precedence constraint Condition 6.2 is a possible anomaly in the order of nodes in a schedule. Consider the schedule of a small task graph in Figure 6.4, where node B is duplicated. Looking at the node order on P_1, one observes that node B starts after node C, even though C depends on B. This is valid according to Condition 6.2, because node B is a duplicated node and the communication e_{BC} is provided by the instance of B on P_2. Hence,

Lemma 4.1, stating that the start order of the nodes in a schedule is a precedence order, is in general not valid for schedules with duplicated nodes.

Complexity   Node duplication is meaningful for those nodes that have more than one leaving edge, that is, for nodes with an out-degree larger than one:

$$\{n_i \in \mathbf{V} : |\{e_{ij} \in \mathbf{E} : n_j \in \mathbf{V}\}| > 1\}. \tag{6.6}$$

Nodes with an out-degree of one or less can also be duplicated, but their duplication alone cannot reduce communication costs. For example, in Figure 6.4, the duplication of node A cannot be beneficial, as only node C receives its data. Imagine node A were duplicated several times. Only that instance of A that provides its data earliest to C had to be retained; all other duplicated nodes of A could be removed.

However, when a node with out-degree larger than one is duplicated, it might be beneficial to recursively duplicate its ancestors, even if their out-degree is only one. To illustrate this, regard the task graph in Figure 6.5(a). Node B is the only node with an out-degree larger than one. Consequently, it is the primary candidate for duplication and Figure 6.5(b) depicts a schedule on three processors where B is duplicated on each of them. While this is beneficial for the start time of node E on P_2, node D could start earlier on P_3 (at $t_s(D) = 6$ instead of $t_s(D) = 8$) without the duplication of B. Yet, the situation changes if node A is also duplicated together with each instance of B as shown in Figure 6.5(c). The resulting schedule length is even optimal, as it is the length of the computation critical path $cp_w = 11$.

Nodes with an out-degree larger than one are the *primary candidates* for duplication. Furthermore, in each duplicated path of nodes (e.g., $p(A \rightarrow B)$ in Figure 6.5), at least one node must have an out-degree larger than one (i.e., B in $p(A \rightarrow B)$); otherwise the duplication of the path cannot be beneficial.

Papadimitriou and Yannakakis [142] showed that the decision problem of scheduling with node duplication continues to be an NP-complete problem, even with unit

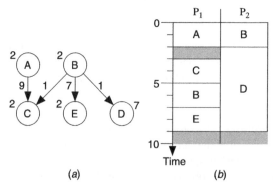

(a)   *(b)*

Figure 6.4. A small task graph (*a*) scheduled on two processors (*b*) with node duplication: node B is duplicated after node C, even though C depends on B.

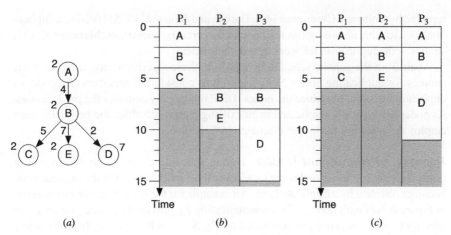

Figure 6.5. A small example task graph (*a*) and two schedules with node duplication; (*b*) only *B* is duplicated; (*c*) *A* and *B* are duplicated.

execution time (UET). Given the above observation, this is not surprising, because task graphs where all nodes have an out-degree of at most one (so called intrees) do not benefit from node duplication and their scheduling is NP-hard. Hence, since this "special case" is NP-hard, the general problem (which includes this case) cannot be easier. More on complexity results for scheduling with node duplication is given in Section 6.4.

Outtrees, on the other hand, strongly benefit from node duplication. In fact, scheduling an outtree on an unlimited number of processors using node duplication is a problem that can be optimally solved in polynomial time (Section 6.4). Hanen and Munier Kordon [86] address the scheduling of outtrees with duplication.

6.2.1 Node Duplication Heuristics

The aim in duplicating a node is to reduce the DRT of its descendant nodes, permitting them to start earlier. So node duplication introduces more computation load into the parallel system in order to decrease the cost of crucial communication. The challenge in scheduling with node duplication is the decision as to which nodes should be duplicated and where.

Node duplication has been employed in a large variety of scheduling approaches. The DSH (duplication scheduling heuristic) by Kruatrachue and Lewis [106] and (E)CPFD by Ahmad and Kwok [4], for instance, are two list scheduling variants with node duplication. Another list scheduling based algorithm with node duplication has been proposed by Hagras and Janeček [83], which is also suitable for heterogeneous processors. Sandnes and Megson [165] use node duplication in a genetic algorithm based heuristic. Clustering algorithms for an unlimited number of processors with node duplication are proposed by Papadimitriou and Yannakakis [142], Liou and Palis [125], and Palis et al. [140]. Another algorithm for unlimited processors has been reported by Colin and Chrétienne [40]. Darbha and Agrawal [49–51] present a duplication heuristic that obtains optimal schedules for certain granularities. Manoharan [134] studies node duplication for work-greedy heuristics.

Apart from the general scheduling approach, these algorithms also differ over the strategy for duplicating nodes. Some algorithms consider all ancestors of a node for duplication in order to reduce the nodes' DRT, others only consider the predecessors. As nodes are typically duplicated in an existing partial schedule, the heuristics must employ the insertion technique discussed in Section 6.1.

Removal of Redundant Nodes   During scheduling with node duplication, it might happen that some nodes are unnecessarily duplicated or that the original node becomes obsolete by a duplicated one. An example for the latter case can be observed in Figure 6.3(c) with node j. The communication e_{jk} provided by node j on P_1 is not used by k on P_2, because j is duplicated on P_2. So j can be removed from P_1, while the schedule remains valid without any effect on its length. For the schedule length to benefit from the idle periods left behind by the removed nodes, a second pass in a scheduling algorithm must be implemented (see Theorem 6.1).

6.3   HETEROGENEOUS PROCESSORS

Up to this point the target system of the scheduling problem always consisted of identical processors. While this is in most cases an accurate modeling of the real parallel system, there are systems that consist of heterogeneous processors. Take, for example, a cluster of PCs (Section 2.1) in which the processor differs from PC to PC. In order to consider the diverse capabilities of the processors in such a system, the scheduling model must be adapted.

Heterogeneity of processors can mean two things:

1. All processors of the system have the same functional capability, but they perform the same task at different speeds.
2. The processors have different functional capabilities; that is, certain tasks can only be performed by certain processors.

This section addresses processor heterogeneity of the first type. The second type of heterogeneity is encountered, for example, in embedded systems. Real time scheduling for embedded systems (Liu [127]) deals with such heterogeneity. The distinction between different services offered by the processors, however, introduces another dimension of conceptual complexity to the scheduling problem. Compromises have to be made for the scheduling algorithms. Introducing heterogeneity in terms of the first type does not change anything fundamental in the scheduling model, as will be seen in the following.

Note that heterogeneity in terms of processing speed also includes processor extensions for certain application domains, such as SSE or AltiVec (Section 2.1). Using such an extension just means that a processor can perform certain tasks faster than others. A processor that does not possess such an extension can still perform the task, just slower. More on this later.

To include heterogeneous processors in task scheduling, the definition of the target parallel system (Definition 4.3) is modified accordingly, without any alteration of its properties.

Definition 6.3 (Target Parallel System—Heterogeneous Processors) *A target parallel system $M_{hetero} = (\mathbf{P}, \omega)$ consists of a set of processors $\mathbf{P}$, whose heterogeneity, in terms of processing speed, is described by the execution time function ω. The processors are connected by a communication network with the properties specified in Definition 4.3.*

Due to the processors' heterogeneity, the execution time of a task is no longer given by its mere node weight, but is a function of the processor.

Definition 6.4 (Execution Time) *Let $G = (\mathbf{V}, \mathbf{E}, w, c)$ be a task graph and $M_{hetero} = (\mathbf{P}, \omega)$ a heterogeneous parallel system. The execution time of $n \in \mathbf{V}$ is the function $\omega : \mathbf{V} \times \mathbf{P} \to \mathbb{Q}^+$.*

Only the finish time of a node is directly affected by the node's execution time. With Definition 6.4, the finish time definition is modified—more precisely is generalized—for heterogeneous systems in the following way.

Definition 6.5 (Node Finish Time—Heterogeneous Processors) *Let S be a schedule for task graph $G = (\mathbf{V}, \mathbf{E}, w, c)$ on a heterogeneous parallel system $M_{hetero} = (\mathbf{P}, \omega)$. The finish time of node n on processor $P \in \mathbf{P}$ is*

$$t_f(n, P) = t_s(n, P) + \omega(n, P). \tag{6.7}$$

It is important to note that Definition 6.5 is a generalization of the corresponding Definition 4.5 made in Section 4.1. If the execution time is set to be the node weight, $\omega(n, P) = w(n)$, Definition 4.5 is obtained.

Furthermore, the definition of the node weight (i.e., the computation cost) also needs an adjustment. Again, it is generalized from Definition 4.4 by representing now the average execution time.

Definition 6.6 (Computation Cost—Heterogeneous Processors) *Let $G = (\mathbf{V}, \mathbf{E}, w, c)$ be a task graph and $M_{hetero} = (\mathbf{P}, \omega)$ a heterogeneous parallel system. The computation cost $w(n)$ of node n is the average time the task represented by n occupies a processor of $\mathbf{P}$ for its execution.*

Since the task graph maintains a node weight also for heterogeneous systems, metrics as node levels or the critical path (Section 4.4) can still be utilized for its analysis. However, since the node weight now reflects the average cost, the interpretation of these metrics is bound to the average case. In general, no safe conclusions can be made on the worst or best case, for example, as done in Lemma 4.4 (CP bound on schedule length) and Lemma 4.6 (level bounds on start time).

A heterogeneous system is said to be *consistent*, or *uniform*, if the relation of the execution times on the different processors is independent of the task. For instance, if processor P executes a task twice as fast as processor Q, it does so for all tasks. In this case, a relative speed $s(P)$, with the average speed set to 1, can be assigned to every processor, so that the execution time of node n on processor P is

$$\omega(n, P) = \frac{w(n)}{s(P)}. \tag{6.8}$$

In an *inconsistent*, or *unrelated*, system the above conclusion is not valid. Processor P might be faster than processor Q with some tasks, but slower with others. In this case, the execution time is given by the corresponding entry in a $|\mathbf{V}| \times |\mathbf{P}|$ cost matrix W:

$$\omega(n_i, P_k) = w_{ik}. \tag{6.9}$$

Once more, the inconsistent model is a generalization including the consistent model. As mentioned before, inconsistency can arise when some of the processors are equipped with acceleration extensions, like SSE or AltiVec, for certain application types (e.g., multimedia). In that case, two processors P and Q, where only P is equipped with a special execution unit, might execute most tasks at the same speed, but some tasks that can benefit from the special unit, are executed faster by P.

In both types of system, consistent or inconsistent, the execution time ω of a node is a function of the node and the processor. For generality, it will be represented as such in the following discussions without any further specification. Of course, in practice, it is often more convenient to regard a heterogeneous system as consistent and to measure the processor speeds with some benchmark.

NP-Completeness Essentially, task scheduling for heterogeneous systems is a straightforward generalization of task scheduling for homogeneous processors, which relates to the execution time of the nodes. Therefore, it is not surprising that scheduling for heterogeneous systems is also an NP-complete problem.

Theorem 6.2 (NP-Completeness—Heterogeneous Processors) *Let $G = (\mathbf{V}, \mathbf{E}, w, c)$ be a task graph and $M_{\text{hetero}} = (\mathbf{P}, \omega)$ a heterogeneous parallel system. The decision problem H-SCHED (G, M_{hetero}) associated with the scheduling problem is as follows. Is there a schedule S for G on M_{hetero} with length $sl(S) \leq T, T \in \mathbb{Q}^+$? H-SCHED (G, M_{hetero}) is NP-complete.*

Proof. First, it is argued that H-SCHED (G, M_{hetero}) belongs to NP; then it is shown that H-SCHED (G, M_{hetero}) is NP-hard by reducing SCHED $(G_{\text{SCHED}}, \mathbf{P}_{\text{SCHED}})$ (Theorem 4.1) in polynomial time to H-SCHED (G, M_{hetero}).

Clearly, for any given solution S of H-SCHED (G, M_{hetero}) it can be verified in polynomial time that S is feasible (Algorithm 5) and $sl(S) \leq T$; hence, H-SCHED (G, M_{hetero}) $\in$ NP.

For any instance of SCHED (G_{SCHED}, $\mathbf{P}_{SCHED}$) an instance of H-SCHED (G, M_{hetero}) is constructed by simply setting $G = G_{SCHED}$, $\mathbf{P} = \mathbf{P}_{SCHED}$, $\omega = w$, and $T = T_{SCHED}$; thus all processors of $\mathbf{P}$ are identical and the node execution time is only a function of the node n: $\omega(n, P) = w(n) \, \forall \, P \in \mathbf{P}$. Obviously, this construction is polynomial in the size of the instance of SCHED (G_{SCHED}, $\mathbf{P}_{SCHED}$). Furthermore, if and only if there is a schedule for the instance of H-SCHED (G, $\mathbf{P}$) that meets the bound T, is there a schedule for the instance SCHED (G_{SCHED}, $\mathbf{P}_{SCHED}$) meeting the bound T_{SCHED}. □

6.3.1 Scheduling

As the modification carried out for heterogeneous systems only affects the finish time calculation of a node, virtually all scheduling heuristics designed for homogeneous systems can be used for heterogeneous systems.

However, in order to produce efficient schedules, a scheduling heuristic should be aware of the heterogeneity of the processors. Yet again, this can often be achieved with a simple generalization. Instead of making decisions based on the start time of a node, the finish time naturally includes the processing capacity of the processor. For example, the possible start time of a node n might be earlier on processor P than on processor Q, but n might finish earlier on Q, due to Q's faster execution. So both the state of the current partial schedule and the heterogeneity of the processors are considered in decisions based on the finish time.

Finish Time Minimization With the above argumentation, the common start time minimization in list scheduling (Section 5.1.1) can be substituted by finish time minimization. The processor selected in each step (see Eq. (5.3)) is then

$$P_{min} \in \mathbf{P} : t_f(n, P_{min}) = \min_{P \in \mathbf{P}} \{\max\{t_{dr}(n, P), t_f(P)\} + \omega(n, P)\}. \tag{6.10}$$

With an accordingly modified procedure for the processor choice in Algorithm 10, a heterogeneous list scheduling heuristic is complete. The same node list construction that is used for homogeneous processors can be employed, because the node weights represent the average costs. Such a list scheduling strategy maintains the greedy characteristic and the same complexity. Moreover, by employing finish time minimization on homogeneous processors, one selects the same processor as one would with start time minimization; hence, it can be used in either case.

Heuristics for Heterogeneous Processors Some scheduling algorithms for heterogeneous processors have been proposed in the past. Heuristics based on the list scheduling approach are reported in Beaumont et al. [18], Liu et al. [126], Menascé et al. [137], Oh and Ha [139], Sinnen and Sousa [177], and Topcuoglu et al. [187, 188].

Also, the DLS algorithm (Sih and Lee [169]) is a dynamic list scheduling heuristic for heterogeneous processors. BSA by Kwok and Ahmad [114] is aware of heterogeneous processors, too. Finally, genetic scheduling algorithms are often designed for heterogeneous processors (e.g., Singh and Youssef [170], Wang et al. [198]) and Woo et al. [205]).

6.4 COMPLEXITY RESULTS

In Chapter 4 the scheduling problem, as stated in Definition 4.13, was defined in quite general terms. One "special case"—scheduling without communication costs—was also studied in Section 4.3. As was mentioned at the time, various other scheduling problems arise, many of which can be treated as special cases derived from the general scheduling problem. In the meantime, the concept of heterogeneous processors was introduced in the previous section. This section surveys and analyzes the NP-hardness of the scheduling problems that are obtained by relaxing, limiting, or generalizing one or more of the general problem's parameters, namely:

- *Processors.* As seen in Section 4.3.2, an unlimited number of processors can make scheduling polynomial-time solvable. Another interesting case to study is whether scheduling a task graph onto a fixed number of processors (in particular, two) is NP-hard.
- *Task Graph Structure.* Some task graph structures are easier to schedule than others. Analyzing the scheduling problem for restricted task graph structures might be valuable for associated real world problems. Moreover, it can give more insights into the general scheduling problem and help with the design of scheduling algorithms.
- *Costs.* Without communication costs, the scheduling problem seems to be less difficult, in particular, for an unlimited number of processors (Section 4.3.2). Other cases of interest relate to the granularity of the task graph or the unity of computation and/or communication costs.

As will be seen, most of these special problems remain NP-hard. The problems are studied in four groups: scheduling *without* communication costs, scheduling *with* communication costs, scheduling with node duplication, and scheduling on heterogeneous processors.

The virtually unlimited number of different scheduling problems makes a classification and designation scheme indispensable in order to maintain an overview and to reveal relations between problems.

6.4.1 $\alpha|\beta|\gamma$ Classification

This section introduces the alpha-beta-gamma ($\alpha|\beta|\gamma$) classification scheme of scheduling problems. It is a simple and extendible scheme that covers more scheduling problems than are discussed in the scope of this book. Originally it was presented by Graham et al. [81] and later extended by Veltman et al. [196], in particular, for scheduling with communication delays. The scheme is typically found in surveys and comparisons of scheduling problems (e.g., Brucker [27], Chrétienne et al. [34], Leung [121]) each of which does its own necessary extension and adaptation.

Each of the three fields specifies one aspect of the scheduling problem. The α field specifies the processor environment, the β field the task characteristics, and the γ field the optimality criterion. The following definitions of the subfields of $\alpha|\beta|\gamma$ are based on the ones found in Brucker [27], Chrétienne et al. [34], Leung [121], and Veltman et al. [196]. Only those subfields are listed here that are relevant for this text and some adaptations and extensions are made, especially for scheduling with communication costs. Many more exist for other scheduling problems, for instance, scheduling with deadlines or with preemption. The interested reader should refer to the above mentioned publications.

Some of the designations used in the scheme are different from those used in this book. For comparability, the scheme's field designations, as found in the above mentioned literature, are utilized in the following. The accompanying explanations should make it clear what is meant in relation to the framework used in this text.

Let $G = (\mathbf{V}, \mathbf{E}, w, c)$ be the task graph to be scheduled and $\circ$ the symbol for an empty field.

α—*Processor Environment* The first field specifies the processor environment and it can have two subfields:

1. *Processor Type.*
   - P. The target system $\mathbf{P}$ consists of identical processors; that is, the execution time (Definition 6.4) of node $n \in \mathbf{V}$ is identical on all processors, $\omega(n, P) = w(n)$, $\forall P \in \mathbf{P}$.
   - Q. The target system $M_{\text{hetero}} = (\mathbf{P}, \omega)$ consists of uniform heterogeneous processors; that is, the execution time of node $n \in \mathbf{V}$ is consistent with the processor speed $s(P)$, $\omega(n, P) = w(n)/s(P)$, $\forall P \in \mathbf{P}$.
2. *Number of Processors.*
   - $\circ$. The number of processors is a variable and is specified as part of the problem instance.
   - m (*positive integer number*). Denotes a fixed number of m processors. Hence, the number of processors m is part of the problem type.
   - ∞. Denotes an unlimited number of processors. Since preemption is not allowed in any of the here discussed problems, this is equivalent to $|\mathbf{P}| \geq |\mathbf{V}|$; that is, there are at least as many processors as tasks.

β—*Task Characteristics* This field specifies the characteristics of the tasks and it can have various subfields, separated by a comma, all of which are optional. You might find different orders of these subfields in other texts, as the order is not consistent across the literature.

1. *Precedence Relations.* This subfield describes the precedence relations between tasks. In other words, it specifies the structure of the task graph G. In can take the following values:

   - o. The tasks are independent; that is, $\mathbf{E} = \emptyset$.
   - *prec.* There are precedence constraints between tasks; hence, $|\mathbf{E}| \geq 1$.
   - {*outtree, intree, tree*}. The task graph G is a rooted tree (*tree*). A rooted tree is a graph with either an in-degree of at most one for each node $n \in \mathbf{V}$ (*outtree*, e.g., Figure 6.6(a)) or an out-degree of at most one for each node $n \in \mathbf{V}$ (*intree*, e.g., Figure 6.6(b)). Note that this definition of tree is general in that a tree can have multiple root nodes, which is usually referred to as forest (Cormen et al. [42]).
   - *op-forest.* The task graph G is an opposing forest, consisting of intrees *and* outtrees.
   - {*fork, join*}. The task graph G is a *fork* or *join* graph. A fork (join) graph is an outtree (intree) with one source (sink) node n_{root}, the root node, where all other nodes are successors (predecessors) of n_{root}, for example, Figure 6.7.
   - *chains.* The task graph G is a tree, where all nodes have an out-degree *and* an in-degree of at most one (e.g., Figure 6.8). That is, the task graph consists of disjoint chains of nodes. Note that it is denoted by *chains* not *chain*, as most scheduling problems are trivial when there is only one chain of nodes.

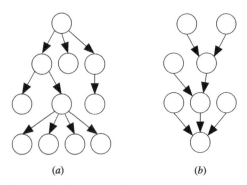

(a)   (b)

Figure 6.6. Examples of (a) outtree and (b) intree.

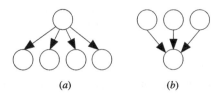

(a)   (b)

Figure 6.7. Examples of (a) fork graph and (b) join graph.

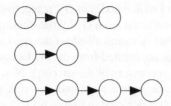

Figure 6.8. Example of chains graph.

- {*harpoon*, *in-harpoon*, *out-harpoon*}. The task graph G is a tree, with a harpoon-like structure. Such a graph is very similar to a fork or join graph. It consists of one root node n_{root} and x chains. In an *out-harpoon* there is an edge from n_{root} to each source node of the x chains, while in an *in-harpoon* there is an edge from each sink node of the x chains to n_{root}. Figure 6.9 illustrates an out-harpoon, where each chain consists of two nodes.

- *fork-join*. The task graph G is a fork-join graph. In a fork-join graph there is one source node n_s and one sink node n_t. All other nodes are successors of n_s and predecessors of n_t and are independent of each other, for example, in Figure 6.10.

- *sp-graph*. The task graph G is a series–parallel graph. A series–parallel graph can be constructed recursively using three basic graphs: (1) a graph consisting of a single node; (2) a single chain graph, that is, $\mathbf{V} = \{n_1, n_2, \ldots, n_l\}$ and $\mathbf{E} = \bigcup_{i=1}^{l-1} e_{i,i+1}$ (the "series" graph); and (3) a fork-join graph (the "parallel" graph), see above. Each of these graphs is a series–parallel graph in itself. Let $G_1 = (\mathbf{V}_1, \mathbf{E}_1)$ and $G_2 = (\mathbf{V}_2, \mathbf{E}_2)$ be two series–parallel graphs, with n_s and

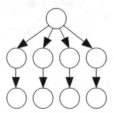

Figure 6.9. Example of out-harpoon graph.

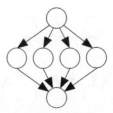

Figure 6.10. Example of fork-join graph.

n_t being the sink and source nodes of G_2, respectively. A new series–parallel graph G can be constructed from G_1 by substituting any node $n_i \in V_1$ with the complete G_2. That is, n_i and all edges that are incident on n_i are removed from G_1. New edges are created from the predecessors of n_i, **pred**(n_i), to n_s and from n_t to the successors of n_i, **succ**(n_i). Note that any series–parallel graph has exactly one source node and one sink node. An example for a series–parallel graph is depicted in Figure 6.11.

- *bipartie*. The task graph G is a bipartie graph. That means V can be partitioned into two subsets V_1 and V_2 such that $e_{ij} \in E$ implies $n_i \in V_1$ and $n_j \in V_2$. In other words, all edges go from the nodes of V_1 to the nodes of V_2 (e.g., Figure 6.12).

- *int-ordered*. The task graph G is interval-ordered. Let each node $n \in V$ be mapped to an interval $[l(n), r(n)]$ on the real line. A task graph is said to be interval-ordered if and only if there exists a node-to-interval mapping with the following property for any two nodes $n_i, n_j \in V$:

$$r(n_i) \leq l(n_j) \iff n_j \in \textbf{desc}(n_i) \tag{6.11}$$

This means that the intervals of any two nodes do not overlap if and only if one node is the descendant of the other (El-Rewini et al. [65], Kwok and Ahmad [113]).

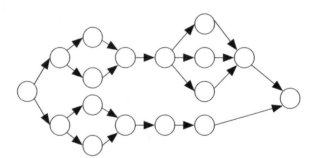

Figure 6.11. Example of series–parallel graph.

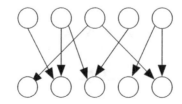

Figure 6.12. Example of bipartie graph.

2. *Computation Costs.* This subfield relates to the computation costs of the tasks.

- o. Each node $n_i \in \mathbf{V}$ is associated with a computation cost $w(n_i)$.
- p_i. If symbol p_i is present, the computation costs of the nodes are restricted in some form. p_i stands for processing requirement (time) of task i; hence, it corresponds directly to $w(n_i)$. Typical restrictions are:
  - $p_i = p$. Each node has the same computation cost, $w(n_i) = w, \forall n_i \in \mathbf{V}$.
  - $p_i = 1$. Each node has unit computation cost, that is, $w(n_i) = 1$, $\forall n_i \in \mathbf{V}$. This is also known as *UET (unit execution time)*. If $p_i = \{1, 2\}$ the computation cost of each node is either 1 or 2.

3. *Communication Costs.* This subfield relates to the communication costs associated with the edges of the task graph.

- o. No communication delays occur. There are no weights associated with the edges of G; that is, $G = (\mathbf{V}, \mathbf{E}, w)$ or, for $G = (\mathbf{V}, \mathbf{E}, w, c)$, $c(e_{ij}) = 0 \, \forall e_{ij} \in \mathbf{E}$.
- c_{ij}. Each edge $e_{ij} \in \mathbf{E}$ is associated with a communication cost $c(e_{ij})$. Restrictions might be specified for the values of $c(e_{ij})$, for instance, as in the next point.
- $c_{ij} \leq p_{\min}$. Each edge has a communication cost that is smaller than or equal to the minimum computation cost of the nodes, $c(e_{ij}) \leq \min_{n \in \mathbf{V}} w(n)$, $\forall e_{ij} \in \mathbf{E}$. In Chrétienne et al. [34] this is referred to as *SCT (small communication time)*. It implies $\min_{n \in \mathbf{V}} w(n) / \max_{e_{ij} \in \mathbf{E}} c(e_{ij}) \geq 1$, which is of course the definition of coarse granularity, $\min_{n \in \mathbf{V}} w(n) / \max_{e_{ij} \in \mathbf{E}} c(e_{ij}) = g(G) \geq 1$ (Section 4.4.3, Definition 4.20). Coarser granularity is specified with $c_{ij} \leq r \cdot p_{\min}$, where r is a constant and $0 < r \leq 1$ (designated r-SCT in Chrétienne et al. [34]).
- c. Each edge has the same communication cost, $c(e_{ij}) = c, \forall e_{ij} \in \mathbf{E}$. If the field is $c = 1$, each edge has *unit communication time (UCT)*, that is, $c(e_{ij}) = 1, \forall e_{ij} \in \mathbf{E}$.

4. *Task Duplication.*

- o. Task duplication is not allowed.
- *dup.* Task duplication is allowed.

*γ—**Optimality Criterion*** The γ field specifies the optimality criterion for the scheduling problem and there are no subfields. All scheduling problems discussed in this book are about finding the shortest schedule length sl, which is represented by the symbol $C_{\max}$ in the γ field. Here C_i stands for the completion time of task n_i; that is, $C_i = t_f(n_i)$, with $C_{\max} = \max_{n_i \in \mathbf{V}} C_i$ being the maximum completion time of all tasks, which is the definition of the schedule length (Definition 4.10).

Examples The following examples illustrate the utilization of the $\alpha | \beta | \gamma$ notation. A short $\alpha | \beta | \gamma$ notation without any optional field, $P || C_{\max}$, refers to the problem of scheduling independent tasks with arbitrary computation costs on $|\mathbf{P}|$ identical processors such that the schedule length is minimized. $P2 | prec, p_i = 1 | C_{\max}$ is the problem of scheduling tasks with precedence constraints and unit computation costs on two identical processors such that the schedule length is minimized. If the two processors are uniformly heterogeneous, this changes to $Q2|prec, p_i = 1|C_{max}$. Finally, $P\infty|tree, c_{ij} \le p_{min}, dup|C_{max}$ denotes the problem of scheduling a task graph with tree structure, arbitrary computation costs, and communication costs that are smaller than or equal to the minimum computation cost (i.e., the graph is coarse grained) on an unlimited number of identical processors. Node duplication is allowed and the objective is again to minimize the schedule length.

Generalizations and Reductions The theory of NP-completeness implies that a general problem cannot be easier, complexity-wise, than any of its special cases. Many of the previously defined subfields restrict the parameters of the task scheduling in some way, making it a subproblem of a general one. In particular, many specific graph structures are listed, which are also related to each other. A fork graph, for instance, is a special form of a tree. If it is proved that a particular scheduling problem is NP-complete for tree structured task graphs, then the scheduling problem is also NP-complete for an arbitrary graph structure. In face of this observation, it is interesting to realize the relations of the different graph structures defined for the β field. Figure 6.13 visualizes these relations. Each arrow points in the direction of the more general structure: that is, the graph at its tail is a special case of the graph at its head.

The above observation is reflected in the way NP-completeness of a problem is usually proved. A common step in such proofs is to reduce a known NP-complete problem in polynomial time to that problem, as is done in all NP-completeness proofs of this text. This is particularly easy if the known problem is a special case, as, for example, in Theorem 6.2. Thus, if NP-completeness is shown for a special case, the NP-completeness of all more general cases can be proved using such reductions. As a consequence, Figure 6.13 can also be interpreted as a reduction graph (Brucker and Knust [29]), although not all task graph structures lead necessarily to an NP-complete scheduling problem.

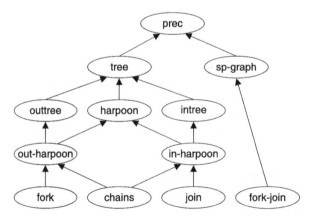

Figure 6.13. Relations between fundamental graph structures.

6.4.2 Without Communication Costs

This section is dedicated to scheduling problems that do not consider communication costs. They are all special cases of the scheduling problem analyzed in Sections 4.3 and 4.3.2. Table 6.1 lists the various scheduling problems, having six columns: (1) $\alpha|\beta|\gamma$ notation of the scheduling problem, (2) number of processors, (3) restrictions on computation costs (in next sections also communication costs), (4) structure of the task graph, (5) NP-hardness or complexity, and (6) reference to corresponding publication and/or section in this book.

As is established in Theorem 4.4, scheduling with an unlimited number of processors, $P\infty|prec|C_{max}$, is polynomial-time solvable.

With a limited number of processors, $P|prec|C_{max}$, the problem is NP-hard (Theorem 4.3) and remains so even for independent tasks $P||C_{max}$, which is closely related to BINPACKING (Garey and Johnson [73]). Also, to schedule task graphs with unit execution time (UET) on a limited number of processors, $P|prec, p_i = 1|C_{max}$, is an NP-hard problem.

However, scheduling without considering communication costs becomes more tractable with UET, as some of the problems with special graph structures are now polynomial-time solvable, for example, $P|p_i = 1|C_{max}$, $P|tree, p_i = 1|C_{max}$. So restricting the graph structure *or* the costs does not help to escape NP-hardness, but doing both makes the problem tractable in some cases.

Scheduling on only two processors proves to be more difficult than one might imagine, as such seemingly simple problems like $P2||C_{max}$ and $P2|chains|C_{max}$ are NP-hard. Nevertheless, scheduling with UET on two processors is polynomial-time solvable for an arbitrarily structured task graph, $P2|prec, p_i = 1|C_{max}$.

The complexity for other fixed numbers of processors m, $Pm|prec|C_{max}$ and also $Pm|prec, p_i = 1|C_{max}$, is yet unknown.

6.4.3 With Communication Costs

This section comes back to scheduling with communication costs—the general scheduling problem of this text. As it encloses the case without communication costs (it is the case where $c(e) = 0, \forall e \in \mathbf{E}$), it can only be more difficult.

Again, Table 6.2 lists the various scheduling problems with the same six column types as in Table 6.1, with the only exception that the costs column now also includes restrictions on the communication costs. As defined in Section 4.4.3, $g(G)$ stands for the granularity of the task graph G, with $g(G) \geq 1$ indicating coarse granularity. In some cases the result remains valid also for weak granularity $g_{weak}(G)$, but this is not distinguished here.

That scheduling with communication costs is more difficult than without is immediately apparent from the first entry. $P\infty|prec, c_{ij}|C_{max}$, that is, the scheduling of an arbitrary task graph on an unlimited number of processors, is NP-hard, while $P\infty|prec|C_{max}$ is polynomial-time solvable. Even with restricted task graph structures, $P\infty|\{tree, harpoon, fork-join\}, c_{ij}|C_{max}$, scheduling remains NP-hard, with the only exception being the primitive fork and join graphs $P\infty|\{fork, join\}, c_{ij}|C_{max}$,

Table 6.1. Complexity Results for Task Scheduling Without Considering Communication Costs

$\alpha\|\beta\|\gamma$	Processors	Costs	Structure	Complexity	Reference
$P\infty\|prec\|C_{max}$	Unlimited			$O(V+E)$	Theorem 4.4 (Section 4.3.2)
$P\|prec\|C_{max}$	$\|P\|$			NP-hard	Karp [99], Theorem 4.3 (Section 4.3.2)
$P\|\|C_{max}$			Independent	NP-hard	Garey and Johnson [72]
$P\|int\text{-}ord\|C_{max}$			Interval–ordered	NP-hard	Papadimitriou and Yannakakis [141]
$P\|prec, p_i = 1\|C_{max}$		$w(n) = 1$		NP-hard	Ullman [192]
$P\|p_i = 1\|C_{max}$			Independent	$O(V)$	Round robin
$P\|tree, p_i = 1\|C_{max}$			Tree	$O(V)$	Hu [93]
$P\|op\text{-}forest, p_i = 1\|C_{max}$			Opposing forest	NP-hard	Garey et al. [74]
$P\|int\text{-}ord, p_i = 1\|C_{max}$			Interval–ordered	$O(V+E)$	Papadimitriou and Yannakakis [141]
$P2\|prec\|C_{max}$	2			NP-hard	
$P2\|\|C_{max}$			Independent	NP-hard	Lenstra et al. [119]
$P2\|chains\|C_{max}$			Chains	NP-hard	Du et al. [56]
$P2\|prec, p_i \in \{1, 2\}\|C_{max}$		$w(n) \in \{1, 2\}$		NP-hard	Ullman [192]
$P2\|prec, p_i = 1\|C_{max}$		$w(n) = 1$		$O(V^2)$	Coffman and Graham [38], Fujii et al. [70], Gabow [71], Sethi [168]

Table 6.2. Complexity Results for Task Scheduling with Considering Communication Costs

$\alpha|\beta|\gamma$	Processors	Costs	Structure	Complexity	Reference
$P\infty|prec, c_{ij}|C_{max}$	Unlimited			NP-hard	Exercise 4.4.3
$P\infty|tree, c_{ij}|C_{max}$			Tree	NP-hard	Chrétienne [33]
$P\infty|harpoon, c_{ij}|C_{max}$			Harpoon	NP-hard	Chrétienne [33]
$P\infty|fork\text{-}join, c_{ij}|C_{max}$			Fork-join	NP-hard	Chrétienne [32], Exercise 4.4.3
$P\infty|\{fork, join\}, c_{ij}|C_{max}$			Fork or join	Polynomial	Chrétienne [31]
$P\infty|prec, c_{ij} \leq r \cdot p_{min}|C_{max}$		$g(G) \geq 1/r,$ $0 < r \leq 1$		NP-hard	Picouleau [149, 150]
$P\infty|tree, c_{ij} \leq p_{min}|C_{max}$		$g(G) \geq 1$	Tree	$O(V)$	Chrétienne [30]
$P\infty|bipartie, c_{ij} \leq p_{min}|C_{max}$			Bipartie	$O(E \log E)$	Chrétienne and Picouleau [35]
$P\infty|sp\text{-}graph, c_{ij} \leq p_{min}|C_{max}$			Series-parallel	Polynomial	Chrétienne and Picouleau [35]
$P\infty|prec, p_i = 1, c \geq 1|C_{max}$		$w(n) = 1,$ $c(e_{ij}) = c \geq 1$		NP-hard	Papadimitriou and Yannakakis [142]
$P\infty|prec, p_i = 1, c = 1|C_{max}$		$w(n) = 1,$ $c(e_{ij}) = 1$		NP-hard	Papadimitriou and Yannakakis [142], Picouleau [149, 150]
$P|prec, c_{ij}|C_{max}$	$|P|$			NP-hard	Theorem 4.1 (Section 4.2.2)
$P|prec, p_i = 1, c = 1|C_{max}$		$w(n) = 1,$ $c(e_{ij}) = 1$		NP-hard	Rayward-Smith [158]
$P|tree, p_i = 1, c = 1|C_{max}$			Tree	NP-hard	Veltman [195], Lenstra et al. [120]
$P|bipartie, p_i = 1, c = 1|C_{max}$			Bipartie	NP-hard	Hoogeveen et al. [90]
$P|int\text{-}ord, p_i = 1, c = 1|C_{max}$			Interval-ordered	$O(E + VP)$	Ali and El-Rewini [10], El-Rewini and Ali [62]
$Pm|tree, p_i = 1, c = 1|C_{max}$	m (fixed)	$w(n) = 1,$ $c(e_{ij}) = 1$	Tree	Polynomial	Varvarigou et al. [194]
$P2|tree, p_i = 1, c = 1|C_{max}$	2	$w(n) = 1,$ $c(e_{ij}) = 1$	Tree	Polynomial	El-Rewini and Ali [61], Lenstra et al. [120] (intree)

which can be scheduled optimally in polynomial time on an unlimited number of processors. Interestingly, the harpoon graph, $P\infty|harpoon, c_{ij}|C_{max}$, which is closely related to the fork and join graphs and only slightly more elaborated, makes scheduling NP-hard.

While any form of coarse granularity, $g(G) \geq 1/r, 0 < r \leq 1$, of the task graph does not make the problem tractable, $P\infty|prec, c_{ij} \leq r \cdot p_{min}|C_{max}$, the combination of coarse granularity and special graph structures does: $P\infty|\{tree, bipartie, sp\text{-}graph\}, c_{ij} \leq p_{min}|C_{max}$. UET and UCT alone, however, do not make scheduling polynomial-time solvable, $P\infty|prec, p_i = 1, c \geq 1|C_{max}$. This situation is similar to scheduling without communication costs: restricting the graph structure *or* the costs does not help, but doing both makes the problem tractable in some cases.

Of course, scheduling with communication costs does not become easier when the number of processors is limited. For example, $P|tree, p_i = 1, c = 1|C_{max}$ is NP-hard, while $P\infty|tree, c_{ij} \leq p_{min}|C_{max}$ (which includes $P\infty|tree, p_i = 1, c = 1|C_{max}$) is not. Only interval-ordered task graphs with UET and UCT can be scheduled optimally on a limited number of processors, $P|int\text{-}ord, p_i = 1, c = 1|C_{max}$, in polynomial time.

When the number of processors is fixed, scheduling a tree structured task graph with UET and UCT is tractable, for example, $P2|tree, p_i = 1, c = 1|C_{max}$.

6.4.4   With Node Duplication

Intuitively, node duplication makes scheduling easier, because the costs of communication can be eliminated by making it local. Hence, the problem comes closer to scheduling without communication costs. Table 6.3 lists various scheduling problems where node duplication is allowed. Of course, node duplication implies communication costs, since duplicating a node is meaningless otherwise.

Needless to say that scheduling with task duplication is NP-hard even on an unlimited number of processors $P\infty|prec, c_{ij}, dup|C_{max}$. On the one hand, task duplication makes the scheduling of an outtree on an unlimited number of processors tractable, $P\infty|outtree, c_{ij}, dup|C_{max}$. On the other hand, it does not help to bring the complexity of scheduling intrees down, $P\infty|intree, c_{ij}, dup|C_{max}$, which consequently remains NP-hard (see also Section 6.2).

Nevertheless, scheduling of coarse grained task graphs on an unlimited number of processors is polynomial-time solvable with task duplication, $P\infty|prec, c_{ij} \leq p_{min}, dup|C_{max}$. One might wonder how this is possible, given that without task duplication, $P\infty|prec, c_{ij} \leq p_{min}|C_{max}$, the problem is NP-hard and that duplication is not beneficial in all cases, in particular, for intrees. Looking back at Section 6.4.3, it can be observed that $P\infty|tree, c_{ij} \leq p_{min}|C_{max}$ is in fact tractable, which includes the scheduling of coarse grained intrees. Hence, the case for which task duplication is not beneficial is already tractable without duplicating nodes.

The scheduling of non-coarse grained task graphs, even with UET and UCT, where UET < UCT, $P\infty|prec, p_i = 1, c > 1, dup|C_{max}$, belongs to the class of NP-hard problems.

Table 6.3. Complexity Results for Task Scheduling with Node Duplication

$\alpha|\beta|\gamma$	Processors	Costs	Structure	Complexity	Reference
$P\infty|prec, c_{ij}, dup|C_{max}$	Unlimited			NP-hard	
$P\infty|outtree, c_{ij}, dup|C_{max}$			Outtree	Polynomial	Chrètienne [33]
$P\infty|intree, c_{ij}, dup|C_{max}$			Intree	NP-hard	Chrètienne [33]
$P\infty|in\text{-}harpoon, c_{ij}, dup|C_{max}$			In-harpoon	NP-hard	Chrètienne [33]
$P\infty|prec, c_{ij} \leq p_{min}, dup|C_{max}$		$g(G) \geq 1$		Polynomial	Colin and Chrètienne [40]
$P\infty|prec, p_i = 1, c > 1, dup|C_{max}$		$w(n) = 1,$ $c(e_{ij}) = c > 1$		NP-hard	Papadimitriou and Yannakakis [142]
$P\infty|prec, p_i = 1, c > 0, dup|C_{max}$		$w(n) = 1,$ $c(e_{ij}) = c > 0$		$O(\mathbf{V}^{c+1})$	Jung et al. [97]
$P|prec, c_{ij}, dup|C_{max}$	$|\mathbf{P}|$			NP-hard	
$P|prec, p_i = 1, c = 1, dup|C_{max}$		$w(n) = 1,$ $c(e_{ij}) = 1$		NP-hard	Chrètienne and Picouleau [35]

169

Table 6.4. Complexity Results for Scheduling on Heterogeneous Processors

$\alpha|\beta|\gamma$	Processors	Costs	Structure	Complexity	Reference
$Q|prec, c_{ij}|C_{\max}$	$|\mathbf{P}|$	$c(e_{ij})$		NP-hard	Theorem 6.2 (Section 6.3)
$Q|prec|C_{\max}$				NP-hard	
$Q|chains,$ $p_i = 1|C_{\max}$		$w(n) = 1$	Chains	NP-hard	Kubiak [107]
$Q2|chains,$ $p_i = 1|C_{\max}$	2	$w(n) = 1$	Chains	$O(\mathbf{V})$	Brucker et al. [28]

With a limited number of processors, task duplication does not change the fate of scheduling—it remains NP-hard, $P|prec, c_{ij}, dup|C_{\max}$, even with UET and UCT, $P|prec, p_i = 1, c = 1, dup|C_{\max}$.

6.4.5 Heterogeneous Processors

There is not much to report for task scheduling on heterogeneous processors. As a generalization of the scheduling problem, it can only be harder. Very few results are known and Table 6.4 lists some of them.

Both scheduling with and without communication costs is NP-hard on a limited number of consistent heterogeneous processors, $Q|prec, c_{ij}|C_{\max}$ and $Q|prec|C_{\max}$.

With heterogeneous processors, even extremely simple task graphs consisting of chains with UET and no communication costs, $Q|chains, p_i = 1|C_{\max}$, cannot be optimally scheduled in polynomial time. On homogeneous processors this is even possible for trees, $P|tree, p_i = 1|C_{\max}$. Only when the number of processors is limited to two, is the problem tractable, $Q2|chains, p_i = 1|C_{\max}$.

It can be concluded that task scheduling remains a difficult problem even for relaxed parameters. Some seemingly simple problems are NP-hard, for example, the scheduling of independent tasks. Optimal solutions in polynomial time are only known for a very few, specific problems. More on scheduling problems, not only limited to task scheduling, and their NP-completeness can be found in Brucker [27], Chrétienne et al. [34], Coffman [37], and Leung [121].

6.5 GENETIC ALGORITHMS

NP-hard problems are often tackled with the application of stochastic search algorithms like simulated annealing or genetic algorithms. Task scheduling is no exception in this respect, and the genetic algorithm (GA) method especially has gained considerable popularity. A GA is a search algorithm that is based on the principles of evolution and natural genetics. A pool of solutions to the problem, the population, creates new generations by breeding and mutation. According to evolutionary theory, only the most suited elements of the population are likely to survive and generate offspring, transmitting their biological inheritance to the next generation (Davis [54], Goldberg [78], Holland [89], Man et al. [133], Reeves and Rowe [161]).

This section studies how a GA can be applied to the task scheduling problem. As the area of genetic algorithms is very broad, only a very brief introduction to its principles can be given. For more information please refer to the corresponding literature, for example, the references just given. The rest of the section then concentrates on how the GA method can be applied to the scheduling problem.

6.5.1 Basics

A GA operates through a simple cycle of stages: creation of a population, evaluation of the elements of the population, selection of the best elements, and reproduction to create a new population. Algorithm 20 outlines a simple GA and the fundamental components are described in the following.

- *Chromosome.* A chromosome represents one solution to the problem, encoded as a problem-specific string.
- *Population.* A pool of chromosomes is referred to as the population. The initial population is typically created randomly, sometimes guided by heuristics.
- *Evaluation.* At each iteration, the quality of each chromosome is evaluated with a *fitness function*.
- *Selection.* Through selection, the GA implements the evolutionary principle "survival of the fittest." The selection operator chooses chromosomes based on their fitness value computed during evaluation.
- *Crossover.* Crossover is the main operator for the creation of new chromosomes. Two "parent" chromosomes are combined to build a new chromosome, which thereby inherits the genetic material of its ancestors.
- *Mutation.* The mutation operator randomly changes parts of a chromosome. It is applied with low probability and mainly serves as a safeguard to avoid the convergence to a locally best solution.

A GA terminates after a specified number of generations, when a time limit is exceeded or when a solution with sufficient quality is found.

Algorithm 20   Outline of a Simple GA
    Create initial population
    Evaluation
    while termination criterion not met **do**
        Selection
        Crossover
        Mutation
        Evaluation
    end while
    Return best solution

6.5.2 Chromosomes

The genetic formulation of a problem begins with the definition of an appropriate chromosome encoding. In order to achieve good performance, the chromosome should be simple, because this permits one to employ simple and fast operators.

For task scheduling, a chromosome represents a solution to the scheduling problem—in other words a schedule (Definition 4.2). A schedule consists of the processor allocation and the start time of each node of the task graph. Theorem 5.1 establishes that it suffices to know the processor allocation and the execution order of the nodes in order to construct a schedule (using the end technique) that is not longer than any other schedule with the same processor allocation and node execution order. Given this observation, two types of chromosome representations for task scheduling can be distinguished (Rebreyend et al. [160]).

Indirect Representation   In the indirect representation, the chromosome string holds information that serves as the input for a heuristic to create a schedule. Two approaches are obvious for the encoding of the chromosome:

- *Processor Allocation.* The chromosome encodes the processor allocation of the nodes. Every chromosome consists of $|\mathbf{V}|$ genes, each of which represents one node of the task graph. Assuming a consecutive numbering of the nodes and processors, starting with 1, gene i corresponds to node $n_i \in \mathbf{V}$. The value of the gene corresponds to the index of the processor, 1 to $|\mathbf{P}|$, to which the node is allocated (Figure 6.14). In GA terminology this is called *value encoding*. Such an encoding is employed, for example, by Benten and Sait [20]. In order to construct a schedule for each chromosome, the node order must be determined by a separate heuristic. Naturally, the techniques for the first phase of list scheduling can be employed (Section 5.1.3). In turn, this means that the same node order can be used for all chromosomes, or that a particular node order is determined for each chromosome, based on the processor allocation, for instance, a node priority based on allocated levels (Section 4.4.2).
- *Node List.* The chromosome encodes a node list. Every chromosome is a permutation of the $|\mathbf{V}|$ nodes of the task graph. Each gene represents a position in the node list and the value of the gene is the index of the node at that position (Figure 6.15). In GA terminology this is called *permutation encoding*. Depending on the operators used in the GA (see below), the node list can be the execution order of the nodes or merely a priority ordering. In the former case, the node list must adhere to the precedence constraints of the nodes. If this holds, a chromosome essentially establishes the first part of list scheduling (Algorithm 9). In the latter case, the position of a node in the list merely specifies its priority. An ordering that adheres to the precedence constraints and the priorities is computed using Algorithm 12. As the chromosome only establishes (directly or indirectly) the node order, the processor allocation must be

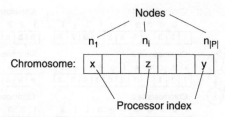

Figure 6.14. Chromosome that encodes processor allocation.

determined by a heuristic. Again, this can be done globally (i.e., identically for all chromosomes) or individually for each chromosome, which naturally can be done with list scheduling (Algorithm 9). Kwok and Ahmad [110] propose to encode the node list in the chromosome, with the nodes always being in precedence order.

Some authors proposed value encoding for the node priorities (e.g., Ahmad and Dhodhi [3], and Dhodi et al. [55], Sandnes and Megson [164, 165]). The chromosome has then the same structure as the processor allocation chromosome—that is, each gene represents a node—while each gene's value is the corresponding node's priority. The actual node order is determined for each chromosome with a procedure such as Algorithm 12. A disadvantage of such an encoding is that nodes can have the same priority.

Figure 6.16 gives examples for the two different encoding approaches. The chromosomes are for the scheduling of the sample task graph (Figure 6.16(a)) on three processors.

Direct Representation In the direct representation, the chromosome encodes the processor allocation *and* the execution order of the nodes. Consequently, the chromosome consists of two parts—the two chromosome approaches used in the indirect representation as discussed earlier. With this information, the nodes' start times are readily computed, using a simple list scheduling (see Theorem 5.1). No heuristic decisions are taken in the construction of the schedule. Wang et al. [198] use such a two-part chromosome—one for the processor allocation and the other for a global node list.

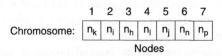

Figure 6.15. Chromosome that encodes node list.

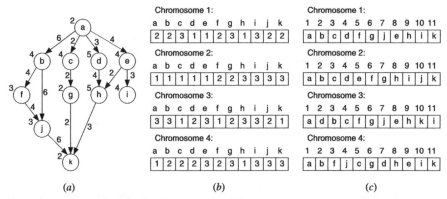

Figure 6.16. Four examples of chromosomes that encode processor allocations (*b*) or node lists (*c*) for the scheduling of the the sample task graph (*a*) on three processors.

Figure 6.17 shows two chromosomes consisting of two parts, one for the processor allocation and one for the global node list. They encode the scheduling of the sample task graph (e.g., Figure 6.16(*a*)) on three processors. In this case, the node lists are in precedence order. Beneath the chromosomes, the corresponding schedules are depicted.

It goes without saying that many other chromosome encodings are possible. Strictly speaking, it is not necessary to know the absolute order of the nodes. It suffices to have partial orders of the nodes allocated to each processor. Such an encoding is reported by Hou et al. [92], which is further improved by Correa et al. [43, 44]. Zomaya et al. [213] employ a similar encoding. Wu et al. [206], use lists of node–processor pairs. The order of the pairs establishes the node order. This encoding can be very useful in combination with node duplication, as a node can be allocated to more than one processor. See also Exercise 6.10, which asks you to suggest a chromosome encoding for task scheduling with node duplication.

Comparison of Representation Types   An essential difference between the direct and indirect representation types of chromosomes is their ability to represent an optimal schedule. As mentioned before, from Theorem 5.1 it is known that the processor allocation and the node order of an optimal schedule are enough to reconstruct a schedule of optimal length. Therefore, it is guaranteed that the optimal solution is situated in the search space of GAs with direct representation.

In contrast, from Section 5.2 it is known that scheduling with a given processor allocation is still NP-hard. So if the chromosome encodes only the processor allocation, the GA based scheduling algorithm is still faced with an NP-hard problem for each chromosome. Heuristic decisions taken for ordering the nodes cannot guarantee that an optimal schedule will be obtained, even if the processor allocation is optimal. Equally, finding the optimal processor allocation for a node ordering given by a chromosome is also NP-hard. To realize this, just consider the

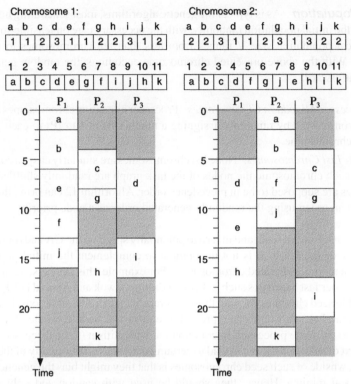

Figure 6.17. Two examples of chromosomes (top) that encode both processor allocation and node list for the scheduling of the sample task graph (Figure 6.16(a)) on three processors, and the schedules they represent (bottom).

scheduling of independent nodes, $P||C_{max}$, which is NP-hard, although the node order is irrelevant. Once again, heuristic decisions taken to allocate the nodes cannot guarantee that an optimal schedule will be obtained, even if the node order is optimal.

It follows that it is not guaranteed for GAs with indirect representation that the optimal solution lies in their search space. In other words, the direct representation permits one to reflect an optimal solution, while the indirect representation sometimes cannot.

This advantage of the direct representation comes with a price, however. First, by having two very different parts, the simplicity of the chromosome is lost. Second, as a direct consequence, more complex crossover and mutation operators must be employed. Essentially, a separate operator must be used for each part, because they are of different type—one is value encoded, the other is permutation encoded. Last but not least, the search space of the GA increases dramatically in comparison to the simpler indirect representation chromosomes. The total number of possible values is the product of the possible values of each part.

Initial Population   As is typical for genetic algorithms, most scheduling heuristics generate the initial population randomly, with the necessary care on feasible solutions. The typical population size N ranges from less than ten chromosomes to about a hundred. With the above presented chromosome encodings, the random generation is straightforward:

- *Processor Allocation Chromosomes.* Processor allocation chromosomes are easily constructed by randomly assigning a number from 1 to $|\mathbf{P}|$ to each gene of the chromosome.
- *Node List Chromosomes.* Node list chromosomes are similarly easy to construct. For each chromosome the nodes of the task graph are randomly shuffled. If the nodes are supposed to be in precedence order, Algorithm 12 can bring them into such an order, using the randomly generated order as a node priority.

To enrich the initial seed, and in turn to potentially speed up the GA and/or improve the quality of the result, it is not uncommon to complement the initial population with nonrandomly generated chromosomes. For example, this can be schedules produced by other fast heuristics such as list scheduling (Kwok and Ahmad [110]). Other conceivable seed chromosomes are those that represent extreme situations. For example, one initial chromosome can represent a sequential schedule, where all tasks are assigned to the same processor. In combination with elitism (to be discussed later in the context of selection), this can lead to certain guarantees on the quality of the result.

The downside of such seed chromosomes is that they might bias the genetic search toward local minima. Hence, they should be used with caution and only a small fraction of the initial population should be nonrandom.

Fitness Function   As the objective in task scheduling is to find the shortest possible schedule, the fitness of a chromosome is directly related to the length of the associated schedule (Hou et al. [92]).

Definition 6.7 (Fitness Function)   *Let c be a chromosome and $\mathcal{S}_c$ the schedule associated with c. The fitness of c is defined as*

$$F(c) = sl_{\max} - sl(\mathcal{S}_c), \qquad (6.12)$$

where $sl_{\max}$ is the maximum schedule length among all chromosomes of the population.

Hence, to evaluate the fitness of a chromosome c, the corresponding schedule $\mathcal{S}_c$ must be constructed. How this is performed depends on the representation type of the chromosome. In case of an indirect representation, this involves the application of a heuristic.

Variations to this definition of the fitness function have been proposed. For instance, in Dhodhi et al. [55] the fitness value is normalized with the sum of all fitness values of the population. In Zomaya et al. [213], a small constant value is added to impede a fitness of $F(c) = 0$, as this is problematic for some selection operators, as will be seen later.

6.5.3 Reproduction

The operators are the instruments with which a GA explores the search space of a problem. Crossover and mutation are reproduction operators that create new chromosomes from existing ones. Their design is closely tied to the chosen chromosome representation. Since a solution represented by a chromosome should be feasible—otherwise the search space becomes even larger—operators must be defined correspondingly.

Crossover The crossover operator is the more significant one of the two. It implements the principle of evolution. New chromosomes are created with this operator by combining two randomly selected parent chromosomes, thereby inheriting the genetic material of its ancestors. How this is done depends on the encoding of the chromosome. In Section 6.5.2 two indirect chromosome encodings are presented, whereby the direct representation is a combination of the two. Since the two encodings differ significantly, two different operators are necessary.

- *Processor Allocation.* For the chromosome encoding of the processor allocation, quite simple crossover operators can be employed. The processor allocation chromosome is a value encoded chromosome, where each gene can assume the same range of values (1 to $|\mathbf{P}|$). Furthermore, the value of one gene has no impact on the possible values of the other genes. Therefore, a simple two-point crossover operator works as follows.

    Given two randomly chosen chromosomes c_1 and c_2, two new chromosomes c_3 and c_4 are generated by swapping a randomly determined gene interval. Let the interval range from i to j. Outside this interval, that is, $[1, i - 1]$ and $[j + 1, |\mathbf{V}|]$, the genes of c_3 have the values of c_1 and inside those of c_2. For c_4 it is the converse.

    Figure 6.18 illustrates this crossover operator. The simpler and more common single-point crossover operator is a special case of this, achieved by setting $j = |\mathbf{V}|$. Figure 6.19 visualizes a single-point crossover operator. The extension of this concept to multiple crossover points is straightforward. Note that the generated new chromosomes are always valid chromosomes; that is, they represent a valid processor allocation of the tasks $\mathbf{V}$.

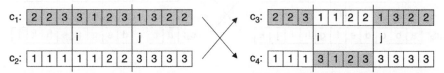

Figure 6.18. Illustration of two-point crossover of processor allocation chromosomes ($i = 4$, $j = 7$).

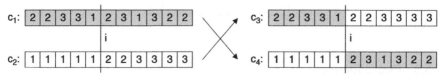

Figure 6.19. Illustration of single-point crossover of processor allocation chromosomes ($i = 6$).

- *Node List.* The chromosome encoding of a node list is a permutation encoding of the node set $\mathbf{V}$. Each element of the permutation, that is, each node $n \in \mathbf{V}$, can appear only once in the chromosome. Thus, the simple crossover operator presented for the processor allocation encoding cannot be employed. To understand this, image two chromosomes c_1 and c_2 in which node f is at position 5 in c_1 and position 6 in c_2. In the case where a single-point crossover operator divides the chromosomes at, for example, position 6, the resulting chromosome c_3 will have node f twice, at position 5 and at position 6, while there will be no f in chromosome c_4. This is depicted in Figure 6.20. The problem is overcome with the following single-point permutation crossover operator.

  Given two randomly chosen chromosomes c_1 and c_2, a crossover point i, $1 \le i < |\mathbf{V}|$, is selected randomly. The genes $[1, i]$ of c_1 and c_2 are copied to the genes $[1, i]$ of the new chromosomes c_3 and c_4, respectively. To fill the remaining genes $[i + 1, |\mathbf{V}|]$ of c_3 (c_4), chromosome c_2 (c_1) is scanned from the first to the last gene and each node that is not yet in c_3 (c_4) is added to the next empty position of c_3 (c_4) in the order that it is discovered. Figure 6.21 illustrates the procedure of this operator.

  Under the condition that the node lists of chromosomes c_1 and c_2 are in precedence order, this operator even guarantees that the node lists of c_3 and c_4 also are. It is easy to see this for the genes $[1, i]$ of both c_3 and c_4 as they are only copied from c_1 and c_2. The remaining genes of c_3 and c_4 are filled in the same relative order in which they appear in c_2 and c_1, respectively. Hence, among themselves, these remaining nodes must also be in precedence order. Furthermore, there cannot be a precedence conflict between the nodes on the left side of the crossover point with those on the right side of the crossover point,

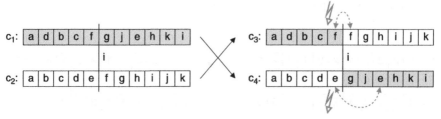

Figure 6.20. Simple single-point crossover as in Figure 6.19 *fails* for node list chromosomes ($i = 6$).

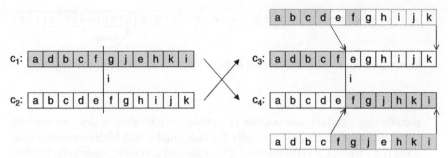

Figure 6.21. Illustration of single-point crossover of node list chromosomes ($i = 6$).

because this separation of the nodes into two groups has not changed from c_1 to c_3 neither from c_2 to c_4 and it adheres to the precedence constraints in c_1 and c_2.

Crossover is usually performed with a probability of 0.5 to 1. If it is not performed, the selected chromosomes c_1 and c_2 are simply copied to the population of the next generation.

Mutation The mutation operator is applied with a much lower probability (about 0.1 or less) than the crossover operator. Its main purpose is to serve as a safeguard to avoid the convergence of the state search to a locally best solution. A new chromosome c_2 is created by copying a randomly picked chromosome c_1 and randomly changing parts of it. Again, a distinction is made between the operators for processor allocation and node list chromosomes.

- *Processor Allocation.* Being a value encoded chromosome, the mutation operator can simply change the values of randomly picked genes. This can be done with any procedure, like increasing or decreasing the current values, or by assigning a randomly determined value. The only condition is that the resulting values are in the range 1–|**P**|. Figure 6.22 illustrates such a simple mutation operator, where two genes are "mutated."

  Another alternative is to swap the values of two randomly picked genes as visualized in Figure 6.23. In terms of processor allocation, this has the effect that the number of nodes on each processor remains constant. While this is good for

Figure 6.22. Illustration of simple mutation operator for processor allocation chromosome.

Figure 6.23. Illustration of swapping mutation operator for processor allocation chromosome.

Figure 6.24. Illustration of swapping mutation operator for node list chromosome.

maintaining a certain load balance in a processor allocation, it does not explore the entire solution space. Especially for task graphs with high communication costs (i.e., high CCRs (Definition 4.23)), reducing communication costs is often crucial. The swapping mutation operator does not support this.

- *Node List.* For chromosomes with permutation encoding, a swapping mutation operator, as just described, is very suitable. The purpose of the mutation operator is to randomly change the node order at some places—swapping two nodes does exactly that (Figure 6.24). In contrast to the previously discussed crossover operator, however, such swapping can bring the nodes out of precedence order. If the node list chromosomes are supposed to be in precedence order, a correcting procedure must be applied after the mutation. As mentioned before, Algorithm 12 can serve this purpose, using the position of the nodes in the mutated chromosome as the node priority.

The direct representation type (as discussed in Section 6.5.2) combines the processor allocation and the node list in one chromosome. Since these parts differ strongly, the simple solution for the operators is to apply the above described ones separately to each part.

Other operators, in combination with appropriate chromosome encodings, have been suggested, which attempt to avoid the creation of invalid chromosomes, especially regarding the node order (e.g., Hou et al. [92], Zomaya et al. [213]).

6.5.4 Selection, Complexity, and Flexibility

Selection   The selection operator of the GA mimics the "survival of the fittest" of evolution. Consequently, it is not simply the selection of the fittest chromosomes from the pool of chromosomes, although fitter chromosomes should have a higher probability of survival. A stochastic approach is needed for the process.

Many selection operators have been suggested for GAs in the literature. As the scheduling problem has a clear and appropriate definition of the fitness function, standard selection methods can be utilized. The following two have been used before.

Tournament   In tournament selection, two chromosomes are randomly picked from the pool. The fitter of the two goes into the population of the next generation, while the losing one is either discarded or put back into the pool. This process is repeated until the initial population size has been reached for the next generation, that is, N tournaments are performed. Such a selection is employed in Sinnen et al. [180].

Roulette Wheel This selection method simulates the behavior of a roulette wheel, where each chromosome has a slot on the wheel. The size of each slot is proportional to the fitness of the corresponding chromosome. In each iteration a position on the roulette wheel is randomly determined and the corresponding chromosome is selected. This process is repeated N times until all chromosomes for the next generation have been selected. In roulette wheel selection, a chromosome c with fitness $F(c) = 0$ has no chance of getting selected. This is not desirable in a GA, especially if such chromosomes are not extremely rare. With Definition 6.7 of the fitness function this can happen though. A simple solution is to add a small value to the fitness value of each chromosome, for example, a very small percentage of the maximum encountered schedule length sl_{max} (Zomaya et al. [213]). Examples for GA based scheduling algorithms that employ this operator are given by Dhodhi et al. [55], Hou et al. [92], and Zomaya et al. [213].

Elitism Elitism is an extension to the selection process that guarantees the survival of the fittest chromosome. With many selection operators (e.g., roulette wheel), the survival of the fittest chromosome is probable, but not guaranteed. With elitism, the simple solution is to just copy the fittest chromosome(s) from one generation to the next.

For GA based task scheduling algorithms, this means that the schedule length of the fittest chromosome upon termination is shorter than or equal to the schedule lengths of the initial chromosomes. By putting a chromosome representing a sequential schedule into the initial population (Section 6.5.2), it is guaranteed that the GA will not produce a solution that is worse than a sequential schedule. This is not as natural as it sounds, given that list scheduling with start time minimization (Theorem 5.2) has no such guarantee. Furthermore, using elitism, GA based task scheduling algorithms can improve schedules produced by other heuristics. It suffices to put chromosomes representing such schedules into the initial population. Elitism then guarantees that the final result will not be worse.

Complexity Usually, the designer of a GA for solving scheduling problems attempts to create operators of low complexity, in particular, lower than the complexity of the evaluation of a chromosome, which is essentially the construction of a schedule. As most GAs use an algorithm based on list scheduling, this complexity is at least $O(\mathbf{V} + \mathbf{E})$, for example, when the processor allocation is encoded in the chromosome. The runtime of a GA is then determined by the population size, N, and the number of generations Gen. Thus, GAs have a complexity of at least $O(N \times Gen \times (\mathbf{V} + \mathbf{E}))$. Typically, the values for the population size and the number of generations are up to a few hundred.

Flexibility GAs are a very flexible instrument for task scheduling. It is quite simple to adapt a GA to scheduling in a special environment, for example, in systems with heterogeneous processors (Singh and Youssef [170], Wang et al. [198], Woo et al. [205]), or in systems with only partially connected processors (Sandnes and Megson [164]). Moreover, scheduling techniques such as node duplication can be included (Sandnes and Megson [165], Tsuchiya et al. [191], Wu et al. [206]). In Dhodhi et al. [55], scheduling of the communications is included in the GA based scheduling algorithm. All this is achieved by correspondingly enhanced chromosomes and the combination of GA with other scheduling heuristics.

The flexibility of GAs is also their shortcoming. GAs have a myriad of parameters which impact on the quality of the solution and the performance. The chromosome representation, the size of the initial population, the probability of crossover and mutation—just to name a few. As a consequence, finding a good GA heuristic involves many cycles of experiment and refinement.

6.6 CONCLUDING REMARKS

This chapter studied advanced aspects of task scheduling. The first sections presented scheduling techniques that are utilized in scheduling algorithms to improve the quality of the produced schedules. The insertion technique tries to use idle slots between already scheduled nodes to improve the processor utilization efficiency. Node duplication, on the other hand, aims at reducing the communication costs of schedules. Following the framework approach of this text, they were analyzed detached from concrete algorithms in a general manner.

Up to that point, everything was based on the scheduling model as introduced in Section 4.1. In Section 6.3 it was extended toward heterogeneous processors. The extension is simple but powerful and most of the discussed scheduling techniques can be employed with little or no modification.

After this, the chapter returned to a topic that had been more or less skipped in Chapter 4—the analysis of special cases of the scheduling problem. Unfortunately, most of them remain NP-hard, as could be seen in the extensive survey of scheduling problems in Section 6.4. The survey employed the $\alpha|\beta|\gamma$ notation for the classification of the various scheduling problems.

As the scheduling problem is in general NP-hard, genetic algorithms have been employed for its solution. This chapter discussed how genetic algorithms can be applied to task scheduling and the implications of different approaches.

Genetic algorithms are employed in many other areas apart from task scheduling. But the GA approach is not the only generic technique that has been applied to task scheduling. Integer linear programming (ILP) is a technique that is utilized for all sorts of scheduling problems. Hanen and Munier [85] formulate an integer linear programming (ILP) based algorithm for task scheduling. Another universal technique, which has been applied recently to task scheduling (Kwok and Ahmad [115]) is A* search (Berman and Paul [22]). In contrast to stochastic search algorithms, A* explores the entire search space systematically and therefore obtains the optimal result. It tries to reduce the runtime by pruning of unpromising state subtrees. For task scheduling, its runtime is nevertheless exponential; therefore, it can only be applied to relatively small graphs.

It goes without saying that there are many other interesting aspects of scheduling. One last example shall be scheduling tools. Many task scheduling algorithms have been implemented in graph based parallelization tools. In the parallelization process of such tools, a graph is generated from an initial program specification. This specification usually has the form of annotated code, where the annotations specify the task graph structure. The graph is then scheduled onto the target parallel system and code is generated according to the schedule and the initial program specification (see also Section 2.3). Examples of such tools are Task Grapher (El-Rewini and Lewis [63]), Parallax (Lewis and El-Rewini [124]), Hypertool (Wu and Gajski [207]), OREGAMI (Lo et al. [130]), CASCH (Ahmad et al. [7]), Meander (Wirtz [201], Sinnen and Sousa [175], Trichina and Oinonen [190]), and PYRROS (Yang and Gerasoulis [209]).

While there are more aspects of task scheduling worth studying, this and the previous chapters covered the fundamentals of task scheduling, basic techniques, and a variety of algorithmic approaches. The next chapters are going to concentrate on an aspect that has been neglected in the literature until recently—the scheduling model. This chapter saw the generalization of the task scheduling model toward heterogeneous processors. This is a generalization of the scheduling model that makes it more realistic for some systems. However, there are other aspects of the scheduling model that do not accurately reflect real systems. As a consequence, schedules can be inaccurate and inefficient on real systems, even if they are optimal under the scheduling model. Therefore, Chapter 7 is going to address the awareness of contention for communication resources in the scheduling model and Chapter 8 will analyze the involvement of the processor in the communication.

6.7 EXERCISES

6.1 In Figure 6.25(*b*) a partial schedule of the sample task graph (Figure 6.25(*a*)) is given. Use list scheduling with start time minimization and the *insertion technique* to schedule the remaining nodes in order c, d, g, i, h, k. Repeat this using the end technique and compare the results.

6.2 Schedule the following task graph on three processors using *node duplication*.
   (a) Which nodes are the primary candidates for duplication?
   (b) Which nodes might be also duplicated as a consequence?

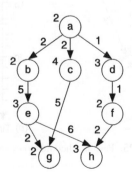

6.3   Repeat Exercise 6.2 for the following task graph to be scheduled on four processors.

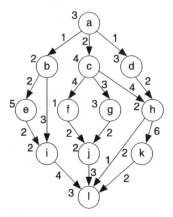

6.4   Schedule the sample task graph (e.g., in Figure 6.25) on three *heterogeneous* processors using list scheduling with *finish* time minimization. The heterogeneity of the processors is consistent and they have the following relative speed values: $s(P_1) = 1$, $s(P_2) = 1.5$, $s(P_3) = 2$. Let the node and edge weights in Figure 6.25 be the average values. Process the nodes in decreasing order of their bottom levels.

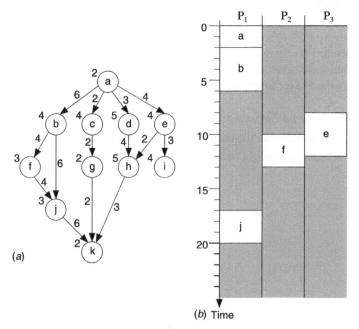

Figure 6.25. Partial schedule (*b*) of sample task graph (*a*).

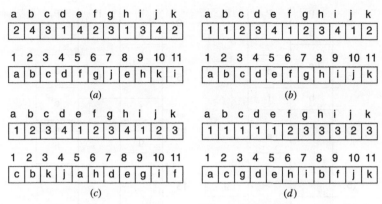

Figure 6.26. Four chromosomes for scheduling the sample task graph (e.g., Figure 6.25) on four processors.

6.5 A given genetic algorithm for task scheduling encodes in each chromosome the processor allocation and the node list (direct representation, Section 6.5.2). Construct the schedules represented by the four chromosomes of Figure 6.26 for scheduling the sample task graph (e.g., Figure 6.25) on four processors. It is not required that the node list is in precedence order; hence, the node order reflects the node priorities.

What is the fitness value of each chromosome (Definition 6.7)?

6.6 Genetic algorithms are very flexible. An example is the fact that the same chromosome encoding as used in Figure 6.26 can be employed for heterogeneous processors. Repeat Exercise 6.5 for four heterogeneous processors. The heterogeneity of the processors is consistent and they have the following relative speed values: $s(P_1) = 1$, $s(P_2) = 2$, $s(P_3) = 2$, $s(P_4) = 1$. Order the chromosomes according to their fitness values. Is this the same order as in Exercise 6.5?

6.7 Figure 6.27 depicts two schedules of Exercise 6.2's task graph on three processors. Specify two chromosomes in direct representation (as in Figure 6.26) that represent these schedules.

6.8 Apply a single-point crossover operator to the chromosomes of Figure 6.26. Perform the crossover between the pairs:

(a) (a,c) with crossover point $i = 6$.

(b) (b,d) with crossover point $i = 5$.

Construct the corresponding schedules of the resulting chromosomes.

6.9 Apply a simple mutation operator to the chromosomes of Figure 6.26. For the processor allocation part of the chromosome, randomly increase or decrease the processor index of one gene, modulo the number of processors $|\mathbf{P}|$. For the node list, swap two randomly determined values.

Construct the corresponding schedules of the resulting chromosomes.

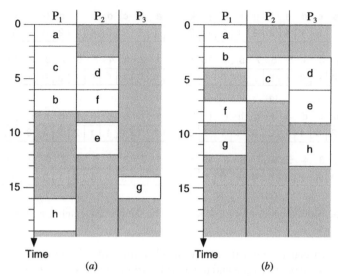

Figure 6.27. Two schedules of Exercise 6.2's task graph on three processors.

6.10 For genetic algorithms and node duplication:

    (a) Why is it more difficult to find an efficient chromosome encoding for task scheduling when node duplication is allowed?

    (b) Suggest an encoding and discuss appropriate operators.

CHAPTER 7

Communication Contention in Scheduling

While task scheduling is an NP-hard problem, its theoretical foundations, as discussed in previous chapters, are easily understood through the high level of abstraction. Computation and communication are strictly separated, whereby the communication is free of contention. Yet, for task scheduling to be more than a mere theoretical exercise, the level of abstraction must be sufficiently close to the reality of parallel systems. During the course of this and the next chapter, it will become clear that this is not the case for the classic model of the target parallel system (Definition 4.3). Therefore, recently proposed models are analyzed that overcome the shortcomings of the classic model, while preserving the theoretical basis of task scheduling.

The critical analysis of the classic model performed in this and the next chapter identifies three of the model's properties, all regarding communication, to be often absent from real systems. Many parallel systems do not possess a dedicated communication subsystem (Property 4); consequently, the processors are involved in communication. The number of resources for communication, such as network interfaces and links, is limited, which contradicts Property 5 of concurrent communication. And finally, only a small number of parallel systems have networks with a fully connected topology, which is Property 6 of the classic model. As a consequence, the following general issues must be investigated:

- The topology of communication network
- The contention for communication resources
- The involvement of processors in communication

This chapter primarily investigates how to handle contention for communication resources in task scheduling. As the network topology has a major impact on contention, the consideration of network topologies, other than fully connected, is a requisite and addressed beforehand. The next chapter will deal with the involvement of the processors in communication. This is done based on the target system model presented in this chapter, so that the final model considers all of the referred issues.

Task Scheduling for Parallel Systems, by Oliver Sinnen
Copyright © 2007 John Wiley & Sons, Inc.

As will be seen, incorporating contention awareness and network topologies into task scheduling leaves the principles of scheduling almost untouched. In contrast, considering the involvement of the processor in communication has a significant impact on how scheduling algorithms work.

The objective of making task scheduling contention aware is the generation of more accurate and efficient schedules. For this objective the following goals are pursued: (1) achieving a more accurate and general representation of real parallel systems; (2) keeping the system model as simple as possible; and (3) allowing the adoption of the scheduling concepts and techniques of the classic model.

Section 7.1 begins the discussion of contention aware scheduling by stating the problem more precisely and reviewing different approaches to its solution. Based on this review, the principles of the approach adopted in this text are described. In order to achieve high generality, the new model should permit one to represent heterogeneity in terms of the communication network as well as in terms of processors. The representation of the communication network, addressed in Section 7.2, is designed accordingly. The essential concept to achieve contention awareness—edge scheduling—is presented and analyzed in Section 7.3. Using edge scheduling, task scheduling becomes contention aware, which is elaborated in Section 7.4, including an analysis of the consequences for the task scheduling framework. Finally, Section 7.5 shows how scheduling algorithms, in particular, list scheduling, are adapted to the new scheduling model.

7.1   CONTENTION AWARENESS

Contention for communication resources arises in parallel systems, since the number of resources is limited. If a resource is occupied by one communication, any other communication requiring the same resource has to wait until it becomes available. In turn, the task depending on the delayed communication is also forced to wait. Thus, conflicts among communications generally result in a higher overall execution time.

The classic model of the target parallel system (Definition 4.3) ignores this circumstance by assuming that all communications can be performed concurrently (Property 5) and that the processors are fully connected (Property 6). However, recent publications have demonstrated the importance of contention awareness for the accuracy and the execution time of schedules (Macey and Zomaya [132], Sinnen and Sousa [171, 176, 178]). At the end of this chapter in Section 7.5.4 these experimental results are reviewed in more detail. Also, in data parallel programming, contention aware scheduling of communications is common (Tam and Wang [183]). Another indicator for the importance of contention is the fact that the LogP programming model (Culler et al. [46, 47]), discussed in Section 2.1.3, explicitly reflects communication contention.

To achieve an awareness of contention, scheduling heuristics must recognize the actualities of the parallel system and consequently adapt the target system model. Only a few scheduling algorithms have been proposed, having a target system view different from the classic model and thereby considering contention. These approaches are discussed in the following. Roughly, contention can be divided into end-point and network contention.

7.1.1 End-Point Contention

End-point contention refers to the contention in the interfaces that connect the processors with the communication network. Only a limited number of communications can pass from the processor into the network and from the network into the processor at one instance in time.

Beaumont et al. [19] extend the classic model by associating a communication port with each processor, called one-port model. At any instance in time, only one incoming and one outgoing communication can be carried out through this port. For example, multiple leaving communications of a node, which are ready as soon as the task finishes execution (node strictness, Definition 3.8), are serialized through this approach. A port can be regarded as the processor's network interface, allowing bidirectional full duplex transfers. Communication between disjoint pairs of processors can still be performed concurrently. Thus, the one-port model captures end-point contention but does not reflect network contention.

A similar, but more general, approach is accomplished with the parameter g of the LogP model (Section 2.1.3). A few scheduling algorithms have been proposed for the LogP model (e.g., Boeres and Rebello [25], Kalinowski et al. [98], Löwe et al. [131]). In these algorithms, the number of concurrent communications that can leave or enter a processor is limited, since the time interval between two consecutive communications must be at least g (see Section 2.1.3). This approach is more general, because the gap can be set to a value smaller than $o + L$ (overhead + latency), thus allowing several communications to overlap. All of the above referenced algorithms neglect the network capacity defined in LogP of at most $\lceil L/g \rceil$ simultaneous message transfers in the network.

End-point contention is only a partial aspect of contention in communication. The limited number of resources within the network (Section 2.2) also leads to conflicts. Furthermore, both of the above approaches suppose completely homogeneous systems and are tied to systems where each processor has one full duplex communication port.

7.1.2 Network Contention

To successfully handle communication contention in scheduling, an accurate model of the network topology is required. Static networks, that is, networks with fixed connections between the units of the system, are commonly represented as undirected graphs. This representation form was informally used during the description of static networks in Section 2.2.1. In the following it is defined formally.

Definition 7.1 (Undirected Topology Graph) *The topology of a static communication network is modeled as an undirected graph $UTG = (\mathbf{P}, \mathbf{L})$. The vertices in $\mathbf{P}$ represent the processors and the undirected edges in $\mathbf{L}$ the communication links* *between them. Link $L_{ij} \in \mathbf{L}$ represents the bidirectional communication link between the processors P_i and P_j and $P_i, P_j \in \mathbf{P}$.*

Examples of such undirected topology graphs are omitted here, but a large number of them can be found in Section 2.2.1. The undirected graph model is a simple and intuitive representation of communication networks. In task scheduling it is employed for various purposes. Wu and Gajski [207] apply it to map scheduled task graphs onto physical processors, trying to balance the network traffic. Other algorithms utilize it for a more accurate determination of communication costs, for example, by counting the "hops" of a communication (i.e., the transitions between links) (e.g., Coli and Palazzari [39], Sandnes and Megson [164]).

Contention Awareness with Undirected Topology Graph   Primarily, the undirected topology graph is the model employed in the few existing network contention aware scheduling heuristics.

El-Rewini and Lewis [63] propose a contention aware scheduling algorithm called MH (mapping heuristic). The target system is represented as an undirected topology graph and a table is associated with each processor. This table maintains for each processor (1) the number of hops on the route, (2) the preferred outgoing link, and (3) the communication delay due to contention. With this information, the communication route between processors is adapted during the execution of the list scheduling based MH. However, the tables are only updated at each start and at each finish of a communication, which is a trade-off between complexity and a real view of the traffic. Furthermore, such a scheduling approach is only meaningful if the target system can be forced to use the routes determined by the algorithm, which is usually not the case (Section 7.2.2).

Concept of Edge Scheduling   A more realistic view of the network traffic is gained when, for each communication, its start and finish times on every utilized communication link are recorded. This corresponds to scheduling the edges on the links, like the nodes are scheduled on the processors, which is illustrated in the Gantt chart of Figure 7.1.

The idea of edge scheduling emerged with the DLS algorithm proposed by Sih and Lee [169]. They proposed representing the target system as an undirected topology

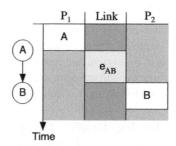

Figure 7.1. The concept of edge scheduling.

graph and employing architecture-dependent routing routines, as opposed to MH. Unfortunately, they do not elaborate on the edge scheduling technique and how it is integrated into classic scheduling.

BSA (bubble scheduling and allocation) is another contention aware algorithm that employs edge scheduling and the undirected topology graph (Kwok and Ahmad [114]). While DLS can be regarded as a dynamic list scheduling algorithm (Section 5.1.2), BSA's approach is quite different, in that it first allocates all nodes to one processor. Subsequently, nodes are migrated to adjacent processors if beneficial. Next, the nodes on the adjacent processors are considered for migration and so on until all processors have been regarded. The routing of the communication is done incrementally with each migration of a node.

BSA has two issues owing to its scheduling strategy, which are analyzed in Sinnen [171]. First, edge scheduling is performed with almost no restrictions. For example, an edge might be scheduled earlier on a link at the end of its route than on a link at the beginning of the route. Obviously, this contradicts causality; hence, it is not a realistic view of the network traffic. Effectively, this approach only captures contention at the last link of a route. Furthermore, the incremental routing can lead to situations where communications are routed in a circle or even use a link twice.

Another task scheduling algorithm that features the scheduling of messages is proposed in Dhodhi et al. [55]. In this genetic algorithm based heuristic, each interprocessor communication is scheduled on the connection between the two communicating processors. The communication network is not modeled, which consequently implies the assumption of a fully connected system.

Virtual Processors Contention for (any type of) resources can also be addressed with the virtual processor concept. This concept is typically found in the domain of real time system scheduling (Liu [127]). Each resource of such a system is modeled as a virtual processor. For interprocessor communication this means that each communication link is a (virtual) communication processor. Correspondingly, the task concept is also abstracted, for example, a communication is considered a job that can only be processed on the (virtual) communication processor.

While this approach is very general, it makes scheduling much more difficult. This is because scheduling has to deal with heterogeneous jobs and heterogeneous (virtual) processors, both in terms of functionality. Task scheduling, as discussed in this text, has only to deal with one type of task. Processor heterogeneity, as introduced in Section 6.3, deals with different processing speeds, not functionality. Due to this conceptual complexity, scheduling heuristics for virtual processors usually have two phases, one for the mapping of the jobs onto the processors and one for the actual ordering/scheduling of the jobs (see also Section 5.2).

This functional heterogeneity of the virtual processor concept is a crucial shortcoming, as many existing task scheduling algorithms cannot be employed. On the other hand, edge scheduling can be integrated into task scheduling with little impact on the general scheduling techniques. Algorithms designed for the classic model can be used almost without any modification as will be seen later. Also, a closer inspection reveals that edge scheduling is in fact similar to the concept of virtual processors: communications (jobs) are scheduled on the links (resources). Yet, edge scheduling is less general and can thereby consider the special circumstances of communication scheduling. For example, local communications should not be scheduled at all, because their costs are negligible.

7.1.3 Integrating End-Point and Network Contention

Edge scheduling, thanks to its strong similarity with task scheduling, has the potential to accurately reflect contention in communication. Therefore, the here adopted approach to contention awareness is based on the edge scheduling concept. The fundamentals of edge scheduling are investigated in Section 7.3. It is aimed at a theory that accurately reflects the real behavior of modern parallel systems, including aspects like routing and causality.

However, edge scheduling on top of the undirected topology graph is limited in several aspects:

- Only half duplex communication links are modeled by the graph.
- Only static networks can be represented.
- End-point contention (Section 7.1.1) is not reflected. A processor connected to multiple links, for example, in a mesh (Section 2.2.1), can simultaneously perform one communication on each link.

Thus, to make edge scheduling an efficient instrument for contention aware scheduling, these issues must be resolved with an improved network model. The next section presents such a model.

7.2 NETWORK MODEL

This section introduces a model for the representation of communication networks of parallel systems in contention aware scheduling (Sinnen and Sousa [172, 179]). The discussion begins with the topology graph that represents the resources of the network. Being a generalization of the undirected graph, the topology graph is capable of capturing both end-point and network contention. The discussion is followed by the analysis of routing, as it determines how the resources are employed.

7.2.1 Topology Graph

The undirected topology graph (Definition 7.1) is a simple and intuitive representation of communication networks. However, as stated in the previous section, it has several shortcomings regarding its utilization for edge scheduling. The discussion and analysis of these issues in the following lead to the proposition of a new topology model. Being a generalization, the new model includes the undirected topology graph but is not limited to static networks.

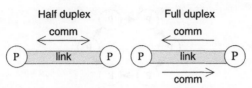

Figure 7.2. Half duplex (left, one communication at a time) and full duplex link (right, two communications—one in each direction).

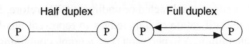

Figure 7.3. Representing half (left) and full duplex links (right) with undirected and directed edges, respectively.

Directed Links The undirected topology graph represents each communication link of the network as one undirected edge. Thus, it is incapable of distinguishing between full and half duplex links. A full duplex point-to-point communication link can transport two communications, or messages, at a time, one in each direction, whereas a half duplex link only transfers one message at a time, independent of its direction. This is illustrated in Figure 7.2.

A distinction between these two communication modes is achieved by representing a full duplex link as two counterdirected edges and a half duplex link as an undirected edge, as shown in Figure 7.3.

An alternative approach would be to use only undirected edges and a special identifier to distinguish between full and half duplex links. Then, however, edge scheduling would have to be aware of the type of link, that is, whether it can schedule two edges with overlapping start–finish intervals (full duplex) or not (half duplex). A better level of abstraction is attained with the utilization of directed edges. The graph encompasses all information about the topology and the communication characteristics of the links. Thus, edge scheduling can be insensible to the edge type.

For comparison, the contention aware scheduling algorithms based on the undirected topology graph (Section 7.1.2) suppose half duplex links, while end-point contention heuristics (Section 7.1.1) assume full duplex links.

Moreover, employing directed edges even admits the representation of unidirectional communication links. They are sometimes encountered in ring-based topologies, as illustrated in Figure 7.4, for example, networks using the SCI (scalable coherent interface) standard (Culler and Singh [48]).

Busses Even when extended with directed edges, the topology model of Definition 7.1 is still limited to static networks. To reflect the important bus topology of dynamic networks (Section 2.2.2) in this model, it must be approximated by a fully connected network (Section 2.2.1). However, for contention scheduling, the fact that the bus is shared among all processors is crucial. Graphs, as defined in Section 3.1, inherently can only represent point-to-point communication links,

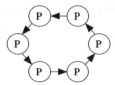

Figure 7.4. Unidirectional ring network, represented with directed edges.

regardless of whether edges are directed or undirected. Therefore, Sinnen and Sousa [176, 178] propose the utilization of hyperedges to represent bus-like topologies in contention aware scheduling. A hypergraph (i.e., a graph containing hyperedges) is a generalization of the undirected graph, where edges—hyperedges—can be incident on more than two vertices (Berge [21]).

Definition 7.2 (Hyperedge and Hypergraph) *Let* **V** *be a finite set of vertices. A hyperedge H is a subset consisting of two or more vertices of* **V**, $H \subseteq \mathbf{V}$, $|H| > 1$. *A hypergraph HG is a pair* $(\mathbf{V}, \mathbf{H})$ *of a finite set of vertices* **V** *and a finite set of hyperedges* **H**.

Note that an undirected edge is a hyperedge incident on exactly two vertices and an undirected graph can be considered a hypergraph containing only such edges. Figure 7.5 visualizes the difference between a 4-vertices undirected graph, fully connected, and a 4-vertices hypergraph, with one hyperedge. In both graphs, every vertex is adjacent to every other vertex.

Switches The above discussed shortcomings of the undirected topology graph and their solutions permit the improved representation of static networks and the incorporation of bus topologies. However, the modeling of dynamic networks in general has not been addressed yet. Since dynamic networks are widely employed in parallel systems (Section 2.2), the approximation of these networks as fully connected is not appropriate.

The essential component of a dynamic network is the switch, which significantly determines the network's characteristic. So while the above improvements target the edges of the topology model, switches are incorporated into the graph by means of a new vertex type—a switch vertex. In other words, a vertex in the topology graph

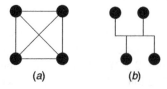

(a)   (b)

Figure 7.5. Undirected graph (*a*) with only point-to-point edges and hypergraph (*b*) with one hyperedge.

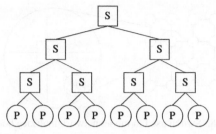

Figure 7.6. Simple binary tree network with processors (P) and switches (S).

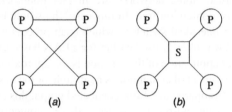

Figure 7.7. LAN modeled as fully connected processors (*a*) or with a switch (*b*).

represents either a processor or a switch. Note that there is still an *implicit* switch within each processor, as was established during the discussion of static networks in Section 2.2.1 (Figure 2.5). With this simple extension, many topologies can be modeled more accurately. For example, in a binary tree network only the leaves are processors; all other vertices are switches (see Figure 7.6).

Another example is a cluster of workstations connected through the switch of a LAN (local area network). In the undirected topology graph, such a network must be represented as fully connected (shown in Figure 7.7(*a*)). Thus, every processor has its own link with every other processor, reducing the modeled impact of contention. In the real system each processor is connected via one link to the switch, which is shared by all communications to and from this processor (Figure 7.7(*b*)).

Moreover, a switch vertex can even be utilized to model the bottleneck caused by the interface that connects a processor to the network, that is, end-point contention (see Section 7.1.1). Imagine one processor in a two-dimensional processor mesh (Section 2.2.1); that is, it has direct links to four neighbors, whose network interface limits the number of communications to one at a time. Deploying a switch as in Figure 7.8 reflects this situation, which, for example, is encountered in the Intel Paragon (Culler and Singh [48]).

Albeit there is already an implicit switch within each processor, as defined for the static network model in Section 2.2.1, Figure 2.5, this switch does not reflect the bottleneck of a network interface, as its form of interaction with the processor is not specified. For what it's worth, it must be assumed as having an unlimited bandwidth, thus being contention free.

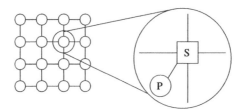

Figure 7.8. Switch used to model network interface bottleneck.

Many dynamic networks, for instance, the butterfly in Figure 2.10, can be described using the switch vertex. Some networks are in fact nonblocking (Cosnard and Trystram [45]), that is, contention free. In this case, the whole network is modeled by encapsulating it in one ideal switch to which each processor is connected with one link. This method was already used earlier for a LAN based cluster, illustrated in Figure 7.7(b). Another application of this method is the approximation of a network, for instance, when the impact of contention is negligible or the description of the network is too complex. Compared to a fully connected graph, this approach still offers the advantage of capturing end-point contention, that is, contention for the network interface. In fact, this approach is almost identical to the one-port model (Beaumont et al. [19]) but has the advantage that it is more flexible, because it can, besides full duplex ports, also represent half duplex ports, using directed or undirected edges, respectively.

Topology Graph   Based on the foregoing considerations, the topology model is defined in the following.

Definition 7.3 (Topology Graph) *The topology of a communication network is modeled as a graph $TG = (\mathbf{N}, \mathbf{P}, \mathbf{D}, \mathbf{H}, b)$, where $\mathbf{N}$ is a finite set of vertices, $\mathbf{P}$ is a subset of $\mathbf{N}$, $\mathbf{P} \subseteq \mathbf{N}$, $\mathbf{D}$ is a finite set of directed edges, and $\mathbf{H}$ is a finite set of hyperedges. A vertex $N \in \mathbf{N}$ is referred to as a* network *vertex, of which two types are distinguished: a vertex $P \in \mathbf{P}$ represents a* processor, *while a vertex $S \in \mathbf{N}, \notin \mathbf{P}$ represents a* switch. *A directed edge $D_{ij} \in \mathbf{D}$ represents a* directed communication link *from network vertex N_i to network vertex N_j, $N_i, N_j \in \mathbf{N}$. A hyperedge $H \in \mathbf{H}$ is a subset of two or more vertices of $\mathbf{N}$, $H \subseteq \mathbf{N}, |H| > 1$, representing a* multidirectional communication link *between the network vertices of H. For convenience, the union of the two edge sets $\mathbf{D}$ and $\mathbf{H}$ is designated link set $\mathbf{L}$, that is, $\mathbf{L} = \mathbf{D} \cup \mathbf{H}$, and an element of this set is denoted by L, $L \in \mathbf{L}$. The weight $b(L)$ associated with a link $L \in \mathbf{L}$, $b : \mathbf{L} \to \mathbb{Q}^{+}$, represents its* relative data rate.

The topology graph is an unusual graph in that it integrates two types of edges—directed edges and hyperedges. Thereby it is able to address the previously discussed shortcomings of the undirected topology graph (Definition 7.1): it reflects directed links, busses, and switches (through the new vertex type). A hyperedge represents a bus or, if it is only incident on two network vertices, a half duplex link. A full duplex link is represented by two counterdirected edges and a unidirectional communication link is modeled by one directed link in the corresponding direction. For edge scheduling, the type of the edge is irrelevant (see Section 7.3); the only relevant consideration is which edges are involved in a communication. Thus, the different edge types are only relevant for the routing of a communication (see Section 7.2.2).

It is assumed that a switch is ideal—that is, there is no contention within the switch. All communications can take every possible route inside the switch—that is, a communication can arrive at any entering edge and leave at any outgoing edge, with the obvious exception of the edge from where it came. Essentially, a switch vertex behaves in the same way as a processor in the undirected topology graph. In fact, a switch can be treated as a processor on which tasks cannot be executed. Or, seen the other way around, a processor is the combination of a switch with a processing unit. This interpretation is used in the undirected topology graph (Figure 2.5, Section 2.2.1). A communication between two nonadjacent processors is routed through other processors, which therefore act as switches.

The topology graph is conservative in that it contains other simpler models. For example, in a network without busses, it reduces to a graph without hyperedges (only directed and undirected edges), or in a static network, all network vertices are processors. In particular, the topology graph is a generalization of the undirected topology graph, which is given with $N = P$ (no switches), $|H| = 2 \forall H \in H$ (only undirected edges, i.e., hyperedges with two vertices) and $D = \emptyset$ (no directed edges).

Also, heterogeneity, in terms of the links' transfer speeds, is captured by the topology graph through the association of relative data rates with the links. In comparison with the introduced processor heterogeneity in Section 6.3, a distinction between consistent and inconsistent heterogeneity does not appear to be sensible. The transfer rate of a link does not depend on the type of data and is thus consistent. In contrast, the type of computation might very well have an impact on the execution speed, which can result in inconsistent heterogeneity. Of course, for homogeneous systems, the relative data rate is set to 1 or simply ignored.

Examples for networks modeled with the topology graph were already given in Figures 7.4 and 7.6–7.8. A topology graph containing directed edges and hyperedges is illustrated in Figure 7.9. It represents 8 dual processor systems, that is, 16 processors, in a dedicated LAN. In each dual system the processors and a network interface (here modeled as a switch) are grouped around one bus. Every dual system is connected to a central switch, representing the LAN, with a full duplex link (the counterdirected edges). Communication between the two processors of each system is performed via the bus, while communication between processors of different systems goes through the central switch. At any time only one of the two processors of a system can communicate, since all communications, local or remote, utilize the bus. Together with a heterogeneous setting of the links' data rates—a system bus is usually faster than a LAN—this topology graph reflects accurately the behavior of such dual processor clusters.

It should be noted that in comparison with a fully connected graph (the only sensible way to model such a network in the undirected topology graph), the number of vertices increases—from 16 to 25—but the number of edges is significantly reduced from 120 (see Eq. (3.2)) to 24.

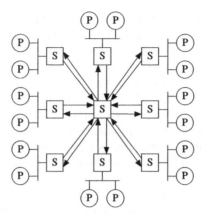

Figure 7.9. Example of topology model: 8 dual processors connected via a switched LAN and full duplex links.

7.2.2   Routing

In Section 7.2.1, an accurate model of the network topology was defined. However, the network's behavior is also affected by the routing of communications. To schedule edges on communication links, it must be known which links are involved in an interprocessor communication and how.

Essentially, this is described by the routing algorithm and the policy of the target system's network. The routing algorithm selects the route through the network, that is, the sequence of links that are employed for the communication, and the routing policy determines how these links are used.

In the next paragraphs the relevant aspects of routing will be reviewed from the perspective of edge scheduling. A general discussion of routing can be found in Culler and Singh [48] and Kumar et al. [108].

Static Versus Adaptive Routing   A routing algorithm is *static*, or *nonadaptive*, if the route taken by a message is determined solely by its source and destination regardless of other traffic in the network. *Adaptive* routing algorithms allow the route for a message/packet to be influenced by traffic it meets along the way. Clearly, it is very complex to model adaptive routing in contention aware scheduling. To accurately simulate the effect of adaptive routing in the topology graph, the exact state of the network at every time instance must be known. Yet, the aim of contention aware scheduling is to avoid contention in the network through an appropriate schedule. Thus, with such a schedule, routes should seldom be adapted. Moreover, adaptive routing is not widely used in parallel machines (one of the few exceptions is the Cray T3E, but not the Cray T3D), due to its higher complexity and possible disadvantages (Culler and Singh [48]).

It is important to understand that static routing does not imply that every message from a given source to a given destination takes the same route. If multiple routes exist between two communication partners, mechanisms like randomization or round robin can choose one of the alternatives, yet not influence the traffic. As these routing techniques are independent of the traffic, they can be integrated into the routing algorithm for the topology graph.

Switching Strategy The next routing aspect to be considered is the switching strategy. Different strategies arise from the circumstance that data to be transfered in a network is split into packets. Two switching strategies can be distinguished: circuit switching and packet switching. In *circuit switching*, the path for the communication is established—by the routing algorithm—at the beginning of the transfer and all data takes the same circuit (i.e., route). With *packet switching* the message is split into packets and routing decisions are made separately for every packet. As a consequence, packets of the same message might be transfered over different routes. In scheduling, a communication associated with an edge is regarded in its entirety and its possible separation into packets is not reflected. Thus, edge scheduling must assume a circuit switched network or—as an alternative interpretation—a static packet switched network, where all packets of a message take the same route between two processors.

Store-and-Forward and Cut-Through Routing If a communication route involves more than one link, the routing of a message can be handled in two different ways: store-and-forward or cut-through routing. In *store-and-forward* routing the data of a communication is stored at every station (in the topology graph this is a processor or switch) on the path until the entire message, or packet, has arrived and is subsequently forwarded to the next station. In *cut-through* routing, a station immediately forwards the data, inflicting only a small routing delay (more on routing delays later), to the next station on the path—the message "cuts through" the station. By showing the transfer of one communication in a linear network of four processors in Figure 7.10, it is illustrated that store-and-forward routing has a much higher latency than cut-through routing, as the transfer time is a function of the number of "hops" (i.e., transitions between links). Store-and-forward routing is used in wide area networks and was used in several early parallel computers, while more modern parallel systems employ cut-through routing, for example, Cray T3D/T3E and IBM SP-2 (Culler and Singh [48]).

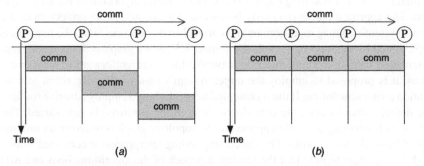

Figure 7.10. Store-and-forward routing (*a*) versus cut-through routing (*b*).

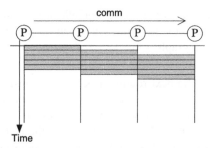

Figure 7.11. Packet based store-and-forward routing is similar to cut-through routing.

Still, for those parallel systems that use store-and-forward, assuming cut-through routing in edge scheduling is often a good approximation. Due to the limited capacity of communication buffers in the stations, store-and-forward routing works packet based. If a message consists of a sufficiently large number of packets, store-and-forward routing behaves approximately like cut-through, with only a packet size dependent delay per hop, as illustrated in Figure 7.11.

Routing Delay per Hop   With every hop that a message or packet takes along its route through the network, a delay might be introduced. In cut-through routing this delay is typically very small, that is, on the order of some network cycles (e.g., Cray T3E has a hop delay of 1 cycle; Culler and Singh [48]). The store-and-forward hop delay is essentially the transfer time of one packet over one link. If the packet is small relative to the message size, it can be regarded as nonsubstantial. For this reason, the hop delay is neglected for simplicity in edge scheduling, but the next section also shows that it can be included in a straightforward manner, if necessary.

Routing Algorithm   The routing algorithm of the network selects the links through which a communication is routed. It is carried out by the involved processors, the switches of the network, or both. In order to obtain as accurate a result as possible, contention aware scheduling ought to reflect the routing algorithm of the target system. The approach that is obviously best suited for this aim is the projection of the target system's routing algorithm onto the model of its network. Since the network is represented by the topology graph (Definition 7.3), and a graph is the common representation form of a communication network, this is straightforward in most cases. Thus, it is proposed to employ the target system's own routing algorithm in contention aware scheduling. If the represented network should employ adaptive routing, the default behavior (i.e., the behavior without traffic conflicts) is reproduced. The job of such a routing algorithm applied to the topology graph is to return an ordered list of links, the route, utilized by the corresponding interprocessor communication.

It is important to note that the routing approach of the algorithms BSA and MH (Section 7.1.2) is exactly the other way around: they determine the route according to their own routing algorithms. Hence, their produced schedules can only be accurate and efficient if the target parallel system can be directed to use the computed routes. However, that is rarely the case in a generic parallel system, especially at the application level (Culler and Singh [48]).

Fortunately, using the target system's own routing algorithm does not imply that for every target system a different routing algorithm must be implemented in scheduling. Most parallel computers employ *minimal routing*, which means they choose the shortest possible path, in terms of number of edges, through the network for every communication. An example is dimension ordered routing in a multidimensional mesh (Culler and Singh [48], Kumar et al. [108]). A message is first routed along one dimension, then along the second dimension, and so on until it reaches its destination.

Given the graph based representation of the network, finding a shortest path can be accomplished with a BFS (breadth first search—Algorithm 1). Thus, the BFS can be used as a generic routing algorithm in the topology graph, which serves, at least, as a good approximation in many cases.

Shortest Path in Topology Graph with BFS   Although BFS is an algorithm for directed and undirected graphs, it can readily be applied to the topology graph. The only graph concept used in BFS is that of adjacency (in the `for` loop, Algorithm 1). As this is already defined for directed and undirected edges (Section 3.1), it only remains to define adjacency for hyperedges.

Definition 7.4 (Adjacency—Hyperedge)   *Let* $\mathbf{V}$ *be a finite set of vertices and* $\mathbf{H}$ *a finite set of hyperedges. A vertex* $u \in \mathbf{V}$, $u \in H$, $H \in \mathbf{H}$ *is adjacent to all vertices* $v \in H\text{-}u$ *and all vertices* $v \in H\text{-}u$ *are adjacent to* u. *The set* $\bigcup_{H \in \mathbf{H}:u \in H} H\text{-}u$ *of all vertices adjacent to* $u \in \mathbf{V}$ *is denoted by* $\mathbf{adj}(u)$.

Now in the topology graph, the total set of all vertices adjacent to a given vertex u is the union of the two adjacent sets induced by the directed edges and the hyperedges. So with this definition of vertex adjacency, the BFS can be applied to the topology graph without any modification and returns a shortest path in terms of number of edges.

Complexity   As routing depends on the algorithm of the target parallel system, there is no general time complexity expression that is valid for all networks. On that account, the routing complexity shall be denoted generically by $O(routing)$.

The algorithm for routing in regular networks is usually linear in the number of network vertices or even of constant time. For example, in a fully connected network it is obviously $O(1)$, as it is in a network with one central switch (Figure 7.7(*b*)). In a topology graph for a mesh network of any dimension (Section 2.2.1), it is at most linear in the number of processors $O(\mathbf{P})$. Whenever it is possible to calculate the routes for a given system once and then to store them; for example, in a table, $O(routing)$ is just the complexity of the length of the route. For example, in a ring network the length of a route (i.e., the number of links) is $O(\mathbf{P})$; hence, $O(routing) \geq O(\mathbf{P})$.

The BFS used to reflect minimal routing, has linear complexity for directed and undirected graphs, $O(V + E)$ (Section 3.1.2). With the above definition of adjacency with hyperedges, this result extends directly to the topology graph. Hence, in the topology graph BFS's complexity is $O(N + L)$. To obtain this complexity it is important that every hyperedge is considered only once and not one time for every incident vertex.

A final word regarding the number of network vertices. In a dynamic network, switches can outnumber the processors. Yet, in practice the number of switches is at most $O(P^2)$—hence $O(N) \leq O(P^2)$—because a higher number of switches would make such a network more expensive than a fully connected network.

Other Aspects   Other aspects of routing, such as flow control, deadlock handling, and buffer sizes (Culler and Singh [48]), are not relevant for edge scheduling. Also, packet dropping due to congestion, for example, in TCP/IP based networks, is not reflected, once more, because the separation into packets is unregarded. In any case, the effect of packet dropping should be significantly reduced when a program is executed according to a contention aware schedule. The contention is already taken care of by the scheduling algorithm and should not arise often during execution.

7.2.3   Scheduling Network Model

Sections 7.2.1 and 7.2.2 analyzed, from the perspective of scheduling, the topologies of communication networks and their routing policies. It remains merely to summarize this analysis to establish a network model based on the topology graph (Sinnen and Sousa [172, 179]).

Definition 7.5 (Network Model)   *A communication network of a parallel system is represented by a topology graph $TG = (N, P, D, H, b)$ (Definition 7.3). The routing algorithm carried out on the topology graph is that of the represented network (or its closest approximation). Furthermore, routing has the following properties:*

1. Static. *Routing in the network is static; the selected route does not depend on the state of the network. If the represented network employs adaptive routing, the default behavior (i.e., the behavior without traffic conflicts) is reproduced.*
2. Circuit Switching. *The communication route between two processors of the network is the same for the entire message.*
3. Cut-through. *A message cuts through the communication stations (processor or switch) on its route; that is, all links of the route are utilized at the same time.*
4. No Routing Delay per Hop. *No delay is inflicted by the transition between links on the communication route.*

Edge scheduling, which is investigated in the next section, employs this model for the representation of the network.

7.3 EDGE SCHEDULING

This section analyzes the theoretical background of the fundamental technique to achieve contention awareness in scheduling: edge scheduling (Sinnen and Sousa [172, 179]). The utilization of edge scheduling in contention aware heuristics and the interaction with task scheduling are treated in the next section.

Edge scheduling is the adoption of the task scheduling principle in the communication domain. In contrast to the classic model of the target system (Definition 4.3), communication resources are treated like processors, in the sense that only one communication can be active on each resource at a time. Thus, edges are scheduled onto the communication links for the time they occupy them.

To schedule an edge on a link, a start time and a finish time are assigned to the edge. The corresponding notations from node scheduling (Section 4.1) are adopted.

Definition 7.6 (Edge Start, Communication and Finish Time on Link) *Let* $G = (\mathbf{V}, \mathbf{E}, w, c)$ *be a task graph and* $TG = (\mathbf{N}, \mathbf{P}, \mathbf{D}, \mathbf{H}, b)$ *a communication network according to Definition 7.5.*

The start time $t_s(e, L)$ *of an edge* $e \in \mathbf{E}$ *on a link* $L \in \mathbf{L}$ $(\mathbf{L} = \mathbf{D} \cup \mathbf{H})$ *is the function* $t_s : \mathbf{E} \times \mathbf{L} \to \mathbb{Q}_0^+$.

The communication time of e on L is

$$\zeta(e, L) = \frac{c(e)}{b(L)}. \tag{7.1}$$

The finish time of e on L is

$$t_f(e, L) = t_s(e, L) + \varsigma(e, L). \tag{7.2}$$

So the time during which link L is occupied with the transfer of the data associated with the edge e is denoted by the communication time $\varsigma(e, L)$. As links might have heterogeneous data rates (Definition 7.3), the communication time is a function of the edge e and the link L. This also brings a change to the interpretation of the edge cost $c(e)$ (Definition 4.4).

Definition 7.7 (Communication Cost—Network Model) *Let* $G = (\mathbf{V}, \mathbf{E}, w, c)$ *be a task graph and* $TG = (\mathbf{N}, \mathbf{P}, \mathbf{D}, \mathbf{H}, b)$ *a communication network.*

The communication cost $c(e)$ *of edge* e *is the* average *time the communication represented by* e *occupies a link of* $\mathbf{L}$ *for its transfer.*

Like a processor, a link is exclusively occupied by one edge.

Condition 7.1 (Link Constraint) *Let* $G = (\mathbf{V}, \mathbf{E}, w, c)$ *be a task graph and* $TG = (\mathbf{N}, \mathbf{P}, \mathbf{D}, \mathbf{H}, b)$ *a communication network. For any two edges* $e, f \in \mathbf{E}$ *scheduled* *on link $L \in \mathbf{L}$:*

$$t_s(e,L) \leq t_f(e,L) \leq t_s(f,L) \leq t_f(f,L)$$

$$or \qquad t_s(f,L) \leq t_f(f,L) \leq t_s(e,L) \leq t_f(e,L). \tag{7.3}$$

Despite its similarity with task scheduling, edge scheduling differs in one important aspect. In general, a communication route between two processors consists of more than one link. An edge is scheduled on each link of the route, while a node is only scheduled on one processor (with the exception of node duplication, Section 6.2). The scheduling of an edge along the links of the route is significantly determined by the routing in the network.

7.3.1 Scheduling Edge on Route

In Section 7.2 the network model of the target system was carefully developed in a way that encapsulates and hides all the details about the network topology from edge scheduling. For any communication between two distinct processors, edge scheduling only sees the communication route in the form of an ordered list of involved links. This list is provided by the routing algorithm of the network model, when it is given the two communicating processors.

Definition 7.8 (Route) *Let $TG = (\mathbf{N}, \mathbf{P}, \mathbf{D}, \mathbf{H}, b)$ be a communication network. For any two distinct processors P_{src} and P_{dst}, $P_{\mathrm{src}}, P_{\mathrm{dst}} \in \mathbf{P}$, the routing algorithm of TG returns a route $R \in TG$ from P_{src} to P_{dst} in the form of an ordered list of links $R = \langle L_1, L_2, \ldots, L_l \rangle$, $L_i \in \mathbf{L}$ for $i = 1, \ldots, l$.*

In Definition 7.8 the route depends only on the source and the destination processor of a communication and not on the traffic state of the network. This conforms to the network model's property of static routing (Definition 7.5). Since the network model also supposes circuit switching, the entire communication is transmitted on the returned route. Hence, the edge e of the communication is scheduled on each link of the route. The type of link (e.g., half or full duplex) is irrelevant to edge scheduling.

Data traverses the links in the order of the route, which must be considered in the scheduling of the edges.

Condition 7.2 (Causality) *Let $G = (\mathbf{V}, \mathbf{E}, w, c)$ be a task graph and $TG = (\mathbf{N}, \mathbf{P}, \mathbf{D}, \mathbf{H}, b)$ a communication network. For the start and finish times of edge $e \in \mathbf{E}$ on the links of the route $R = \langle L_1, L_2, \ldots, L_l \rangle$, $R \in TG$,*

$$t_f(e, L_{k-1}) \leq t_f(e, L_k), \tag{7.4}$$

$$t_s(e, L_1) \leq t_s(e, L_k), \tag{7.5}$$

for $1 < k \leq l$.

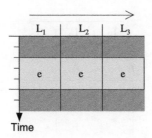

Figure 7.12. Edge scheduling on a route without contention.

Inequality (7.4) of the causality condition reflects the fact that communication does not finish earlier on a link than on the link's predecessor. For homogeneous links, inequality (7.5) is implicitly fulfilled by Eq. (7.4), but for heterogeneous links this condition is important as will be seen later.

The edge scheduling must further comply with the other routing properties of the network model in Definition 7.5, namely, the cut-through routing and the neglect of hop delays. Together, these two properties signify that all links of the route are utilized simultaneously. Hence, an edge must be scheduled at the same time on every link. Figure 7.12 illustrates this for a route consisting of three homogeneous links on which edge e is scheduled.

This example shows the trivial case of edge scheduling, where all links are homogeneous and idle. When contention comes into play (i.e., when a communication meets other communications along the route), edge scheduling must be more sophisticated. The scheduling times of an edge on the different links are strongly connected through the causality condition and the routing properties of the network model. If an edge is delayed on one link, the scheduling times of the edge on the subsequent links are also affected. Two different approaches must be examined for the determination of the scheduling times on a route with contention.

Consider the same example as in Figure 7.12, with the modification that link L_2 is occupied with the communication of another edge, say, e_{xy}, at the time e is scheduled.

Nonaligned Approach One solution to the scheduling of e is depicted in Figure 7.13(a). On link L_2, e is delayed until the link is available; thus, $t_s(e, L_2) = t_f(e_{xy}, L_2)$. To adhere to the causality condition, e is also delayed on link L_3; it cannot finish on the last link until it has finished on the previous link, that is, $t_s(e, L_3) = t_s(e, L_2)$ and $t_f(e, L_3) = t_f(e, L_2)$. On L_1, e is scheduled at the same time as without contention (Figure 7.12). This approach is referred to as nonaligned.

Aligned Approach Alternatively, e can be scheduled later on all links; that is, it starts on all links after edge e_{xy} finishes on link L_2, $t_s(e, L_1) = t_s(e, L_2) = t_s(e, L_3) = t_f(e_{xy}, L_2)$, even on the first L_1. At first, this aligned scheduling of the edges, illustrated in Figure 7.13(b), seems to better reflect cut-through routing, where communication takes place on all involved links at the same time. Scheduling the edge at different times, as done in the first approach (Figure 7.13(a)), seems to imply the storage of

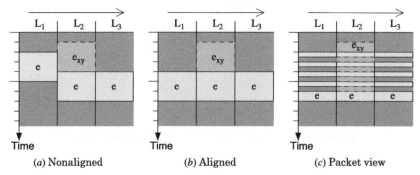

(a) Nonaligned   (b) Aligned   (c) Packet view

Figure 7.13. Edge scheduling on a route with contention.

at least parts of the communication at the network vertices—a behavior similar to store-and-forward routing.

Packet View However, communication in parallel systems is packet based (Section 7.2.2) and the real process of the message transfer is better illustrated by the Gantt chart of Figure 7.13(c) (assuming the usual fair sharing of the resources). Each of the symbolized packets performs cut-through routing and link L_2 is shared between both edges.

For several reasons, it is not the intention to go down to the packet level in edge scheduling: (1) the scheduling model would become more complicated; (2) new system parameters, like the packet size and arbitration policies, would be introduced; and (3) the scheduling of an edge would affect previously scheduled edges (e.g., in Figure 7.13(c), e_{xy} is affected by the scheduling of e).

But the packet view of the situation in Figure 7.13(c) illustrates that the nonaligned scheduling (Figure 7.13(a)) can be interpreted as a message based view of the situation in Figure 7.13(c). Nonaligned scheduling holds an accurate view of the communication time spent in the network. The communication lasts from the start of edge e on the first link L_1 until it finishes on link L_3, exactly the same time interval as in the packet view of Figure 7.13(c). In contrast, aligned scheduling delays the start of the e on the first link L_1 and therefore does not reflect the same delay within the network as in the packet view. In both approaches, the total occupation time of the links is identical to the packet view, even though approximated as an indivisible communication.

Moreover, scheduling the edges aligned on a route means that idle intervals have to be found that are aligned across the entire route. In consequence, this approach is likely to produce many idle time slots on the links, since many solutions that are feasible under the causality condition (e.g., the solution of Figure 7.13(a)), cannot be employed. Also, employment of the insertion technique (Section 6.1) in edge scheduling could be virtually useless, as it can be expected that only a few of such idle channels exist. Nonaligned scheduling allows the successive treatment of each link, since the edge's start and finish times are only constrained by the scheduling on the foregoing links.

In conclusion, the nonaligned approach appears to better reflect the behavior of communication networks and its utilization in edge scheduling is adopted. Before edge scheduling is formally defined, the nonaligned approach must yet be analyzed for heterogeneous links.

Heterogeneous Links The causality condition must also be obeyed on heterogeneous links, that is, links with different data rates. Figure 7.14(a) illustrates again the scheduling of an edge on three links, whereby link L_2 is now twice as fast as the other links ($b(L_2) = 2b(L_1) = 2b(L_3)$). Due to Eq. (7.4) of the causality Condition 7.2, e cannot finish earlier on L_2 than on L_1 and therefore the start of e on L_2 is delayed accordingly. As noted before, this delay seems to suggest a storage between the first and second link, but in fact it is the best approximation of the real, packet based communication illustrated in Figure 7.14(b). This explains also why e starts and finishes on L_3 at the same times as on L_1, which is in concordance with Eq. (7.5).

Another possibility, which seems more intuitive at first, is to start e on L_3 later, namely, at the same time it starts on L_2. This would correspond to the seemingly more intuitive $t_s(e, L_{k-1}) \leq t_s(e, L_k)$ as the causality equation (7.5). Then, however, e would finish on L_3 later than on L_2 and L_1. Consequently, this implies that having a faster link between two other links *retards* the communication. This is certainly not realistic and also does not correspond to the packet view in Figure 7.14(b). Thus, the causality Condition 7.2 as defined ensures a realistic scheduling of the edge in this example.

The next example further emphasizes the importance of inequality (7.5) of the causality Condition 7.2 for heterogeneous links. Consider the same example as before, now with the last link slower than the two others ($b(L_3) = b(L_1)/2 = b(L_2)/4$), which is illustrated in Figure 7.14(c). Without Eq. (7.5), e could start earlier on L_3 than on L_1: it only had to finish not earlier than on L_2. Evidently, the principle of causality would be violated, because the communication would be active on L_3 before it even entered into the network. Due to the lower data rate of L_3, the correct behavior, as shown in Figure 7.14(c), is not implicitly enforced by inequality (7.4) as in all other examples. Note that the communication is not delayed by the contention on L_2, because even without it, e could not start earlier on L_3. This is a realistic reflection of the behavior

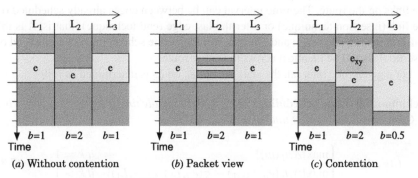

(a) Without contention   (b) Packet view   (c) Contention

Figure 7.14. Edge scheduling on a heterogeneous route.

of a real network, where a fast link (L_2) can compensate contention in relation to slower links on the route.

7.3.2 The Edge Scheduling

In the following, the equations are formulated for edge scheduling. As stated earlier, the nonaligned approach allows determination of the start and finish times successively for each link on the route. It is not necessary to consider all links at the same time to find an idle time channel.

On the first link, the edge's scheduling is only constrained by the finish time of its origin node on the source processor of the communication. On all subsequent links, the edge has to comply with the causality condition.

As for the scheduling of the nodes (Condition 6.1), a scheduling condition can be formulated that integrates both the end technique (Definition 5.2) and the insertion technique (Definition 6.1).

Condition 7.3 (Edge Scheduling Condition) *Let $G = (\mathbf{V}, \mathbf{E}, w, c)$ be a task graph, $TG = (\mathbf{N}, \mathbf{P}, \mathbf{D}, \mathbf{H}, b)$ a communication network, and $R = \langle L_1, L_2, \ldots, L_l \rangle$ the route for the communication of edge $e_{ij} \in \mathbf{E}$, $n_i, n_j \in \mathbf{V}$. Let $[A, B]$, $A, B \in [0, \infty]$, be an idle time interval on L_k, $1 \leq k \leq l$, that is, an interval in which no other edge is transferred on L_k. Edge e_{ij} can be scheduled on L_k within $[A, B]$ if*

$$B - A \geq \varsigma(e_{ij}, L_k), \tag{7.6}$$

$$B \geq \begin{cases} t_f(n_i) + \varsigma(e_{ij}, L_1) & \text{if } k = 1 \\ \max\{t_f(e_{ij}, L_{k-1}), t_s(e_{ij}, L_1) + \varsigma(e_{ij}, L_k)\} & \text{if } k > 1 \end{cases}. \tag{7.7}$$

Inequality (7.6) ensures that the time interval is large enough for e_{ij}. Furthermore, inequality (7.7) guarantees that the constraint of the origin node's finish time or the causality Condition 7.2 can be fulfilled for e_{ij} within this interval. On the first link $(k = 1)$ of the route R, e_{ij}'s scheduling is only constrained by the finish time of its origin node n_i on the source processor of the communication. On all subsequent links $(k > 1)$ of R, the scheduling of e_{ij} has to comply with the causality Condition 7.2.

A time interval, obeying Condition 7.3, can be searched successively for each of the links on the route. The time interval can be between edges already scheduled on the link (insertion technique) or after the last edge (end technique); that is, A is the finish time of the last edge and $B = \infty$. Obviously, the scheduling technique (end or insertion) should be identical across the entire route.

For a given time interval, the start time of e_{ij} on L_k is determined as follows.

Definition 7.9 (Edge Scheduling) *Let $[A, B]$ be an idle time interval on L_k adhering to Condition 7.3. The start time of e_{ij} on L_k is*

$$t_s(e_{ij}, L_k) = \begin{cases} \max\{A, t_f(n_i)\} & \text{if } k = 1 \\ \max\{A, t_f(e_{ij}, L_{k-1}) - \varsigma(e_{ij}, L_k), t_s(e_{ij}, L_1)\} & \text{if } k > 1 \end{cases}. \tag{7.8}$$

So edge e_{ij} is scheduled as early within the idle interval as the causality condition or the finish time of the origin node admits. This corresponds exactly to the nonaligned approach discussed in Section 7.3.1.

A last remark regarding hop delays: it should be easy to see that the introduction of such a hop delay is straightforward. A simple addition of a delay value to the calculation of the start times on the links L_k, $k > 1$, is sufficient. Of course, the conditions on the idle time interval must be adapted, too (Condition 7.3). Moreover, if the hop delay equals the time spent on the previous link, even store-and-forward routing can be simulated.

7.4 CONTENTION AWARE SCHEDULING

The network model and edge scheduling have been investigated in the previous sections. Based on this, contention aware task scheduling is analyzed in this section. The basics are presented next, followed by the proposal of contention aware scheduling heuristics.

7.4.1 Basics

The new, more realistic, target system model is defined as follows.

Definition 7.10 (Target Parallel System—Contention Model) *A target parallel system $M_{TG} = (TG, \omega)$ consists of a set of possibly heterogeneous processors* **P** *connected by the communication network $TG = (\mathbf{N}, \mathbf{P}, \mathbf{D}, \mathbf{H}, b)$, according to Definition 7.5. The processor heterogeneity, in terms of processing speed, is described by the execution time function ω. This system has Properties 1 to 4 of the classic model of Definition 4.3:*

1. *Dedicated system*
2. *Dedicated processor*
3. *Cost-free local communication*
4. *Communication subsystem*

Two of the classic model's properties are substituted by the detailed network model: concurrent communication (Property 5) and a fully connected network topology (Property 6).

With the abandonment of the assumption of concurrent communication, task scheduling must consider the contention for communication resources. For this purpose, edge scheduling is employed. The actual scheduling of the edges on the links was discussed in Section 7.3. It remains to establish a connection with task scheduling.

The integration of edge scheduling into task scheduling is simple, due to the careful formulation of the scheduling problem in Section 4.2. It is only necessary to redefine the total edge finish time (Definition 4.6).

Definition 7.11 (Edge Finish Time—Contention Model)   *Let $G = (\mathbf{V}, \mathbf{E}, w, c)$ be a DAG and $M_{TG} = ((\mathbf{N}, \mathbf{P}, \mathbf{D}, \mathbf{H}, b), \omega)$ a parallel system. The finish time of $e_{ij} \in \mathbf{E}$, $n_i, n_j \in \mathbf{V}$, communicated from processor P_{src} to P_{dst}, $P_{\text{src}}, P_{\text{dst}} \in \mathbf{P}$, is*

$$t_f(e_{ij}, P_{\text{src}}, P_{\text{dst}}) = \begin{cases} t_f(n_i, P_{\text{src}}) & \text{if } P_{\text{src}} = P_{\text{dst}} \\ t_f(e_{ij}, L_l) & \text{otherwise} \end{cases} \qquad (7.9)$$

where L_l is the last link of the route $R = \langle L_1, L_2, \ldots, L_l \rangle$ from P_{src} to P_{dst}, if $P_{\text{src}} \neq P_{\text{dst}}$.

So the edge's (global) finish time is its finish time on the last link on the route, unless the communication occurs between nodes scheduled on the same processor (Property 3), where it remains the finish time of the origin node.

The scheduling of the nodes is not concerned with the internals of edge scheduling. Everything that is relevant to node scheduling is encapsulated in the finish time of an edge. As before, a node must not start earlier than the incoming edge finishes communication. Figure 7.15 illustrates the precedence constraint with edge scheduling. It displays an integrated Gantt chart of the node and edge schedule of a two-node task graph on two processors connected by one link. The edge e_{AB} starts on the link after node A has finished (inequality (7.8), $k = 1$), and the execution of B is started immediately after the communication has terminated on the link (Eqs. (4.5) and (7.9)).

The time interval spanned by the edge's start time on the first link $t_s(e, L_1)$ and its finish time on the last link $t_f(e, L_l)$ of the route R corresponds to the communication delay (Definition 4.4) in the classic model. The interval is identical with the delay in the case of no contention. Furthermore, in this case an edge is scheduled on every link at the same time (cut-through routing) and the edge weight is defined to be the average time a link is occupied with the transfer (Definition 7.7). Hence, the edge weight in a task graph for the classic model is identical to the edge weight in a task graph for the contention model. Thus, the new system model has no impact on the

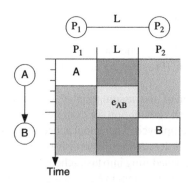

Figure 7.15. Precedence constraint with edge scheduling.

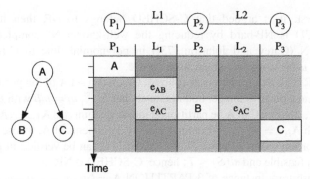

Figure 7.16. Contention aware scheduling: scheduling order of edges is relevant.

creation of the task graph. It only differs for heterogeneous target systems; but this case is not addressed in the classic model, anyway.

Contention Awareness When an edge is scheduled along its route, it might conflict with other, already scheduled edges on one or more links. As a result, the start of the edge is delayed on these links, which eventually results in a delayed total finish time of the edge. Thus, the contention is exposed in the extended transfer time of the edge. This delays the execution start of the destination node and can in the end result in a longer schedule. For instance, in Figure 7.16, showing the integrated schedule of the depicted task graph and homogeneous target system, edge e_{AC} is delayed on link L_1 due to contention, because e_{AB} already occupies L_1. This also delays the start of node C on P_3, as the communication arrives later. In the classic model, both communications are transferred at the same time so that node C starts at the same time as B. A scheduling heuristic "sees" the contention effect through the node's later DRT on the processors to which communication is affected by contention.

An interesting consequence of edge scheduling is that the order in which edges are scheduled is relevant. For example, the order of the outgoing edges of a node has an influence on the DRTs of the node's successors. This is desired, as it reflects the behavior of real systems. If edge e_{AC} in Figure 7.16 was scheduled before e_{AB}, the start times between node B and C would be swapped—that is, node C would start before node B.

7.4.2 NP-Completeness

Intuitively, scheduling is more complicated when considering contention, and in fact, the problem remains NP-complete (Dutot et al. [57]).

Theorem 7.1 (NP-Completeness—Contention Model) *Let $G = (\mathbf{V}, \mathbf{E}, w, c)$ be a DAG and $M_{TG} = ((\mathbf{N}, \mathbf{P}, \mathbf{D}, \mathbf{H}, b), \omega)$ a parallel system. The decision problem C-SCHED (G, M_{TG}) associated with the scheduling problem is as follows. Is there a schedule S for G on M_{TG} with length $sl(S) \leq T, T \in \mathbb{Q}^+$? C-SCHED (G, M_{TG}) is NP-complete in the strong sense.*

Proof. First, it is argued that C-SCHED belongs to NP, then it is shown that C-SCHED is NP-hard by reducing the well-known NP-complete problem 3-PARTITION (Garey and Johnson [73]) in polynomial time to C-SCHED. 3-PARTITION is NP-complete in the strong sense.

The 3-PARTITION problem is as follows. Given is a set $\mathbf{A}$ of $3m$ positive integer numbers a_i and a positive integer bound B such that $\sum_{i=1}^{3m} a_i = mB$ with $B/4 < a_i < B/2$ for $i = 1, \ldots, 3m$. Can $\mathbf{A}$ be partitioned into m disjoint sets $\mathbf{A}_1, \ldots, \mathbf{A}_m$ (triplets) such that each $\mathbf{A}_i$, $i = 1, \ldots, m$, contains exactly 3 elements of $\mathbf{A}$, whose sum is B?

Clearly, for any given solution $\mathcal{S}$ of C-SCHED it can be verified in polynomial time that $\mathcal{S}$ is feasible and $sl(\mathcal{S}) \leq T$; hence, C-SCHED $\in$ NP.

From an arbitrary instance of 3-PARTITION $\mathbf{A} = \{a_1, a_2, \ldots, a_{3m}\}$, an instance of C-SCHED is constructed in the following way.

Task Graph G. The constructed task graph G is a fork graph as illustrated in Figure 7.17(a). It consists of one parent node n_x and $3m + 1$ child nodes $n_0, n_1, \ldots, n_{3m}$. There is an edge e_i directed from n_x to every child node n_i, $0 \leq i \leq 3m$. The weights assigned to the nodes are $w(n_x) = 1$, $w(n_0) = B + 1$, and $w(n_i) = 1$ for $1 \leq i \leq 3m$. The edge weights are $c(e_0) = 1$ and $c(e_i) = a_i$ for $1 \leq i \leq 3m$.

Target System M_{TG}. The constructed target system $M_{TG} = ((\mathbf{N}, \mathbf{P}, \mathbf{D}, \mathbf{H}, b), \omega)$, illustrated in Figure 7.17(b), consists of $|\mathbf{P}| = m + 1$ identical, fully connected processors. Each processor P_i is connected to each other processor P_j through a half duplex link L_{ij}, which is represented by an undirected edge (a hyperedge incident on two vertices: $L_{ij} = H_{ij} = \{P_i, P_j\}$). Hence formally, $\mathbf{N} = \mathbf{P}$, $\mathbf{D} = \emptyset$, $\mathbf{H} = \bigcup_{i=1}^{m} \bigcup_{j=i+1}^{m+1} L_{ij}$, and, as all processors and links are identical, $\omega(n, P) = w(n) \; \forall P \in \mathbf{P}$ and $b(L_{ij}) = 1 \; \forall L_{ij} \in \mathbf{H}$.

Time Bound T. The time bound is set to $T = B + 2$.

Clearly, the construction of the instance of C-SCHED is polynomial in the size of the instance of 3-PARTITION.

It is now shown how a schedule $\mathcal{S}$ is derived for C-SCHED from an arbitrary instance of 3-PARTITION $\mathbf{A} = \{a_1, a_2, \ldots, a_{3m}\}$, which admits a solution

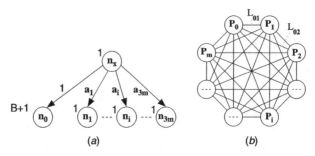

(a)   (b)

Figure 7.17. The constructed fork DAG (*a*) and target system (*b*).

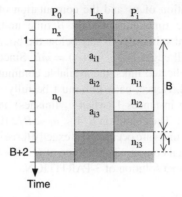

Figure 7.18. Constructed schedule of $n_x, n_0, n_{i1}, n_{i2}, n_{i3}$ and e_{i1}, e_{i2}, e_{i3} on P_0, P_i and L_{0i}.

to 3-PARTITION: let $A_1, \ldots, A_m$ (triplets) be m disjoint sets such that each A_i, $i = 1, \ldots, m$, contains exactly 3 elements of A, whose sum is B.

Nodes n_x and n_0 are allocated to the same processor, which shall be called P_0. The remaining nodes $n_1, \ldots, n_{3m}$ are allocated in triplets to the processors $P_1, \ldots, P_m$. Let a_{i1}, a_{i2}, a_{i3} be the elements of A_i. The nodes n_{i1}, n_{i2}, n_{i3}, corresponding to the elements of triplet A_i, are allocated to processor P_i. Their entering edges e_{i1}, e_{i2}, e_{i3}, respectively, are scheduled on L_{0i} as early as possible in any order. The resulting schedule is illustrated for P_0, P_i and L_{0i} in Figure 7.18.

What is the length of this schedule? The time to execute n_x and n_0 on P_0 is $w(n_x) + w(n_o) = B + 2 = T$. On each link L_{0i} the three edges corresponding to the triplet A_i take B time units for their communication (Figure 7.18). The first communication starts as early as possible, that is, after 1 time unit when n_x finishes. After the last communication has finished, only one node remains to be executed on processor P_i; the other two nodes are already executed during the communication (Figure 7.18). This is guaranteed by the fact that a_i is a positive integer and $w(n_i) = 1$ for $1 \leq i \leq 3m$. The execution of this last node takes 1 time unit and, hence, each processor P_i finishes at $1 + B + 1 = T$.

Thus, a feasible schedule S was derived, whose schedule length matches the time bound T; hence, it is a solution to the constructed C-SCHED instance.

Conversely, assume that the instance of C-SCHED admits a solution, given by the feasible schedule S with $sl(S) \leq T$. It will now be shown that S is necessarily of the same kind as the schedule constructed previously.

Nodes n_x and n_0 must be scheduled on the same processor, say, P_0, since otherwise the communication e_0 is remote and takes one time unit. As a consequence, the earliest finish time of n_0 would be $w(n_x) + c(e_0) + w(n_0) = B + 3$, which is larger than the bound T.

Since processor P_0 is fully occupied with these two nodes until $B + 2$, all remaining nodes have to be executed on the other processors. That means that all corresponding communications are remote.

Between the computation of n_x and the computation of the last node on each processor (both take one time unit) exactly B time units are available for the communication on the links, in order to stay below the bound $T = B + 2$. The total communication time of all edges is $\sum_{i=1}^{3m} c(e_i) = mB$. Since there are m processors connected to P_0 by m dedicated links, the available communication time B per link must be completely used. Hence, each link must be fully occupied with the communication between time instances 1 (when n_x finishes) and $B + 1$ (when the last node must start execution). The condition $B/4 < a_i < B/2$ (i.e., $B/4 < c(e_i) < B/2$) enforces that these B time units are occupied by exactly three edges; any other number of edges cannot have a total communication time of B. This distribution of the edges on the links corresponds to a solution of 3-PARTITION. □

$\alpha|\beta|\gamma$ **Notation**   To characterize the problem of scheduling under the contention model with the $\alpha|\beta|\gamma$ classification (Section 6.4.1), the following extensions are proposed for the α and β fields.

The distinguishing aspect from the problems discussed in Section 6.4 is the contention awareness achieved through edge scheduling. This is symbolized by the extension -*sched* in the β field, immediately after the specification of the communication costs (e.g., c_{ij}-*sched*).

Furthermore, the complexity of a problem also depends on the topology graph of the parallel system. Hence, the α field is extended with a third subfield that specifies the communication network, similar to the description of the precedence relations of the task graph:

- o. The processors of the parallel system are fully connected.
- *net.* The parallel system has a communication network, modeled by a topology graph $TG = (\mathbf{N}, \mathbf{P}, \mathbf{D}, \mathbf{H}, b)$.
- *snet.* The parallel system has a static communication network, modeled as an undirected graph $UTG = (\mathbf{P}, \mathbf{L})$.
- *star.* The parallel system consists of a star network, where each processor is connected to a central switch via two counterdirected links (full duplex). This network corresponds to the one-port model (Beaumont et al. [19]), as discussed in Section 7.1.1.
- *star-h.* The parallel system consists of a star network, where each processor is connected to a central switch via an undirected edge (half duplex), as illustrated in Figure 7.7(b).

With these extensions, the general problem of scheduling under the contention model is specified by $P, net|c_{ij}$-*sched*$|C_{max}$ for homogeneous processors.

Just before Theorem 7.1 and its proof, it was mentioned that scheduling with contention is intuitively more complicated than scheduling under the classic model. The proof provides evidence. It showed the NP-completeness of the special case $P|fork, c_{ij}$-*sched*$|C_{max}$. While the corresponding problem under the classic model, $P|fork, c_{ij}|C_{max}$, is also NP-complete (Section 6.4.3), the scheduling of the particular problem utilized in the forgoing proof is not. One can construct an optimal schedule in polynomial time for the fork graph of Figure 7.17 on the equivalent classic target system $M_{\text{classic}} = (\mathbf{P}, \omega)$. $M_{\text{classic}} = (\mathbf{P}, \omega)$ corresponds to M_{TG} by consisting of $|\mathbf{P}| = m + 1$ identical processors, that is, $\omega(n, P) = w(n) \forall P \in \mathbf{P}$. The construction of the schedule is described in the following.

With the same argument as in the previous proof, nodes n_x and n_0 must be scheduled on the same processor, say, P_0. Now all other nodes are scheduled on the remaining processors, by putting any three nodes on each processor. Let the nodes scheduled on P_i be n_j, n_k, n_l, $1 \le j, k, l \le 3m$. Since there is no contention for the communication resources, all data sent from n_x to n_j, n_k, n_l is transmitted at the same time immediately after n_x finishes. In other words, the edges e_j, e_k, e_l are communicated from P_0 to P_i at the same time. With $c(e_i) = a_i < B/2$, $1 \le i \le 3m$, the total communication time of these edges is less than $B/2$. It follows for the finish time of processor P_i that $f_t(P_i) < w(n_x) + B/2 + w(n_j) + w(n_k) + w(n_l) = 4 + B/2$ (Figure 7.19).

It holds that $4 + B/2 \le B + 2$ for $B \ge 4$. Furthemore, in order that B and a_i, $i = 1, \ldots, 3m$, comply with the constraints of the 3-PARTITION problem, B must be an integer $B \ge 3$ (otherwise there can be no solution to the 3-PARTITION instance). If $B = 3$ then $a_i = 1, i = 1, \ldots, 3m$, in order to comply with $B/4 < a_i < B/2$ and the fact that a_i is a positive integer; but then the finish time of processor P_i is $t_f(P_i) = 5 = B + 2 = T$. Hence, the constructed schedule under the classic model (without contention) has the schedule length T and is therefore optimal. Furthermore, the construction clearly takes only polynomial time.

A similar complexity result is obtained by Beaumont et al. [19], where the NP-completeness of scheduling under the one-port model is proved by demonstrating the NP-completeness of $P\infty, star|fork, c_{ij}\text{-}sched|C_{\max}$. Under the classic model, the corresponding problem $P\infty|fork, c_{ij}|C_{\max}$ is tractable (Section 6.4.3). Finally, Sinnen and Sousa [172, 179] show that $P\infty, star\text{-}h|fork, c_{ij}\text{-}sched|C_{\max}$ is also NP-complete.

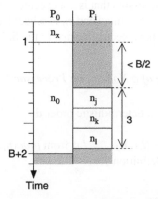

Figure 7.19. Schedule of n_x, n_0, n_j, n_k, n_l under classic model, that is, without contention.

7.5 HEURISTICS

Scheduling under the contention model essentially requires only one change compared to the classic model: an edge is scheduled on the network links in order to determine the delay of the corresponding interprocessor communication. In this section it is investigated how task scheduling heuristics can be adapted to the contention model. This is done exemplarily with list scheduling in its general form—static or dynamic (Section 5.1). Having contention aware list scheduling is crucial, due to list scheduling's significant role in task scheduling, and allows one to transform many of the existing heuristics into contention aware algorithms.

Other algorithms that are not list scheduling based can easily be enhanced with contention awareness based on the same procedures as described in the following. For example, in Sinnen [172] BSA (Section 7.1.2) was adapted to contention awareness as discussed in this chapter. Moreover, Section 7.5.3 briefly outlines how clustering can be adapted to the contention model.

The only obvious exception is node duplication. To integrate this technique, the duplication of the respective communications of duplicated nodes must first be investigated.

7.5.1 List Scheduling

List scheduling—static (Algorithm 10) or dynamic (Algorithm 11)—is only affected by the contention model and the associated edge scheduling in two aspects: the determination of the DRT and the scheduling of a node.

Scheduling a Node   At each step of list scheduling, a partial feasible schedule is extended with one free node. In the contention model, a feasible schedule also comprises the schedule for all edges whose communication is transferred between processors. Thus, the scheduling of a free node implies the preceding scheduling of its entering edges. Algorithm 21 describes how a node n_j is scheduled in the contention model. For each of n_j's predecessors that is not executed on the same processor as n_j, the communication route is determined and e_{ij} is scheduled on this route. Only when all remote communications have been allocated to the links, is the node scheduled.

Algorithm 21   Scheduling of a Node n_j on Processor P in Contention Model
Require: n_j is a free node
  for each $n_i \in$ **pred**(n_j) in a deterministic order **do**
    if $proc(n_i) \neq P$ **then**
      determine route $R = \langle L_1, L_2, \ldots, L_l \rangle$ from $proc(n_i)$ to P
      schedule e_{ij} on R (Definition 7.9)
    end if
  end for
  schedule n_j on P

Note that each leaving edge e_{ij} of a node n_i is only scheduled during the same algorithm in which its destination node n_j is scheduled. There is no alternative, since for edge scheduling $proc(n_j)$ must be known, which is only determined when n_j is scheduled. Thus, the leaving edges are automatically scheduled in the scheduling order of their destination nodes, that is, the nodes' priority order.

Determining DRT Most scheduling heuristics determine the DRT of a node not only for one but for several processors, before the decision is taken about which processor the node is to be scheduled on. List scheduling with start time minimization (Section 5.1) is a prominent example. In the contention model, the DRT is calculated by temporarily scheduling the entering edges of a node. This is necessary to obtain the correct view of the communication times of the edges. Algorithm 22 expresses the procedure in the contention model. Basically, the edges are scheduled as if the node was to be scheduled on the respective processor. While doing so, the DRT is calculated and before the procedure terminates, the scheduled edges are removed from the links.

Algorithm 22   Determine $t_{dr}(n_j, P)$ in Contention Model
Require: n_j is a free node
    $t_{dr} \leftarrow 0$
    for each $n_i \in \mathbf{pred}(n_j)$ in a deterministic order **do**
        $t_f(e_{ij}) \leftarrow t_f(n_i)$
        if $proc(n_i) \neq P$ **then**
            determine route $R = \langle L_1, L_2, \ldots, L_l \rangle$ from $proc(n_i)$ to P
            schedule e_{ij} on R (Definition 7.9)
            $t_f(e_{ij}) \leftarrow t_f(e_{ij}, L_l)$
        end if
        $t_{dr} \leftarrow \max\{t_{dr}, t_f(e_{ij})\}$
    end for
    remove edges $\{e_{ij} \in \mathbf{E} : n_i \in \mathbf{pred}(n_j) \wedge proc(n_i) \neq P\}$ from links
    return t_{dr}

Scheduling Order of Edges As discussed in Section 7.4, and illustrated in Figure 7.16, the scheduling order of edges has an impact on the DRTs of the destination nodes. For a scheduling algorithm it is crucial that the determination of the DRT is deterministic. A DRT that varies between repeated calculations for the same node and processor might even lead to an infeasible schedule. For this reason, Algorithms 21 and 22 require a deterministic processing order of the edges (`for` loop).

For example, the edges might be processed in the order of their indices. Of course, an attempt can be undertaken to find an order that positively influences the DRTs of the destination nodes, and thereby the schedule length. For example, the edges can be ordered according to the start time of their origin nodes or by the size of the communication. Yet, sorting the edges increases the complexity of the list scheduling heuristic, albeit not significantly. Node priorities and the complexity of list scheduling are discussed later.

Heuristics With the discussed two algorithms for the DRT calculation (Algorithm 22) and the scheduling of a node (Algorithm 21), the basic structure of list scheduling need not be altered for the contention model. The actual scheduling of a node or an edge—in Algorithms 21 and 22 this is performed by the statements "schedule ..."—can be done either with the end technique or the insertion technique. The scheduling condition (Condition 6.1) and the edge scheduling condition (Condition 7.3) regulate the choice of an appropriate time interval for node and edge scheduling, respectively.

An Example For the illustration of contention aware list scheduling, the sample task graph (e.g., in Figure 6.25) is scheduled on a homogeneous linear network of three identical processors, as depicted in Figure 7.20. The chosen node order is identical with the order of the list scheduling example in Figure 5.1, namely, $a, b, c, d, e, f, g,$ i, h, j, k. Nodes $a, b,$ and c are scheduled as in the classic model, but the scheduling of d is affected by contention on L_1. Edge e_{ad} would arrive at time unit 7 at P_3, since L_1 is occupied with e_{ac} until time unit 4. Thus, d is scheduled earlier on P_1 and in

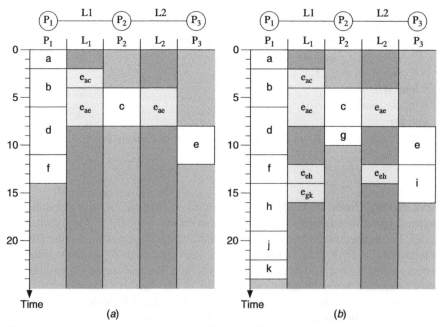

Figure 7.20. Example for contention aware list scheduling with start time minimization: intermediate (*a*) and final schedule (*b*). Node order of the sample task graph (e.g., in Figure 6.25) is $a, b, c, d, e, f, g, i, h, j, k$.

compensation e is scheduled on P_3 (vice versa in the classic model (Figure 5.1)). Also, f is scheduled on P_1 to avoid contention on L_1. Figure 7.20(a) shows the intermediate schedule at this point. Then, g and i are scheduled on the processors of their parent nodes. The following node h is decisive for the rest of the schedule: its earliest start time is on P_1, but its entering edge e_{eh} now blocks L_1 for the entering communication of j, which is quite large. Thus, j and k must be scheduled on P_1. The final schedule length (24) (Figure 7.20(b)) is not longer than in the classic model (Figure 5.1), but here more nodes are executed on P_1 and less interprocessor communications are performed. Hence, list scheduling achieves in this example the desired contention awareness in its scheduling decisions. In general, the schedule lengths under the contention model and the classic model are not directly comparable. Contention aware scheduling produces normally longer schedules, but shorter execution times on real machines. The experiments in Sinnen and Sousa [172, 176] verify this.

Complexity The complexity of contention aware list scheduling is similar to list scheduling under the classic model (Section 5.1). Essentially, only the calculation of the DRT is more complex than in the classic model. For example, the complexity of the second part of simple list scheduling (Algorithm 9) with start time minimization and the end technique is as follows. To calculate the start time of a node n, the data ready time $t_{dr}(n)$ is computed, which involves the scheduling of n's entering edges. The scheduling of an edge includes the determination of the route and its scheduling on each link of the route. This is described by $O(routing)$ (see Section 7.2.2) for each edge. If the route can be determined in $O(1)$ time (calculated or looked up), then $O(routing)$ is just the complexity of the length of the route, as the edge must be scheduled on each link of the route. So for all nodes on all processors, calculating the DRT amortizes to $O(\mathbf{P}EO(routing))$. The start time of every node is computed on every processor, that is, $O(\mathbf{P}V)$ times; hence, the total complexity is $O(\mathbf{P}(V + EO(routing)))$. In comparison, the complexity under the classic model is $O(\mathbf{P}(V + E))$ and with the insertion technique $O(V^2 + \mathbf{P}E)$ (Section 6.1). Using the insertion technique in contention aware list scheduling has a complexity of $O(V^2 + \mathbf{P}E^2O(routing))$, as the calculation of an edge's start time changes from $O(1)$ to $O(E)$, since there are at most $O(E)$ edges on each link. As discussed in Section 7.2.2, for many systems $O(routing)$ is at most linear in the number of processors $|\mathbf{P}|$ and links $|\mathbf{L}|$.

Sorting the entering edges of each node according to some heuristic amortizes to $O(E \log E)$ for all nodes. Thus, the complexity of the second part of list scheduling is then $O(\mathbf{P}(V + EO(routing)) + E \log E)$ using the end technique and remains $O(V^2 + \mathbf{P}E^2O(routing))$ using the insertion technique.

7.5.2 Priority Schemes—Task Graph Properties

So far, only the second part of list scheduling has been considered. But the question arises if the priority schemes used for the classic model are appropriate for the contention model. This question mainly refers to priority metrics based on the task graph properties (e.g., node levels) because dynamic metrics (e.g., the earliest start time of a node) are fully adapted to the contention model with the modified calculation of the DRT.

As analyzed in Section 7.4.1, the creation of the task graph remains unmodified for the contention model. Hence, all metrics like node levels or critical path can still be employed. However, in the contention model, the edge weights reflect the best case that no contention occurs. Thus, metrics involving edge weights (e.g., the bottom level) are inaccurate in two situations: (1) when a communication is performed locally, because it then has zero cost (as in the classic model); and (2) when a communication is delayed due to contention. As a result, metrics based on communication costs are less accurate. In particular, bounds on the schedule length, as established in Section 4.4, lose their validity. On the other hand, communication becomes even more important due to contention and must be considered in the ordering of the nodes.

Various node priority schemes for the first part of contention aware list scheduling are analyzed and compared in Sinnen and Sousa [177], among which the *bottom level* provides the best results. Experiments show that this combination—list scheduling with bottom-level node order—outperforms all other compared contention aware algorithms (Sinnen and Sousa [176, 178]).

7.5.3 Clustering

The major hurdle for the utilization of clustering (Section 5.3) under the contention model is the fact that clustering relies on an unlimited number of clusters (processors). As is well known by now, the classic model assumes a fully connected network (Property 6 of Definition 4.3), which is completely symmetric with an identical distance (of one) between any pair of processors. Consequently, the labeling of the clusters and their position within this fully connected network has no influence on the communication behavior and is therefore irrelevant for the clustering.

Under the contention model, however, the target system consists of a limited number of processors and is modeled by a topology graph. In general, the topology graph is not completely symmetric nor is the distance between all processor pairs identical. Hence, it does matter for the communication behavior where a processor is positioned within the network. How can such a topology graph be expanded in a meaningful way to an unlimited number of processors (i.e., at least $|V|$ processors) for the clustering phase? The positioning of the clusters in this network should not influence the implicit schedule lengths of each clustering step.

For this reason, a solution to this problem can only be an approximation of the actual target system. During the clustering phase, a symmetric topology graph with an identical distance between every processor pair must be employed. Both the *fully connected network* and the *star network* (see Figure 7.7(b) and Section 7.4.2) fulfill this requirement. Employing the corresponding topology graphs with $|V|$ processors makes clustering aware of contention, especially with the star network, which fully captures end-point contention. With such contention awareness, merging of clusters might be beneficial, when it would not be under the classic model.

Of course, in the third step of a clustering based heuristic (Algorithm 14)—after the mapping of the clusters to the real processors in the second step—the topology graph representing the real system should be employed. This third step is list scheduling based (Section 5.4), which has just been discussed in Section 7.5.1.

Clustering Approaches What follows is a brief discussion of how the three different approaches to clustering, as analyzed in Section 5.3, are employable under the contention model.

- *Linear clustering* (Section 5.3.2) does not require any modification whatsoever. There are no decisions taken in the linear clustering Algorithm 16 based on the schedule length. The algorithm only analyzes properties of the task graph; hence, the underlying scheduling model is irrelevant.
- *Single edge clustering* (Section 5.3.3) can be applied under the contention model with an approximated topology graph (see earlier discussion). The only part that requires some attention is the line "Schedule C_i using heuristic" in Algorithm 17, that is, the construction of the implicit schedule. Essentially, this is scheduling with a given processor allocation—a list scheduling that was already covered in Section 7.5.1.
- *List scheduling as clustering* (Section 5.3.4), as the name implies, is based on list scheduling, whose utilization under the contention model is analyzed in Section 7.5.1. It can be applied with an approximated topology graph as discussed earlier.

7.5.4 Experimental Results

It is a well-known fact that contention in the communication networks of parallel systems has a significant influence on the overall system performance. However, there are only a few experimental studies on the impact of contention in scheduling.

In Macey and Zomaya [132], schedules produced under the classic model are analyzed by simulating their execution in a system with link contention. The authors conclude that the assumption of concurrent communication in the classic model (Property 5 of Definition 4.3) is completely unrealistic; hence, it is imperative to consider contention in task scheduling.

Contention aware scheduling aims at producing schedules that are more efficient than those produced under the classic model. To attain this goal, the contention model attempts to reflect parallel systems more accurately than the classic model. But how accurate are the classic model and the contention model? And how efficient are the schedules produced under the two models?

Answers to these questions can only be found with an experimental evaluation on real parallel systems. Sinnen and Sousa [171, 176, 178] proposed a methodology with which they subsequently performed such an experimental evaluation. The methodology begins with the common procedure of scheduling algorithm comparisons: a large set of task graphs is scheduled by different heuristics—under the classic model and the contention model—on various target systems.

The employed target systems cover a wide spectrum of parallel architectures: (1) a cluster of 16 PCs (Telford [184]); (2) an 8-processor shared-memory SMP system, Sun Enterprise 3500 (Culler and Singh [48]); and (3) a 344-processor massively parallel system, Cray T3E-900 (Booth et al. [26]). These systems possess different network topologies and are modeled with corresponding topology graphs.

In the next step, code is generated that corresponds to the task graphs and their schedules, using the C language and MPI (message passing interface [138]), the standard for message passing on parallel systems.

This code is then executed on the real parallel systems and the accuracy, the speedup, and the efficiency of a schedule and its model are evaluated. Let $pt(S)$ be the execution time—the parallel time—of the code generated for schedule S and task graph $G = (\mathbf{V}, \mathbf{E}, w, c)$ on a target system. The *accuracy* of S is the ratio of the real execution time on the parallel system to the schedule length (i.e., the estimated execution time) $acc(S) = pt(S)/sl(S)$. The *speedup* of S is the ratio of sequential time to execution time $sup(S) = seq(G)/pt(S)$. Finally, the *efficiency* of S is the ratio of speedup to processor number $eff(S) = sup(S)/|\mathbf{P}| = seq(G)/(pt(S) \times |\mathbf{P}|)$.

The experimental evaluation exposed that scheduling under the classic model results in very inaccurate and inefficient schedules for real parallel systems. Contention aware scheduling significantly improves both the accuracy and the efficiency of the produced schedules. Still, the results are improvable.

It is observed that results get worse—under both models—as the amount of communication in the task graph increases. In terms of the CCR (Definition 4.23), the classic model is extremely inaccurate for medium ($CCR = 1$) and high ($CCR = 10$) communications.

Note that the efficiency improvements from the employment of the contention model are superior to the typical difference between various scheduling algorithms under the classic model (Ahmad et al. [6], Khan et al. [103], Kwok and Ahmad [111]).

Nonetheless, contention aware scheduling, even though significantly more accurate and efficient than scheduling under the classic model, does not produce results that really satisfy. As analyzed by Sinnen and Sousa [171, 176, 178], this is in part due to inappropriate topology graphs employed in the experiments, which do not exploit the full potential of the topology graph of Definition 7.3. It is argued that especially the switch vertex and the hyperedge are important instruments of the topology graph which should be used for an accurate reflection of communication networks.

Another possible explanation for the still improvable results is that communication contention is not the only aspect that is inaccurately reflected in the classic model. Even for the contention model, a relation between the increase of communication (CCR) and the degradation of accuracy can be observed in the presented experimental results. This indicates another deficiency of the scheduling models regarding communication. Like the classic model (Definition 4.3), the contention model (Definition 7.10) supposes a dedicated communication subsystem to exist in the target system (Property 4). With the assumed subsystem, computation can overlap with communication, because the processor is not involved in communication. However, the analysis of the three parallel systems shows that none of them possesses such a dedicated communication subsystem. On all three systems the processors are involved, in one way or the other, in interprocessor communication. Consequently, to further improve the accuracy and efficiency of scheduling, the involvement of the processor should be investigated. This is addressed in the next chapter.

7.6 CONCLUDING REMARKS

This chapter was dedicated to communication contention in task scheduling. Driven by the insight that the classic model is inappropriate for many parallel systems, a more realistic system model was studied. This model abandons the idealizing assumptions of completely concurrent communications and fully connected networks. Task scheduling under this model attains its awareness of contention by scheduling the edges onto the communication links. Owing to its similarity with the scheduling of the nodes, edge scheduling is intuitive and powerful.

The analyzed contention aware scheduling unifies the consideration of both types of contention—end-point and network contention—in one model. This is accomplished through an extended network model and its representation of the topology. At the same time, the details of the network are hidden from edge scheduling, which only sees the route for each communication.

It was shown that only a few modifications are necessary to adapt the techniques of classic scheduling for contention aware scheduling. In particular, the discussion covered how list scheduling, in all its forms, is employed under the new model.

In conclusion, there is almost no reason against favoring the contention model over the classic model in practice. The few changes that are necessary to make task scheduling algorithms contention aware usually result in only a small increase of the runtime complexity. The complexity difference between different scheduling heuristics for the classic model is often much larger. Yet, various experiments demonstrated that contention aware scheduling rewards with significantly increased accuracy and shorter execution times.

The only drawback is that the implementation of a contention aware algorithm is more involved. Its utilization also requires the modeling of the target system as a topology graph. In order to facilitate this task, one can start with a simple topology graph. Very appropriate in this context is the star topology, which corresponds to the one-port model and efficiently captures end-point contention. Later, the topology model can be enhanced, (i.e., made more accurate) in order to further improve the scheduling results.

Unfortunately, contention aware scheduling is not as accurate and efficient as one would like it to be. As discussed, this suggests that the scheduling model must be investigated further, especially with respect to the involvement of the processor in communication. This issue will be addressed in the next chapter.

7.7 EXERCISES

7.1 Draw the network of the following parallel systems as a topology graph (Definition 7.3):

    (a) A cluster of PCs composed of three subnetworks. In each subnetwork 8 processors are connected to one central switch. The networks are linked through a direct connection from each switch to every other switch.

    (b) A system of 9 processing nodes, which are connected through a 3×3 cyclic grid (Section 2.2.1). Each processing node consists of one interface to the grid and two processors, all connected via a single bus.

7.2 Describe how to model a butterfly network of 8 processors (depicted in Figure 2.11) as a topology graph. Consider:

    (a) The edge types to be employed, that is, directed, undirected, or hyperedge.

    (b) Can the communication routes in such a topology graph be determined with a shortest-path algorithm or is a specific routing algorithm necessary? Give examples for routes between processors.

    (c) Draw the topology graph.

7.3 Visit the online *Overview of Recent Supercomputers* by van der Steen and Dongarra [193] on the site of the TOP500 Supercomputer Sites [186] and choose one modern system. Model this system with a topology graph (Definition 7.3).

7.4 Figure 7.9 displays the topology graph for 8 dual processors connected via a switch:

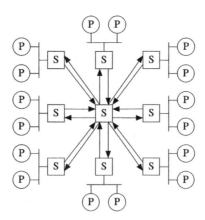

    (a) Find examples for four processor pairs that can communicate concurrently, and indicate which links are occupied by the communication.

    (b) Find examples for processor pairs that cannot communicate concurrently and indicate the links where the communication conflicts.

    (c) Repeat this exercise for the topology graphs constructed in the previous exercises.

7.5 Schedule the following task graph

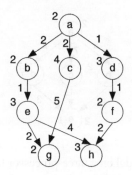

under the contention model on four processors connected as a ring, employing half duplex links:

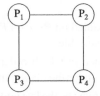

The nodes are allocated to the processors as specified in the following table:

Node	a	b	c	d	e	f	g	h
Processor	1	3	2	4	2	1	3	4

Schedule the nodes in alphabetic order as early as possible.

7.6 Repeat the scheduling done in Exercise 7.5, now on a heterogeneous linear network

where link L_{34} is twice as fast as the other two links; that is, $b(L_{34}) = 2b(L_{13}) = 2b(L_{42})$.

How does the resulting schedule change for full duplex links (i.e., each undirected edge of the linear network is substituted by two counterdirected edges)?

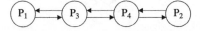

7.7   Schedule the following fork-join task graph

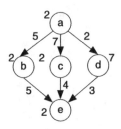

under the contention model on a three-processor linear network

Nodes a and b shall be scheduled on P_1, c on P_2, and d and e on P_3. Their start times should be as early as possible.

(a) Does it make a difference in which order nodes c and d are scheduled? Compare to scheduling under the classic model.

(b) Node e, as the exit node, must be the last node to be scheduled. Does the order in which its incoming communications are scheduled make a difference for the schedule length?

(c) Describe in general terms the situation when the scheduling order of incoming communications makes a difference. Suggest an ordering strategy.

7.8   Perform list scheduling with start time minimization under the contention model with the following task graph on a four-processor ring (see Exercise 7.5). Order the nodes in decreasing bottom-level order.

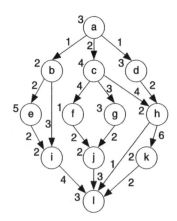

7.9 Section 7.5 studies how scheduling algorithms can be employed under the contention model. For the cases discussed this is relatively simple. One exception seems to be the utilization of node duplication (Section 6.2) under the contention model.

Describe the challenges that node duplication imposes under the contention model. Try to find examples for these challenges.

CHAPTER 8

Processor Involvement in Communication

The inclusion of communication contention in task scheduling yields more accurate and more efficient schedules compared with those produced under the classic model. Yet, the experimental results referenced in Section 7.5.4 suggest that even the contention model does not represent a parallel system accurately enough. An examination of the results (Sinnen and Sousa [172, 176]) reveals a clear relation between the increase of communication (CCR) and the decrease of accuracy of the produced schedules. A possible explanation is that the contention model is still unrealistic regarding communication.

In the prelude of the previous chapter, three very idealizing assumptions of the classic model (Definition 4.3) were queried. Two of them—the assumption of contention-free communication (Property 5) and the assumption of a fully connected network (Property 6)—have already been addressed in Chapter 7. Property 4, which supposes a dedicated communication subsystem to be present in the target system, is examined in this chapter. With the assumed subsystem, computation can overlap with communication, because the processor is not involved in communication.

However, many parallel systems do not possess such a subsystem (Culler and Singh [48]). Therefore, in many systems the processors are involved, in one way or another, in interprocessor communication. Furthermore, involvement of the processor also serializes communication, even if the network interfaces are capable of performing multiple message transfers at the same time, since a processor can only handle one communication at a time. For example, a processor can usually only perform one memory copy at a time. Thus, considering the processors' involvement in task scheduling is of importance as it serializes the communication and, more importantly, prevents the overlap of computation and communication.

In fact, the involvement of the processors in communication is an aspect that has been considered before. The LogP model (Section 2.1.3) dedicates the parameter o to describe the computational overhead inflicted on a processor for each communication. Per definition of the model, the processor cannot execute other operations while occupied with this overhead. Furthermore, the system model used by Sarkar [167], for which a clustering algorithm is proposed, includes the computational overhead of communication. Falsafi and Wood [66] compare the performance of a parallel system with and without the utilization of a dedicated communication subsystem.

This chapter investigates involvement of the processor in communication, its impact on task scheduling, and how it can be considered in the latter. A system model for scheduling that considers the involvement of the processor in communication is developed based on the contention model. As a result, the new model is general and unifies the other scheduling models. Since scheduling under the new model requires adjustment of existing techniques, it is shown how this is done for list scheduling and a simple genetic algorithm based heuristic.

The chapter begins in Section 8.1 by classifying interprocessor communication into three types, which differ based on the involvement of participating processors. Subsequently, the classification is refined by analyzing the characteristics of the involvement. To integrate the processor involvement into scheduling, the scheduling model must be adapted, which is elaborated in Section 8.2. Section 8.3 discusses the general approaches for scheduling under the new model, which differ from those followed under the contention model. Finally, Section 8.4 puts the pieces together by formulating heuristics for scheduling under the involvement–contention model. This includes a short review of experimental results that demonstrate the fundamentally improved accuracy and the significantly reduced execution times of the schedules produced under this model.

8.1 PROCESSOR INVOLVEMENT—TYPES AND CHARACTERISTICS

8.1.1 Involvement Types

Regarding the involvement of the processor, interprocessor communication can be divided into three basic types: *two-sided*, *one-sided*, and *third-party* (Sinnen [172], Sinnen et al. [180]). For illustration of these types, consider the following situation: the output of a task A on processor P_1 is transferred via the communication network to a task B on processor P_2. Figure 8.1 visualizes the involvement of the processors in this simple situation for the three types. In the following paragraphs, the network is treated as a general and abstract entity.

Two-Sided   In two-sided interprocessor communication both the source and the destination processor are involved in the communication (Figure 8.1(a)). In which kinds of parallel systems does this occur? It can be the case for message passing as well as for shared-memory architectures. On message passing systems, the involvement of the processors consists of preparing the message and sending it to and receiving it from the network. For example, in a PC cluster (Section 2.1.2), the TCP/IP based communication over the LAN involves both processors. The sending processor must divide a message into packets and wrap them into TCP/IP protocol envelopes before setting up the network card for the transfer. On the receiving side, the processor is involved in the unwrapping and assembling of the packets into the original message

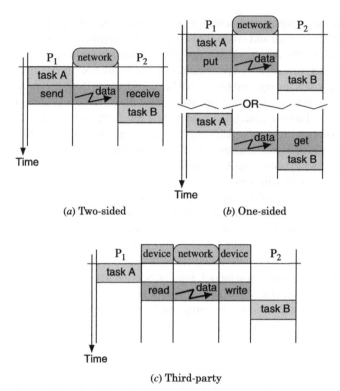

Figure 8.1. Types of interprocessor communication: (*a*) two-sided—both processors are involved ("send" and "receive"); (*b*) one-sided—only one processor is involved, either source or destination processor; (*c*) third-party—a special communication device takes care of the communication leaving the processor uninvolved.

(Culler and Singh [48]). In shared-memory systems, communication can be two-sided, if the interprocessor communication is performed via a common buffer. The source processor writes the data into the common buffer and the destination processor reads it from there. For example, the implementation of MPI on the SGI Origin 2000 works this way (White and Bova [199]). In general, the involvement of the source and the destination processor can have different extents.

One-Sided   Communication is said to be one-sided, if only one of the two participating processors is involved (Figure 8.1(*b*)). Thus, this type of communication is limited to shared-memory systems: either the source processor writes the data into its destination location (shared-memory "put") or the destination processor reads the data from its source location (shared-memory "get"). For one-sided communication it is essential that the active processor, that is, the processor that performs the communication, has direct access to the data location. Note that the shared-memory communication does not have to be transparent. Special devices or routines might be used by the active processor, but the memory access must be performed without the involvement of the passive processor. For example, Cray T3E uses special registers for remote memory access (Culler and Singh [48]).

Third-Party In third-party interprocessor communication, the data transfer is performed by dedicated communication devices, as illustrated in Figure 8.1(c). The communication devices must be able to directly access the local and the remote memory without involvement of the processor (i.e., it must possess some kind of a direct memory access (DMA) engine). The processor only informs the communication device of the memory area it wants transferred and the rest of the communication is performed by the device. Thus, the communication overhead on the processor takes approximately constant time, independent of the amount of data to be transferred. All devices together comprise the communication subsystem of the parallel computer, which autonomously handles all interprocessor communication. Examples for machines that possess such subsystems are the Meiko CS-2 (Alexandrov et al. [9]) or the IBM SP-2 [96]. The IBM Blue Gene/L (van der Steen and Dongarra [193]) has two identical PowerPC processors per module, but it can be configured to employ one of them as either a communication processor or a normal processor for computation. In the former case, one processor acts like a third-party communication device for the other processor.

It is important to realize that a processor engaged in communication cannot continue with computation. The charts of Figure 8.1 illustrate this by placing the involvements ("send/receive," "get/put") onto the time lines of the processors. On the contrary, in third-party communication it is the device, not the processor, that is occupied during the communication process.

Task scheduling, under both the classic model and the contention model, assumes the third-party type of interprocessor communication (Property 4 of Definitions 4.3 and 7.10). Consequently, the produced schedules overlap computation with communication; that is, a processor executes tasks while data is entering and leaving. However, this overlap can only exist on systems with a dedicated communication subsystem. On other systems, those that perform one- or two-sided communication, the produced schedules are unrealistic and thereby inappropriate. The experiments referenced in Section 7.5.4 demonstrated the negative consequences on the accuracy and the execution times of the schedules.

Among the communication types an order can be established regarding the degree of the processor involvement: two-sided, one-sided, and third-party. While the type of communication is restricted by the hardware capabilities of the target system, the software layer employed for interprocessor communication can increase the degree of involvement. For example, in a shared-memory system, communication can be one sided, but the software layer might use a common buffer, which turns it into two-sided communication (see two-sided above). This effect is not uncommon, as shown by the analysis of MPI implementations on common parallel systems by White and Bova [199]. It becomes important, since MPI is the standard for message passing and message passing is the dominant programming paradigm for parallel systems. Furthermore, almost all of today's parallel applications are written using a higher level

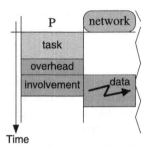

Figure 8.2. Decomposing of the processor's participation in communication into overhead and direct involvement.

programming paradigm in order to allow portability (e.g., MPI, PVM, OpenMP, HPF), which might be subject to the same effect. Task scheduling should of course reflect the effective type of involvement, taking into account the software layer employed in the code generation.

8.1.2 Involvement Characteristics

The first rough classification of interprocessor communication is established by its separation into three different types. For a more precise characterization, however, the involvement of the processor must be examined further. In the following it is shown how the processor's involvement can be decomposed into overhead and direct involvement.

Overhead The first component of the processor's involvement is the communication setup or *overhead*. From the initiation of the communication process until the first data item is put onto the network, the processor is engaged in preparing the communication. An overhead is in general also imposed on the destination processor after the data has arrived until it can be used by the receiving task.

The overhead consists mainly of the path through the communication layers (e.g., MPI [138], Active Messages (von Eicken et al. [197]), TCP/IP) and therefore becomes larger with the abstraction level of the communication environment. Typical tasks performed by these environments are packing and unpacking of the data, buffering, format conversions, error control, and so on. The communication hardware can also impose an overhead for control of the communication device, especially with third-party communication.

Note that the overhead does not arise for communication between tasks executed on the same processor. Therefore, it cannot be made part of the computation reflected by the origin and the destination tasks of the communication.

The importance of the overhead depends on the granularity of the task graph. For fine grained tasks, the overhead might have a significant influence on the program execution.

Direct Involvement After the communication has been prepared by the processor during the overhead, any further participation in communication is the direct involvement of the processor. It is characterized by the circumstance that data is already in transit on the communication network. Figure 8.2 focuses on the behavior of the origin processor in interprocessor communication, but the same principles are also valid for the destination processor. It features both the overhead and the direct involvement of the processor and thereby illustrates their differences. The direct involvement begins simultaneously with the data transit on the network, subsequent to the overhead. From now on, the term *involvement* means the direct involvement of the processor and the term *overhead* is used for the pre- and postphases discussed earlier. An example for direct involvement is the memory copy performed by a processor in shared-memory systems. The involvement only ends when the store command for the last data word has been executed. If the amount of data is small, the time of the processor involvement is short in comparison to the overhead, in particular, when only one data item is transferred.

With the notions of overhead and involvement, the generalized treatment of the three types of communication is possible. The overhead is a common part of interprocessor communication, independent of the type of communication. As mentioned earlier, the overhead is typically dominated by the communication layer and not by the underlying communication hardware. Thus, third-party communication is affected like two-sided and one-sided communication. In contrast, the involvement largely depends on the type of communication.

Length of Overhead and Involvement The distinction between communication types becomes obsolete if the communication is described in terms of overhead and involvement. Therefore, it is assumed that overhead and involvement are imposed on both the source and the destination processor. Then, one communication type differs from another merely by the length of the two components.

Overhead The overhead is usually of constant time on both processors, albeit of possibly different length. As explained earlier, the overhead is mainly due to the path through the communication layer, which does not change significantly for different data transfers. In other words, the overhead is usually independent of the data size. In some environments, however, data might be copied into and from a buffer, which is of course an operation taking time proportional to the data size. Examples are some MPI implementations as described by White and Bova [199].

Involvement The length of the involvement is primarily determined by the type of communication. Logically, this time is zero on some processors, namely, on both in third-party communication and on one—either the sending or the receiving processor—in one-sided communication. Figure 8.3 illustrates overhead and involvement in examples of all three types of communication. The involvement is omitted in Figure 8.3(a) and on P_2 in Figure 8.3(b), since it is zero due to the communication type.

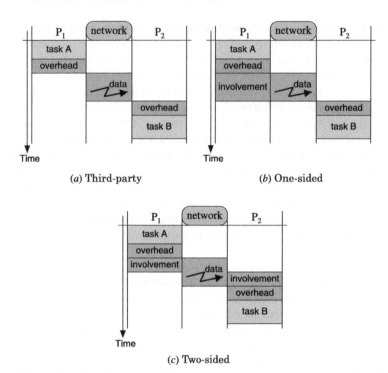

Figure 8.3. Overhead and involvement in the three communication types: (*a*) no involvement achieved with DMA (zero-copy network devices); (*b*) shared-memory "put"; (*c*) only partial involvement, for example, shared-memory using a common buffer or communication devices without DMA.

The third-party communication as shown in Figure 8.3(*a*) reflects systems, where the communication subsystem possesses a DMA engine so that the processor does not need to copy the data. Nevertheless, the processor must call communication functions that initialize the network device and, on the receiving side, an interrupt must inform the processor about the incoming data, hence the overheads. Such interprocessor communication needs *zero-copy* capable devices, for example, the IBM SP-2 network adapter [96] supports zero-copy communication.

Figure 8.3(*b*) depicts one-sided communication in a shared-memory system. The source processor is the active one and copies the data into the destination location ("put"). It is involved during almost the entire communication—only the last word is transferred after the end of its involvement. For sufficiently long messages the difference between involvement and network usage can be neglected. The passive destination processor is of course not involved at all, yet it might be subject to an overhead, for example, when it becomes aware of the incoming data.

As already insinuated in Figure 8.2, involvement of the processor does not always last for the duration of the entire communication. This is the case in the example of two-sided communication shown in Figure 8.3(*c*). The example reflects shared-memory communication via a common buffer. Both processors are only involved in communication during one-half of the network activity: the source processor while writing into the common buffer and the destination processor while reading from the buffer.

Another case of partial involvement can be given on a system with a communication device or network adapter that performs the data transfer over the network, but the processors have to copy the data to and from it. The time the processor is involved is a function of the data size, but the processor is not engaged during the entire communication. Note that the communication device is considered part of the communication network.

Packet Based Communication   The clear separation of overhead and involvement is not so obvious for packet based communication. For instance, consider two-sided interprocessor communication as in the last example and assume that the communication is packet based. At the beginning of a communication the sending processor goes through the communication layer, which is represented by the overhead. If the data has at most the size of one packet, it is copied to the network adapter, which then sends the message to its destination. This behavior can very well be captured with the notions of overhead and involvement. However, a long message must be split into several packets. During the transfer of each packet on the network, the processor, is not involved and can continue computation. But the processor experiences a new, although smaller, overhead and involvement for submitting the next packet to the adapter, after the transfer of the previous one finished. The situation is similar on the receiving processor, which participates in the reception and assembling of the packets.

Like the treatment of packet based communication in edge scheduling (Section 7.3.1), the packet based communication can be approximated with a single overhead and a single involvement. Figure 8.4 illustrates this for the previously described example. The various overheads and involvements (Figure 8.4(a)) of all packets are unified into one overhead and one involvement (Figure 8.4(b)). Of course, the period of network activity is identical in both views. Note that the single overhead in Figure 8.4(b) only corresponds to the initial overhead in the packet view (Figure 8.4(a)). The small incremental overheads are accredited to the involvement. This is sensible, as those overheads happen concurrently with the network activity and can thus be considered involvement. If instead they were accredited to the initial overhead, the network activity would be shifted to a later time interval, since its start would be delayed correspondingly.

The notions of overhead and involvement are very flexible as they allow the unified description of all three types of interprocessor communication, even when the processors' participation is only partial. At the same time they are intuitive and, as will be seen in the next sections, easily integrate with edge scheduling. The separation into overhead and involvement is also more general than the approach taken by the LogP model, as will be shown next. After that, the notions of overhead and involvement are employed in the task scheduling strategy that considers the involvement of the processor in communication.

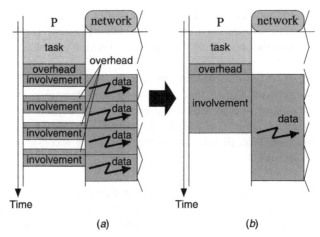

Figure 8.4. Overhead and involvement for a packet based communication: (*a*) packet view and (*b*) the approximation as one overhead and one involvement.

8.1.3   Relation to LogP and Its Variants

The LogP model (Section 2.1.3) was motivated by a problem similar to the one investigated in this chapter: the popular PRAM model for algorithm design and complexity analysis is not sufficiently realistic for modern parallel systems. Therefore, it is interesting to compare the discussed notions of overhead and scheduling with the approach of LogP, and some of its variants, since LogP also recognizes that the processor often participates in communication.

LogP   LogP (Culler et al. [46, 47]) differs in one important aspect from the discussed notions of overhead and involvement. The three parameters L, o, and g are specified assuming a message of some fixed short size M (Section 2.1.3). So while LogP's approach is equivalent to the notions of overhead and involvement for such a short message M—o corresponds to the overhead and the involvement is negligible or can be considered part of the overhead—it is quite different for a long message. In LogP, a long message of size N must be modeled as a series of $\lceil N/M \rceil$ short messages. Figure 8.5 illustrates the sending of a message of size $6 \times M$; the reception on the destination processor works correspondingly. Since LogP does not distinguish between overhead and involvement, both are represented by the several o's of the long message. Thus, the communication cost imposed on the processor is modeled as being linear in the message size. Consequently, the LogP model does not capture the nature of the large overhead of communication layers such as MPI or even the smaller, but still significant, overhead (Culler et al. [47]) of communication environments such as Active Messages. This overhead is typically only paid for once for every communication, independent of the message size. This is especially an issue for third-party communication, because the overhead is typically of constant length independent of message size.

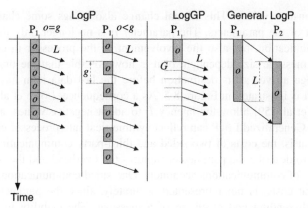

Figure 8.5. Sending a long message of size $6 \times M$ in LogP, LogGP, and generalized LogP model.

LogP was proposed as a model of parallel computing for algorithm design and evaluation. Its basis on small messages reflects that aim and in fact Culler et al. [47] argue that the overhead of high level communication layers is computation and should be modeled as such. From the task scheduling point of view, every additional computation due to communication is a cost that only has to be paid for interprocessor communication and is not inherent to the execution of tasks. A precise modeling of this overhead is therefore significant for scheduling.

A further limit of the LogP model is its basis on two-sided communication. Both processors are involved and both are imposed the cost o for every small message. That is, one-sided communication cannot be modeled accurately in LogP.

LogGP Alexandrov et al. [9] address one limitation of LogP by incorporating long messages into the model. Their new model, called LogGP, is a simple extension of LogP introducing only one additional parameter, G, which is the gap per byte for *long messages*. The reciprocal of G characterizes the available per-processor communication bandwidth in case of a long message. For a short message of the implicit fixed size M, LogGP is identical to LogP. For a long message, however, o is only paid once. After this time o, the first message M is sent into the network. Subsequent messages of size M take G time each to get out. Figure 8.5 depicts the sending of a long message in LogGP. The processor is only busy during o; the rest of the time it can perform computation, that is it can overlap computation with communication. Hence, the long message extension of LogGP represents third-party communication. LogGP is therefore not realistic for systems that do not possess a dedicated communication subsystem. Furthermore, the network capacity of LogP (at most $\lceil L/g \rceil$ messages) is not redefined for LogGP.

Generalized LogP Löwe et al. [131] and Eisenbiegler et al. [59] generalize the LogP model by making the parameters L, o, and g functions of the message size instead of constant values. This formal change also brings some changes to the interpretation of the parameters. The parameter o is no longer only the overhead of the communication, but also the involvement of the processor in communication. With a message size dependent o, it is now possible that the first data item of the message arrives at its destination before the last data item has been sent. This situation is illustrated in Figure 8.5. As a consequence, Löwe et al. [131] and Eisenbiegler et al. [59] allow the latency L to adopt negative values, as shown in the example. Generalized LogP can reflect symmetrical interprocessor communication costs, that is, the costs of two-sided and third-party communication. Its basis on functions to describe the parameters captures the overhead and the involvement of these kinds of communications accurately. One-sided communication, which has unsymmetrical costs, is not represented accurately, since the parameters are the same for both sending and receiving of a message. The inability to distinguish between overhead and involvement makes it also impossible to analyze contention for network resources. In particular, the network capacity of the LogP model— only $\lceil L/g \rceil$ messages can be concurrently in transit—is not defined for generalized LogP.

From the above comparison it follows that the notions of overhead and involvement as introduced in this section are more flexible than LogP and its variants. While LogGP and generalized LogP address some of the shortcomings of LogP, the separation of overhead and involvement is more general, for example, it also covers one-sided communication. Moreover, this separation allows the new scheduling model, introduced in the next section, to be based on the contention model. It therefore inherits the flexible and powerful modeling of end-point and network contention. In scheduling based on LogP and its variants (e.g., Boeres and Rebello [25], Kalinowski et al. [98], Löwe et al. [131]), network contention is completely ignored.

LogP is a model for algorithm design and complexity analysis. As such, a certain simplicity is essential, for example, an identical and constant o on both sides of the communication. The overhead and involvement introduced in this section are not subject to such restriction. The possible increase of conceptual complexity is justified, since these notions are used for scheduling performed by a computer algorithm and not for algorithm design performed by a human.

8.2   INVOLVEMENT SCHEDULING

The notions of overhead and involvement discussed in the last section are the key concepts to enhance task scheduling toward the awareness of processor involvement in communication (Sinnen [172], Sinnen et al. [180]). In the first step, a new target system model is defined.

Definition 8.1 (Target Parallel System—Involvement–Contention Model) *A target parallel system $M = (TG, \omega, o, i)$ consists of a set of possibly heterogeneous processors $\mathbf{P}$ connected by the communication network $TG = (\mathbf{N}, \mathbf{P}, \mathbf{D}, \mathbf{H}, b)$,* *according to Definition 7.5. This system has Properties 1 to 3 of the classic model of Definition 4.3:*

1. *Dedicated system*
2. *Dedicated processor*
3. *Cost-free local communication*

So in comparison with the contention model (Definition 7.10), the involvement–contention model departs from the assumption of a dedicated communication subsystem (Property 4). Instead, the role of the processors in communication is described by the new components o—for overhead—and i—for (direct) involvement.

Definition 8.2 (Overhead and Involvement)   *Let* $G = (V, E, w, c)$ *be a task graph and* $M = ((N, P, D, H, b), \omega, o, i)$ *a parallel system. Let* $R = \langle L_1, L_2, \ldots, L_l \rangle$ *be the route for the communication of edge* $e \in E$ *from* $P_{\mathrm{src}} \in P$ *to* $P_{\mathrm{dst}} \in P$, $P_{\mathrm{src}} \neq P_{\mathrm{dst}}$.

Overhead. $o_s : E \times P \to \mathbb{Q}_0^+$ *and* $o_r : E \times P \to \mathbb{Q}_0^+$ *are the communication overhead functions (the subscript s stands for send and r stands for receive). $o_s(e, P_{\mathrm{src}})$ is the computational overhead, that is, the execution time, incurred by processor P_{src} for preparing the transfer of the communication associated with edge e and $o_r(e, P_{\mathrm{dst}})$ is the overhead incurred by processor P_{dst} after receiving e.*

Involvement. $i_s : E \times L_s \to \mathbb{Q}_0^+$ *and* $i_r : E \times L_r \to \mathbb{Q}_0^+$ *are the communication involvement functions, where* L_s *is the set of links that leave processors,* $L_s = \{H \in H : H \cap P \neq \emptyset\} \cup \{D_{ij} \in D : N_i \in P\}$, *and* L_r *is the set of all links that enter processors,* $L_r = \{H \in H : H \cap P \neq \emptyset\} \cup \{D_{ij} \in D : N_j \in P\}$. $i_s(e, L_1)$ *is the computational involvement, that is, execution time, incurred by processor P_{src} during the transfer of edge e and $i_r(e, L_l)$ is the computational involvement incurred by P_{dst} during the transfer of e.*

This is the general definition of overhead and involvement for arbitrary, possibly heterogeneous systems. Therefore, the overhead is made a function of the processor and the involvement a function of the utilized communication link. As discussed in the previous section, the overhead depends largely on the employed communication environment and is thereby normally unaffected by the utilized communication resources. In contrast, the involvement depends to a large extent on the capabilities of the utilized communication resources. Hence, the processor involvement is characterized by the outgoing or incoming link utilized for a communication. Figure 8.6 illustrates this for a parallel system, where each processor is incident on several links. A communication between the two depicted processors inflicts the involvement associated with the undirected link between them, while a communication into and from the not shown rest of the network inflicts another involvement, associated with the directed edges that are utilized for these communications.

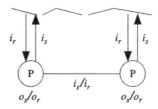

Figure 8.6. Extract of a parallel system represented by the involvement–contention model.

Of course, if the overhead changes with the link employed for a communication, it can also be defined as a function of the link, as done for the involvement.

With the distinction between the sending (o_s, i_s) and the receiving side (o_r, i_r) of communication, all three types of communication—third-party, one-sided, and two-sided—can be precisely represented. The corresponding functions are simply defined accordingly, for example, $i_s(e, L) = i_r(e, L) = 0$ for involvement-free third-party communication.

Homogeneous Systems   For homogeneous systems or systems that have homogeneous parts, the definition of overhead and involvement can be simplified.

In the case where the entire system has only one link per processor (that encompasses full duplex links represented by two counterdirected edges), i_s and i_r can be defined as a function of the processor, instead of as a function of the link, that is, $i_{s,r}(e, L) \Rightarrow i_{s,r}(e, P)$.

If interprocessor communication is symmetric, that is, the overhead and the involvement on both sides of the communication have the same extent (which is possible in third-party and two-sided communications), the distinction between the sending and receiving side can be abolished, resulting in only one function o and one function i, that is, $i_s(e, L) = i_r(e, L) = i(e, L)$ and $o_s(e, P) = o_r(e, P) = o(e, P)$.

Furthermore, in homogeneous systems and considering communication links, the involvement functions can be defined globally, that is, $i_{s,r}(e, L) = i_{s,r}(e)$. The same is true for the overhead functions in the case of homogeneous processors, that is, $o_{s,r}(e, P) = o_{s,r}(e)$.

8.2.1   Scheduling Edges on the Processors

In contention aware scheduling (Section 7.4), the edges of the task graph are scheduled onto the links of the communication network, like the nodes are scheduled on the processor. Incorporating overhead and involvement into contention aware task scheduling is accomplished by extending edge scheduling so that edges are also scheduled on the processors. To do so, the first step is to define the start and finish times of an edge on a processor.

Definition 8.3 (Edge Start and Finish Times on Processor)   *Let $G = (\mathbf{V}, \mathbf{E}, w, c)$ be a task graph and $M = ((\mathbf{N}, \mathbf{P}, \mathbf{D}, \mathbf{H}, b), \omega, o, i)$ a parallel system.*

The start time $t_s(e, P)$ of an edge $e \in \mathbf{E}$ on a processor $P \in \mathbf{P}$ is the function $t_s : \mathbf{E} \times \mathbf{P} \to \mathbb{Q}^+$.

Let $R = \langle L_1, L_2, \dots, L_l \rangle$ be the route for the communication of edge $e \in \mathbf{E}$ from $P_{\mathrm{src}} \in \mathbf{P}$ to $P_{\mathrm{dst}} \in \mathbf{P}$, $P_{\mathrm{src}} \neq P_{\mathrm{dst}}$. The finish time of e on P_{src} is

$$t_f(e, P_{\mathrm{src}}) = t_s(e, P_{\mathrm{src}}) + o_s(e, P_{\mathrm{src}}) + i_s(e, L_1) \tag{8.1}$$

and on P_{dst} is

$$t_f(e, P_{\mathrm{dst}}) = t_s(e, P_{\mathrm{dst}}) + o_r(e, P_{\mathrm{dst}}) + i_r(e, L_l). \tag{8.2}$$

Figure 8.7 illustrates scheduling under the involvement–contention model. The edge e_{AB} of the simple task graph is not only scheduled on the two links between the communicating processors, but also on both processors in order to reflect the incurred overhead and their involvement. Consequently, the execution time of an edge on a processor is the sum of the overhead and the involvement (see Eqs. (8.1) and (8.2)).

Clearly, for a sensible scheduling of the edges on the processors, some conditions must be formulated.

Like a node, an edge scheduled on a processor represents computation, precisely the computation necessary for the communication of the edge. Thus, the exclusive processor allocation Condition 4.1 also applies to edges scheduled on the processors. Remember, this is a requirement imposed by Property 2 of the system model (Definition 8.1).

An edge is only scheduled once on a processor, from which it follows that a separation between the overhead and the involvement is not possible. This is already ensured by the definition of the finish time of an edge on a processor (Eqs. (8.1) and (8.2)). Even for systems where the separation of overhead and involvement might

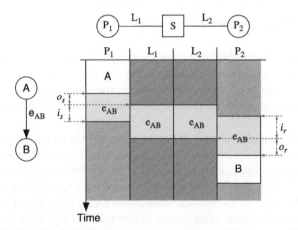

Figure 8.7. Scheduling under the involvement–contention model: edges are also scheduled on the processors. S is a switch or other processor.

occur, the improvement on accuracy by allowing the separation would presumably not outweigh the more complicated scheduling.

For a meaningful and feasible schedule, the scheduling of the edges on the processors must obey Condition 8.1.

Condition 8.1 (Causality in Involvement Scheduling) *Let* $G = (\mathbf{V}, \mathbf{E}, w, c)$ *be a task graph and* $M = ((\mathbf{N}, \mathbf{P}, \mathbf{D}, \mathbf{H}, b), \omega, o, i)$ *a parallel system. Let* $R = \langle L_1, L_2, \ldots, L_l \rangle$ *be the route for the communication of edge* $e_{ij} \in \mathbf{E}$, $n_i, n_j \in \mathbf{V}$, *from* $P_{src} \in \mathbf{P}$ *to* $P_{dst} \in \mathbf{P}$, $P_{src} \neq P_{dst}$.

To assure the node strictness (Definition 3.8) of n_i

$$t_s(e_{ij}, P_{src}) \geq t_f(n_i, P_{src}). \tag{8.3}$$

Edge e_{ij} *can be transferred on the first link* L_1 *only after the overhead is completed on the source processor* P_{src}:

$$t_s(e_{ij}, L_1) \geq t_s(e_{ij}, P_{src}) + o_s(e_{ij}, P_{src}). \tag{8.4}$$

To assure the causality of the direct involvement on the destination processor P_{dst},

$$t_s(e_{ij}, P_{dst}) \geq t_f(e_{ij}, L_l) - i_r(e_{ij}, L_l). \tag{8.5}$$

The three inequalities can be observed in effect in Figure 8.7. Edge e_{AB} starts on P_1 after the origin node A finishes (inequality (8.3)). On the first link L_1, e_{AB} starts after the overhead finishes on P_1 (inequality (8.4)), at which time the involvement of P_1 begins. And last, e_{AB} starts on P_2 so that the involvement finishes at the same time as e_{ij} on L_2 (inequality (8.5)).

Discussion of Condition 8.1

Order of Overhead–Involvement For simplicity, Condition 8.1 fixes the temporal order between overhead and involvement in the way it has been assumed so far: first the overhead then the involvement on the source processor (established by Eq. (8.4)) and the other way around on the destination processor (established by Eq. (8.5)).

Freedom of Node–Edge Order It is not required by Condition 8.1 that an edge e_{ij} is scheduled on the source processor immediately after the termination of the origin node n_i. If n_i has multiple leaving edges, this is not possible anyway—only one edge can start immediately after n_i—but other edges, not leaving n_i, or even nodes might be scheduled between n_i and e_{ij}. The same holds for e_{ij} and the receiving node n_j, whereby e_{ij} must of course finish before n_j starts. This freedom in the scheduling order of the nodes and edges on the processors grants a high degree of flexibility to scheduling algorithms. Communications can be delayed, while executing other nodes

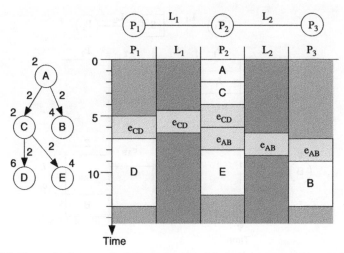

Figure 8.8. Example for a schedule under the involvement–contention model, where a delayed communication (e_{AB}) results in a shorter schedule length. $O_{s,r}(e,P) = 0.5, i_{s,r}(e,L) = 0.75\varsigma(e,L)$.

in the meantime, to obtain more efficient schedules. Figure 8.8 shows an example where the delayed scheduling of edge e_{AB} reduces the schedule length. If it were scheduled before C on P_2, the start of the communication e_{CD} would be delayed by two time units and in turn the start of D.

It is essential that the code generated from such a schedule respects the defined order of tasks and communications. For comparison, under the classic model it is assumed that all communications are sent immediately after the origin node finishes (Section 4.1). Even though the order of communications is established in contention aware scheduling, the relation to the node finish times is not clearly defined, as nodes are executed concurrently with the communications. Thus, only scheduling under the involvement–contention model admits the precise definition of the task and communication order.

Of course, the data to be sent must remain unmodified until it is sent. This might require additional buffering, for example, if an array holding the data is reutilized by other tasks on the sending processor. Yet, for nonblocking communications additional buffering is necessary anyway. In nonblocking communication the processor that initiates a communication does not have to actively wait (blocking wait) until its communication partner becomes available; that is, it can execute other tasks in the meanwhile.

Nonaligned Scheduling Condition 8.1 does not strictly adhere to the notion of the direct processor involvement introduced in the previous section. There, it is stated that a processor is involved in communication at the same time the communication is sent to or received from the network. But inequalities (8.4) and (8.5) also allow a later start

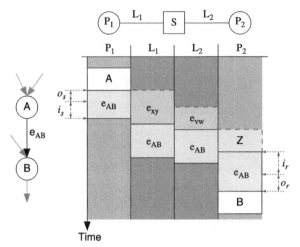

Figure 8.9. Nonaligned scheduling of the edges on the processors and links; chart and task graph are only extracts.

of the edge on the first link and an earlier start on the last link, so that the involvement is scheduled earlier (on the source processor) or later (on the destination processor) than the network activity. An extract of a corresponding schedule is depicted in Figure 8.9. Due to contention, the start of edge e_{AB} is delayed on the links and on P_2 so that the alignment with the involvement on the processors is not given.

This relaxation corresponds to the nonaligned scheduling approach in edge scheduling (Section 7.3.1). An enforced aligned scheduling, where the involvement part of the edge on the processors is aligned with the edge on the links, would result in the same issues as for the aligned edge scheduling approach (Section 7.3.1). Many idle time periods would be created on the processors, and the scheduling of an edge, on the processors and links, became more complex. The discussion of scheduling algorithms for the involvement–contention model in the next section makes this clearer.

Scheduling   As for the edge scheduling on the links (Section 7.3.2), a scheduling condition is formulated for the correct choice of an idle time interval into which an edge can be scheduled on a processor, with either the end or the insertion technique (Sections 5.1 and 6.1).

Condition 8.2 (Edge Scheduling Condition on a Processor)   *Let $G = (\mathbf{V}, \mathbf{E}, w, c)$ be a task graph and $M = ((\mathbf{N}, \mathbf{P}, \mathbf{D}, \mathbf{H}, b), \omega, o, i)$ a parallel system. Let $R = \langle L_1, L_2, \ldots, L_l \rangle$ be the route for the communication of edge $e_{ij} \in \mathbf{E}$, $n_i, n_j \in \mathbf{V}$, from $P_{\text{src}} \in \mathbf{P}$ to $P_{\text{dst}} \in \mathbf{P}$, $P_{\text{src}} \neq P_{\text{dst}}$. Let $[A, B]$, $A, B \in [0, \infty]$, be an idle time interval on P, $P \in \{P_{\text{src}}, P_{\text{dst}}\}$, that is, an interval in which no other edge or node is scheduled* *on P. Edge e_{ij} can be scheduled on P within $[A, B]$ if*

$$B - A \geq \begin{cases} o_s(e_{ij}, P_{\text{src}}) + i_s(e_{ij}, L_1) & \text{if } P = P_{\text{src}} \\ o_r(e_{ij}, P_{\text{dst}}) + i_r(e_{ij}, L_l) & \text{if } P = P_{\text{dst}} \end{cases}, \qquad (8.6)$$

$$B \geq \begin{cases} t_f(n_i, P_{\text{src}}) + o_s(e_{ij}, P_{\text{src}}) + i_s(e_{ij}, L_1) & \text{if } P = P_{\text{src}} \\ t_f(e_{ij}, L_l) + o_r(e_{ij}, P_{\text{dst}}) & \text{if } P = P_{\text{dst}} \end{cases}. \qquad (8.7)$$

This condition corresponds to Condition 7.3 for edge scheduling on the links. It ensures that the time interval $[A, B]$ adheres to the inequalities (8.3) and (8.5) of the causality Condition 8.1. For a given time interval, the start time of e_{ij} on P_{src} and P_{dst} is determined as follows.

Definition 8.4 (Edge Scheduling on a Processor) *Let $[A, B]$ be an idle time interval on P adhering to Condition 8.2. The start time of e_{ij} on P is*

$$t_s(e_{ij}, P) = \begin{cases} \max\{A, t_f(n_i)\} & \text{if } P = P_{\text{src}} \\ \max\{A, t_f(e_{ij}, L_l) - i_r(e_{ij}, L_l)\} & \text{if } P = P_{\text{dst}} \end{cases}. \qquad (8.8)$$

So, as with the scheduling on the links (Definition 7.9), the edge is scheduled as early as possible within the limits of the interval. Of course, the choice of the interval should follow the same policy on the links and on the processors; that is, either the end or insertion technique should be used.

The Overhead and Involvement Functions In all definitions and conditions related to scheduling under the involvement–contention model, overhead and involvement are simply treated as functions for the sake of generality. Nevertheless, these functions have a certain nature for most parallel systems, which is studied next. Generally, the overhead and the involvement functions are at most linear in the communication volume for probably all parallel systems.

The overhead is of constant time for many parallel systems and their communication environment. As mentioned before, the overhead is normally the path through the communication layer. Thus,

$$o_s(e, P) = o_s(P), \quad o_r(e, P) = o_r(P), \qquad (8.9)$$

that is, the overhead merely depends on the processor speed, so it is constant for homogeneous systems. Sometimes, however, the data to be transferred is copied between buffers during the overhead by which the overhead becomes a linear function of the communication volume.

The involvement is by definition not longer than the network activity. On the sending side, the involvement stops at the latest when all data has been sent across the first link, and on the receiving side, the involvement does not start before the communication begins on the last link. Hence, the involvement must be smaller than the communication time on the respective link:

$$i_s(e, L_1) \leq \varsigma(e, L_1), \tag{8.10}$$

$$i_r(e, L_l) \leq \varsigma(e, L_l). \tag{8.11}$$

Since the involvement represents the actual data transfer performed by the processor, this time is typically proportional to the communication time of the first/last link:

$$i_s(e, L_1) = C_s \cdot \varsigma(e, L_1), \tag{8.12}$$

$$i_r(e, L_l) = C_r \cdot \varsigma(e, L_l), \tag{8.13}$$

where $C_s, C_r \in [0, 1]$. For example, $C_s = C_r = 0.5$ for the two-sided memory copy as shown in Figure 8.3(c).

Consequently, in many common parallel systems, the overhead and the involvement are simply characterized by constants. Furthermore, these constants are independent of the task graph and they only need to be determined once to describe the target system for the scheduling of any task graph.

On the other hand, the functions might also be used to describe in detail the behavior of interprocessor communication. For instance, consider a parallel system, where the overhead of third-party communication is quite high. For few data words, other means of communication (e.g., memory copy by the processor) might be cheaper than setting up the communication device. In this case, there is a threshold data size for which it is worthwhile to employ the communication device. The overhead and involvement functions can be formulated accordingly. Similar things can happen on the protocol level of the communication environment, where an optimized procedure is employed for short messages (White and Bova [199]).

Note that the size of the involvement does not depend on the contention in the network. The assumption is that if the processor has to wait to send or receive a communication due to contention, this wait is passive or nonblocking, which means it can perform other operations in the meantime.

8.2.2 Node and Edge Scheduling

Few alterations are imposed by the new model on the edge scheduling on the links and on the scheduling of the nodes.

Edge Scheduling on Links While inequalities (8.3) and (8.5) of Condition 8.1 are newly introduced for the scheduling of the edges on processors, Eq. (8.4) substitutes a constraint of edge scheduling. The edge scheduling Condition 7.3 assures that an edge starts on the first link of a route after the origin node has finished. With Eq.

(8.4) the start of the edge on the first link must be further delayed by at least the time of the communication overhead. Hence, the first case ($k = 1$) of inequality (7.7) of Condition 7.3, $B \geq t_f(n_i) + \varsigma(e_{ij}, L_1)$, becomes

$$B \geq t_s(e_{ij}, P_{\text{src}}) + o_s(e_{ij}, P_{\text{src}}) + \varsigma(e_{ij}, L_1). \tag{8.14}$$

The calculation of the edge's start time on the first link (Definition 7.9) must be modified accordingly. Thus, the first case ($k = 1$) of Eq. (7.8), $t_s(e_{ij}, L_1) = \max\{A, t_f(n_i)\}$, becomes

$$t_s(e_{ij}, L_1) = \max\{A, t_s(e_{ij}, P_{\text{src}}) + o_s(e_{ij}, P_{\text{src}})\} \tag{8.15}$$

The rest of the edge scheduling procedure is completely unaffected by the scheduling of the edges on the processors and remains unmodified.

Node Scheduling   To adapt the scheduling of the nodes to the new model, it is only necessary to redefine the total finish time of the edge, which was originally defined for the classic model in Definition 4.6 and redefined for the contention model in Definition 7.11.

Definition 8.5 (Edge Finish Time)   *Let* $G = (V, E, w, c)$ *be a task graph and* $M = ((N, P, D, H, b), \omega, o, i)$ *a parallel system. The finish time of* $e_{ij} \in E$, $n_i, n_j \in V$, *communicated from processor* P_{src} *to* P_{dst}, $P_{\text{src}}, P_{\text{dst}} \in P$, *is*

$$t_f(e_{ij}, P_{\text{src}}, P_{\text{dst}}) = \begin{cases} t_f(n_i, P_{\text{src}}) & \text{if } P_{\text{src}} = P_{\text{dst}} \\ t_f(e_{ij}, P_{\text{dst}}) & \text{otherwise} \end{cases} . \tag{8.16}$$

As with the edge scheduling discussed earlier, the rest of the node scheduling procedure is completely unaffected by the scheduling of the edges on the processors and remains unmodified.

8.2.3 Task Graph

For scheduling under the contention model, the communication cost (i.e., the edge weight) was defined to be the average time a link is occupied with the communication associated with the edge (Definition 7.7). As elaborated in Section 7.4.1, this definition is compatible with the edge weight under the classic model, which represents the communication delay. Scheduling under the involvement–contention model is based on edge scheduling and adopts the communication cost definition of edge scheduling. Thus, it remains unmodified.

Yet, in practice, some caution is indicated. Under the contention model, the time an edge spends on a link corresponds to the communication delay. In consequence, pre- and postprocessing parts are either ignored or accredited to the communication cost represented by the edge weight. Under the involvement–contention model, the communication delay is also composed of the overheads inflicted on the communicating processors; hence, these overheads are not included in the edge weights (see below). Thus, task graph weights determined for different models are not necessarily identical.

Path Length   Most metrics used for analyzing a task graph and its nodes are based on the path length, for example, node levels or the critical path. As stated in Section 4.4, the length of a path can be interpreted as its execution time, when all communications are interprocessor communications. Under the involvement–contention model, the total communication time, or communication delay, does not merely consist of the edge weight, but also of the overheads incurred by the sending and receiving processors. Thus, the path length must be adapted in order to still represent the execution time of the path.

Definition 8.6 (Path Length with Overhead)   *Let $G = (V, E, w, c)$ be a task graph and $M = ((N, P, D, H, b), \omega, o, i)$ a parallel system. Let $\bar{o}_{s,r}(e) = \sum_{P \in \mathbf{P}} o_{s,r}(e, P)/|\mathbf{P}|$ be the average overhead. The length of a path p in G is*

$$len(p) = \sum_{n \in p, \mathbf{V}} w(n) + \sum_{e \in p, \mathbf{E}} (\bar{o}_s(e) + c(e) + \bar{o}_r(e)). \tag{8.17}$$

Of course, nothing changes for the computation path length, where all communications are assumed to be local. All metrics based on path lengths (e.g., node levels) are used as before with the new definition. Thus, to establish node priorities, the schemes from the classic model and the contention model can be used directly.

In general, communication is more important in the involvement-contention model than in the other models. In particular, the involvement of the processors is a crucial aspect of communication. But the involvement does not "lie" on the path and is therefore not considered in its length.

8.2.4 NP-Completeness

As can be expected, scheduling under the involvement–contention model remains an NP-hard problem. This is easy to see, as the involvement model is based on the contention model, which is NP-hard. Therefore, the proof of the following NP-completeness theorem employs the straightforward reduction from the decision problem C-SCHED(G, M_{TG}) associated with the scheduling under the contention model (Theorem 7.1).

Theorem 8.1 (NP-Completeness—Involvement-Contention Model)   *Let $G = (V, E, w, c)$ be a task graph and $M = ((N, P, D, H, b), \omega, o, i)$ a parallel system.*

The decision problem INVOLVEMENT-CONTENTION-SCHED *(G, M), shortened to* IC-SCHED, *associated with the scheduling problem is as follows. Is there a schedule S for G on M with length $sl(S) \leq T, T \in \mathbb{Q}^+$? IC-SCHED (G, M) is NP-complete.*

Proof. First, it is argued that IC-SCHED belongs to NP; then it is shown that IC-SCHED is NP-hard by reducing the NP-complete problem IC-SCHED (G_C, M_{TG}) (Theorem 7.1) in polynomial time to IC-SCHED.

Clearly, for any given solution S of IC-SCHED it can be verified in polynomial time that S is feasible (Algorithm 5, adapted for edge scheduling on links and processors) and $sl(S) \leq T$; hence, IC-SCHED $\in$ NP.

For any instance of C-SCHED, an instance of IC-SCHED is constructed by simply setting $G = G_C$, $M = (M_{TG}, o, i)$, $o_{s,r} = 0$, $i_{s,r} = 0$, and $T = T_{C-\text{SCHED}}$; thus, the overhead and the involvement take no time. Obviously, this construction is polynomial in the size of the instance of C-SCHED. Furthermore, if and only if there is a schedule for the instance of IC-SCHED that meets the bound T, is there a schedule for the instance C-SCHED meeting the bound $T_{C-\text{SCHED}}$. $\square$

$\alpha|\beta|\gamma$ **Notation** In terms of the $\alpha|\beta|\gamma$ notation (Section 6.4.1), the problem of scheduling under the involvement–contention model can be characterized with a simple extension of the notation used for the contention model (Section 7.4.2).

The essential aspect of the contention model is the fact that edges are scheduled. In Section 7.4.2 it was suggested to denote this by *-sched* in the β field, immediately after the specification of the communication costs (e.g., c_{ij}-*sched*). To indicate on which network the edges need to be scheduled, the α field was extended with a third subfield describing the network topology (e.g., *net* or *star*).

In the involvement–contention model, the edges are also scheduled and the β field is therefore left unchanged. What changes is the fact that edges also need to be scheduled on the processors. To indicate this, the α field is extended, with fields that indicate the presence of overhead and direct involvement.

The fourth subfield relates to the overhead and can take the following values:

- o. Communication inflicts no computational overhead on the sending or receiving processor.
- $o_{s,r}$. Communication inflicts computational overhead on the sending and on the receiving processor. Restrictions might be specified for the overhead in an obvious way as in the following examples.

    $\{o_s, o_r\}$. Communication inflicts computational overhead only on the sending (o_s) or on the receiving processor (o_r).

    o. Communication inflicts identical computational overhead on the sending and receiving processors, that is, $o_s = o_r = o$.

The fifth subfield relates to the (direct) involvement and can take the following values:

- o. Communication inflicts no computational involvement on the sending or receiving processor. This implies that there is a dedicated communication subsystem.
- $i_{s,r}$. Communication inflicts computational involvement on the sending and receiving processors. Restrictions might be specified for the involvement in an obvious way as in the following examples.
    - $\{i_s, i_r\}$. Communication inflicts computational involvement only on the sending (i_s) or on the receiving processor (i_r). In other words, the communication type is one-sided.
    - i. Communication inflicts identical computational involvement on the sending and receiving processors, that is, $i_s = i_r = i$.

So, unless both the fourth and the fifth field are empty, edges are also scheduled on the processors. With these notations, the general problem of scheduling under the involvement–contention model is specified by $P, net, o_{s,r}, i_{s,r} | prec, c_{ij}\text{-}sched | C_{\max}$ for a limited number (α's second subfield is empty) of homogeneous processors (P).

8.3 ALGORITHMIC APPROACHES

The integration of edge scheduling into task scheduling in order to achieve contention awareness had little to no effect on the fundamental scheduling methods, as was shown in Chapter 7. In contrast to scheduling on the links, the scheduling of the edges on the processors, which seems at first a simple extension, has a strong impact on the operating mode of scheduling algorithms (Sinnen [172], Sinnen et al. [180]).

Essentially, the problem is that at the time a free node n is scheduled, it is generally unknown to where its successor nodes will be scheduled. It is not even known if the corresponding outgoing communications will be local or remote. Thus, no decision can be taken whether to schedule n's leaving edges on its processor or not. Later, at the time a successor is scheduled, the period of time directly after node n might have been occupied with other nodes. Hence, there is no space left for the scheduling of the corresponding edge.

The general issue behind the described problem is not specific to a certain heuristic, for example, list scheduling, but it applies to all scheduling algorithms under the involvement–contention model. Also, scheduling under the LogP model faces the same problem with the scheduling of o for each communication (Kalinowski et al. [98]). Ideally, the processor allocations of the nodes are known *before* the scheduling.

This problem does not arise in contention scheduling, because communication can overlap with computation; that is, an edge can be scheduled on a link at the same time a node is executed on the incident processor.

Two different approaches to handle the described issue in scheduling under the involvement-contention model, which have reasonable complexity, can be distinguished: (1) direct scheduling and (2) scheduling based on a given processor allocation (Sinnen [172], Sinnen et al. [180]).

8.3.1 Direct Scheduling

Direct scheduling means that the processor allocation and the start/finish time attribution of a node are done in one single step. This is the dominant technique of most scheduling heuristics discussed so far. Thus, the direct scheduling approach is basically an attempt to adapt the known scheduling techniques to the involvement–contention model.

The essential question is *when* to schedule the edges on the processors. It is clear that in direct scheduling (i.e., a one-pass scheduling), the nodes considered for scheduling must be free. Otherwise, the start time of a node cannot be determined and it has to be set in a later step of the algorithm.

In the first attempt, the approach of list scheduling under the contention model is considered; that is, edges are scheduled at the same time as their destination nodes.

Scheduling Entering Edges   Scheduling an edge together with its free destination node is surely the easiest approach. At the time an edge is scheduled, the source and the destination processor are known and the route can be determined. An edge can be scheduled in one go on the source processor, the links, and the destination processor. Unfortunately, in many cases, this approach generates inefficient schedules.

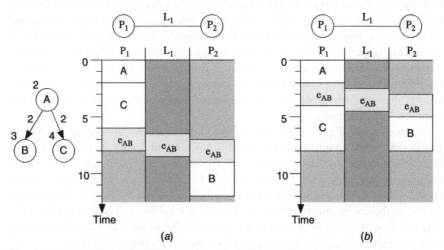

Figure 8.10. In direct scheduling (*a*), edges are scheduled with their destination nodes; (*b*) the optimal schedule. $o_{s,r}(e, P) = 0.5$, $i_{s,r}(e, L) = 0.75\varsigma(e, L)$.

Consider the example of Figure 8.10(a), where the nodes of the depicted task graph were scheduled in the order A, C, B, which is their bottom-level order. As communication e_{AB} was scheduled together with node B, the time period after node A was already occupied by C. A much shorter schedule length is achieved, when e_{AB} starts immediately after A as shown in Figure 8.10(b).

Even though there are cases where the delayed scheduling of communication can be of benefit (see Figure 8.8), in most situations the parallelization of the task graph is hindered with this approach. The start time of a node is delayed not only by the total communication time of an entering edge but also by the execution times of the nodes scheduled between the origin node and the entering edge on the origin node's processor (in the example of Figure 8.10(a) this is node C between A and e_{AB} on P_1). In fact, a list scheduling with start time minimization based on this approach would schedule all nodes in Figure 8.10 on P_1, as the parallelization is never beneficial: the start/finish time of each considered node is always earliest on P_1. Moreover, this even generalizes to any fork task graph: list scheduling with start time minimization would produce a sequential schedule!

Provisionally Scheduling Leaving Edges   The direct application of the scheduling method from contention scheduling is inadequate under the new model. Consequently, it is necessary to investigate how edges can be scheduled earlier. Of course, the problem remains that at the time a free node is scheduled, it is not known to where its successors will be scheduled. Nevertheless, the scheduling of the leaving edges must be prepared in some way.

The most viable solution is to reserve an appropriate time interval after a node for the later scheduling of the leaving edges. This must be done in a worst case fashion, which means the interval must be large enough to accommodate all leaving edges. A straightforward manner is to schedule all leaving edges on the source processor, directly after the origin node. The scheduling of the edges on the links and the destination processors is not possible at that time, since the destination processors, and with them the routes, are undetermined. Fortunately, this is also not necessary, as the scheduling on the links and the destination processor can take place when the destination node is scheduled, in the way it is done under the contention model. If the destination node is scheduled on the same processor as the origin node, the corresponding edge, which was provisionally scheduled with the origin node, is simply removed from that processor.

As an example, Figure 8.11 depicts three Gantt charts illustrating the scheduling process of the task graph on the left side. First, A is scheduled on P_1, together with its three leaving edges (shown in Figure 8.11(a)); hence, the worst case that B, C, and D are going to be scheduled on P_2 is assumed. Indeed, node B is scheduled on P_2, which includes the preceding scheduling of e_{AB} on the link and on P_2. Next, C is scheduled on P_1. So the communication between A and C is local and e_{AC} is removed from P_2. The situation at this point is shown in Figure 8.11(b). Finally, D is scheduled on P_2 with the respective scheduling of e_{AD} on the link and P_2 (Figure 8.11(c)).

On heterogeneous systems, the described provisional scheduling of an edge on its source processor must consider that the involvement depends on the first link of the

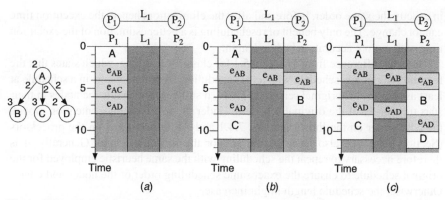

Figure 8.11. Direct scheduling: edges are scheduled on source processor together with their origin node; (a, b) charts illustrate intermediate schedules, while (c) shows the final schedule of the depicted task graph. $o_{s,r}(e,P) = 0.5$, $i_{s,r}(e,L) = 0.75\varsigma(e,L)$.

utilized route. Again, as the route is unknown at the time of the scheduling, the worst case must be assumed and the finish time is calculated as defined in the following.

Definition 8.7 (Provisional Finish Time of Edge on Source Processor) *Let* $G = (\mathbf{V}, \mathbf{E}, w, c)$ *be a task graph and* $M = ((\mathbf{N}, \mathbf{P}, \mathbf{D}, \mathbf{H}, b), \omega, o, i)$ *a parallel system. The provisional finish time of edge* $e_{ij} \in \mathbf{E}$ *on its source processor* $P = proc(n_i)$, $P \in \mathbf{P}$, *is*

$$t_f(e_{ij}, P) = t_s(e_{ij}, P) + o_s(e_{ij}, P) + i_{s,\max}(e_{ij}, P), \tag{8.18}$$

where

$$i_{s,\max}(e_{ij}, P) = \max_{L \in \mathbf{L}: L \text{ leaving } P} \{i_s(e_{ij}, L)\}. \tag{8.19}$$

When the destination node n_j is scheduled, the finish time must be reduced, if applicable, to the correct value.

With the reservation of a time interval for the outgoing edges on the processor, the rest of scheduling can be performed as under the contention model. The clear disadvantage of this approach are the gaps left behind by removed edges, which make a schedule less efficient. In order to relieve this shortcoming, two techniques can help to eliminate or even avoid the gaps.

Gap Elimination—Schedule Compaction In a completed schedule, gaps can be eliminated by repeating the scheduling procedure. The nodes and their edges must be scheduled in the exact same order as in the first run and the processor allocation is taken from the completed schedule. This makes the provisional scheduling of edges needless and the gaps are avoided.

It is important to realize that this technique can reduce the schedule length, but it has no impact on the execution time of the schedule. All nodes and edges are scheduled in exactly the same order, before and after the elimination; hence, the execution time cannot change. The only benefit of rescheduling is a better estimation of the execution time with a more accurate schedule length.

Note the difference from Theorem 5.1 for classic scheduling, which states that the rescheduling of any schedule with a list scheduling algorithm results in a schedule at least as short as the original schedule. There, it suffices to schedule the nodes in their start time order of the original schedule. Under the involvement–contention model, the edge order on the links (Sections 7.4.1 and 7.5.1) as well as on the processors (Section 8.2.1, see also later) is relevant for the schedule length. Generally, it is therefore necessary to repeat the scheduling with the same heuristic employed for the original schedule to ensure the exact same scheduling order of the nodes and edges. Otherwise, the schedule length might increase.

Inserting into Gaps The insertion technique can be applied to use the emerging gaps during the scheduling. This has the advantage that scheduling decisions might be different when a node or an edge can be inserted into a gap and therefore start earlier. Note that use of the insertion technique in a straightforward manner is only possible because the causality Condition 8.1 for involvement scheduling gives the freedom of the node and edge order (Section 8.2.1). Inserting a node or an edge into a gap is very likely to separate edges from their origin or destination nodes, as, for example, in Figure 8.8.

Scheduling Order of Edges As in contention scheduling (Sections 7.4.1 and 7.5.1), the order in which the edges are scheduled is relevant and might lead to different schedule lengths. Here, the scheduling of the edges is divided into the scheduling of the entering edges on the links and the destination processors and the leaving edges provisionally on the source processors. Scheduling the entering edges is very similar to the scheduling under the contention model and the same considerations apply to their order (Sections 7.4.1 and 7.5.1). A natural order of the leaving edges is the scheduling order of their destination nodes, as it follows the same rational as the minimization of the node's start and finish times (Section 5.1). Of course, the ordering of the edges slightly increases the complexity of a scheduling algorithm. It is very important for most algorithms that the scheduling order of the edges is deterministic (Section 7.5.1).

In Section 8.4.1, a list scheduling algorithm will be introduced, employing the technique of reserving time intervals for the leaving edges. With list scheduling, utilization of the end or insertion technique is no problem at all.

8.3.2 Scheduling with Given Processor Allocation

The second approach to involvement scheduling assumes a processor allocation, or mapping, to be given at the time the actual scheduling procedure starts. Since it is known for every node to where its successors will be scheduled, there is no need for the provisional scheduling of the edges on the processors.

Scheduling an Edge The scheduling of an edge e_{ij} can be divided into three parts: scheduling on the source processor P_{src}, on the links of the route R, and on the destination processor P_{dst}. On the source processor, an edge must be scheduled together with its origin node n_i, as the foregoing considerations in the context of the direct scheduling showed. Of course, when n_i is scheduled, only those leaving edges, whose destination nodes will be executed on a processor other than P_{src}, are scheduled on P_{src}. The scheduling on the links and on the destination processor can take place with either the origin node n_i or the destination node n_j. Hence, there are three alternatives for the scheduling of an edge e_{ij}.

1. *Schedule e_{ij} on P_{src} with n_i, on R, P_{dst} with n_j.* This alternative is identical to the approach of direct scheduling, where the edge e_{ij} is scheduled on the links of R and on the destination processor P_{dst}, when its destination node n_j is scheduled. Figure 8.12 illustrates this alternative, where the nodes A and C of the depicted task graph are allocated to processor P_1, and B and D are allocated to P_2, and they are scheduled in the order A, B, C, D. Figure 8.12(a) shows the intermediate schedule after scheduling B and Figure 8.12(b) shows the final schedule.

2. *Schedule e_{ij} on P_{src}, R with n_i, on P_{dst} with n_j.* Here, e_{ij} is scheduled not only on P_{src}, but also on the links, when n_i is scheduled. This way, the edges are scheduled on the links of R in the scheduling order of their origin nodes, while in the first alternative the edges are scheduled on the links in the order of their destination nodes. Figure 8.13 demonstrates this difference, by repeating the example of Figure 8.12, now scheduling e_{ij} on the links with the origin node n_i. So e_{AD} starts earlier than e_{BC}; in Figure 8.12 it is the other way around.

3. *Schedule e_{ij} on P_{src}, R, P_{dst} with n_i.* Edge e_{ij} is completely scheduled together with its origin node n_i. This alternative is likely to produce the best scheduling

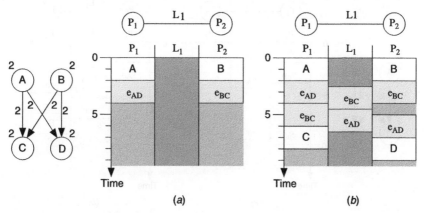

Figure 8.12. Scheduling with given allocation: edge e_{ij} scheduled on P_{src} with n_i, on R, P_{dst} with n_j, as in direct scheduling. Node order is A, B, C, D; $o_{s,r}(e, P) = 0.5$, $i_{s,r}(e, L) = 0.75\varsigma(e, L)$.

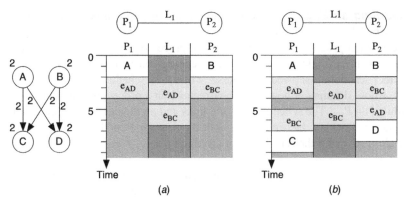

Figure 8.13. Scheduling with given allocation: edge e_{ij} scheduled on P_{src}, R with n_i, on P_{dst} with n_j. Node order is A, B, C, D; $o_{s,r}(e, P) = 0.5$, $i_{s,r}(e, L) = 0.75\varsigma(e, L)$.

alignment of the edge on the source processor, the links and the destination processor, as the scheduling is done in a single step. Note that the scheduling of an edge on the processors and links does not require alignment; therefore, it is called the nonaligned approach, but it is obvious that large differences between the scheduling times on the links and the processors are not realistic. Hence, in terms of the scheduling of the edge, this alternative is probably the most accurate. Unfortunately, it has a similar disadvantage as the direct scheduling of the entering edges: the scheduling of the edges on their destination processors might prevent the efficient scheduling of the nodes. Figure 8.14 illustrates the previous example, now with the complete scheduling of the edges on all resources with their origin nodes. As the edge e_{AD} is scheduled before

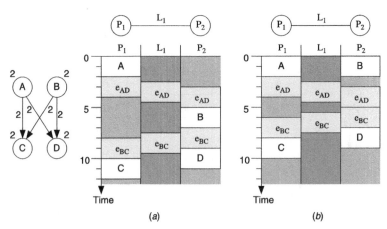

Figure 8.14. Scheduling with given allocation: edge e_{ij} scheduled on P_{src}, R, P_{dst} with n_i; (a) without insertion; (b) with insertion. Node order is A, B, C, D; $o_{s,r}(e, P) = 0.5$, $i_{s,r}(e, L) = 0.75\varsigma(e, L)$.

node B, B can only start after e_{AD} on P_2 (Figure 8.14(a)), unless the insertion technique is employed (Figure 8.14(b)). The early scheduling of the edges on their destination processors rapidly increases their finish times, leaving large idle gaps. So the conjoint scheduling of an edge on all resources is only sensible with the insertion technique.

There is no clear advantage of the first over the second alternative or vice versa. Which one is better (i.e., more realistic) depends on the way the communication is realized in the target parallel system, whether it is initiated by the receiving (first alternative) or by the sending side (second alternative).

In all three alternatives, the scheduling order of the edges is relevant and the same considerations apply as for direct scheduling (Section 8.3.1).

Section 8.4.2 looks at two-phase scheduling heuristics, where the first phase establishes the processor allocation and the second phase constructs the actual schedule with a simple list scheduling heuristic, based on the foregoing considerations.

8.4 HEURISTICS

This section formulates heuristics for scheduling under the involvement–contention model. It starts with the venerable list scheduling and then discusses two-phase heuristics, where the scheduling in the second phase disposes of a processor allocation determined in the first phase.

8.4.1 List Scheduling

As under the contention model (Section 7.5.1), it suffices to reformulate the determination of the DRT and the scheduling procedure of a node in order to adapt list scheduling for the involvement–contention model. The two procedures are modified using the direct scheduling approach discussed in Section 8.3.1.

Scheduling of Node   Algorithm 23 shows the procedure for the scheduling of a node in accordance with the direct scheduling approach. As under the contention model, the actual scheduling of a node n is preceded by the scheduling of its entering edges on the links and the destination processor P, including the possibly necessary correction of the edges' finish times on the source processors (lines 6–13). If the origin node of an edge is scheduled on P, the provisionally scheduled edge is removed from P first (lines 1–5), since the communication is then local. It is important that those edges are removed before the scheduling of the entering edges coming from remote processors, as the latter might already use the freed time intervals. After the scheduling of n on P (line 14), each of its leaving edges is scheduled on P in order to reserve the corresponding time periods (lines 15–17). As discussed in Section 8.3.1, the finish time of each edge is calculated for the worst case, that is, the longest possible involvement (Definition 8.7).

Algorithm 23   Scheduling of Node n_j on Processor P in Involvement–Contention Model

Require: n_j is a free node

  1: **for** each $n_i \in$ **pred**(n_j) **do**
  2:    **if** $proc(n_i) = P$ **then**
  3:       remove e_{ij} from P
  4:    **end if**
  5: **end for**
  6: **for** each $n_i \in$ **pred**(n_j) in a deterministic order **do**
  7:    **if** $proc(n_i) \neq P$ **then**
  8:       determine route $R = \langle L_1, L_2, \ldots, L_l \rangle$ from $proc(n_i)$ to P
  9:       correct $t_f(e_{ij}, proc(n_i))$
10:      schedule e_{ij} on R (Definition 7.9)
11:      schedule e_{ij} on P (Definition 8.3 and 8.4)
12:    **end if**
13: **end for**
14: schedule n_j on P
15: **for** each $n_k \in$ **succ**(n_j) in a deterministic order **do** $\triangleright$ *reserve space for leaving edges*
16:    schedule e_{jk} on P with worst case finish time (Definition 8.7)
17: **end for**

Determining DRT   The calculation of the DRT, formulated by Algorithm 24, is performed in the same fashion as the scheduling of a node. It is noteworthy that provisionally scheduled edges, whose origin nodes are scheduled on the processor P for which the DRT is calculated (i.e., communication is local), are first removed (lines 1–5) and then replaced at the end (line 19). Also, the possibly necessary correction of the incoming edges' finish times on the source processors is calculated, but the schedule is not changed (line 11).

End and Insertion Technique   As before, both techniques, end and insertion, can be employed with list scheduling. Under the involvement–contention model, the insertion technique is more indicated, since the removing of provisionally scheduled edges leaves gaps, which should be filled by other nodes or edges. However, inserting a node or an edge into a gap is very likely to separate edges from their origin or destination nodes. If this is not supported by the code generation or is simply not desired, the end technique should be employed, where the order of nodes and edges is strict.

    The scheduling order of the edges is required to be deterministic (see the corresponding `for` loops in Algorithms 23 and 24), just as under the contention model (Section 7.5.1).

Algorithm 24   Determine $t_{dr}(n_j, P)$ in Involvement–Contention Model

Require: n_j is a free node
1: **for** each $n_i \in \textbf{pred}(n_j)$ **do**
2:   **if** $proc(n_i) = P$ **then**
3:     remove e_{ij} from P
4:   **end if**
5: **end for**
6: $t_{dr} \leftarrow 0$
7: **for** each $n_i \in \textbf{pred}(n_j)$ in a deterministic order **do**
8:   $t_f(e_{ij}) \leftarrow t_f(n_i)$
9:   **if** $proc(n_i) \neq P$ **then**
10:     determine route $R = \langle L_1, L_2, \ldots, L_l \rangle$ from $proc(n_i)$ to P
11:     calculate $t_f(e_{ij}, proc(n_i))$ for route R
12:     schedule e_{ij} on R (Definition 7.9)
13:     schedule e_{ij} on P (Definitions 8.3 and 8.4)
14:     $t_f(e_{ij}) \leftarrow t_f(e_{ij}, P)$
15:   **end if**
16:   $t_{dr} \leftarrow \max\{t_{dr}, t_f(e_{ij})\}$
17: **end for**
18: remove edges $\{e_{ij} \in \mathbf{E} : n_i \in \textbf{pred}(n_j) \land proc(n_i) \neq P\}$ from links and P
19: replace provisionally scheduled edges $\{e_{ij} \in \mathbf{E} : n_i \in \textbf{pred}(n_j) \land proc(n_i) \neq P\}$
    on P
20: return t_{dr}

An Example   As in previous chapters, the list scheduling for the involvement–contention model is illustrated with an example. Since scheduling the sample task graph of Figure 3.15 under the involvement–contention model is too extensive to be instructive, the smaller task graph displayed in Figure 8.15 is utilized for the example.

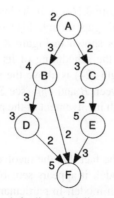

Figure 8.15. Task graph for list scheduling example in Figure 8.16.

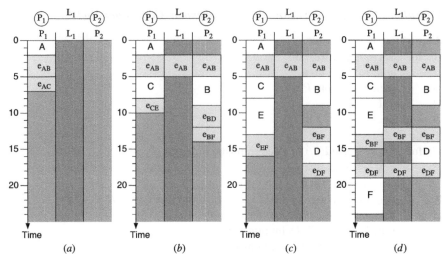

Figure 8.16. Example of list scheduling under involvement–contention model; (a)–(c) intermediate schedules, (d) final schedule. Task graph of Figure 8.15 is scheduled in the order A, B, C, E, D, F; $o_{s,r}(e, P) = 0$, $i_s(e, L_1) = \varsigma(e, L_1)$, $i_r(e, L_l) = \varsigma(e, L_l)$.

Its nodes are scheduled in the order A, B, C, E, D, F by a list scheduling heuristic with start time minimization and the end technique on two identical processors connected by a half duplex link. For simplicity, the overhead is assumed to be zero, $o_{s,r}(e, P) = 0$, and both processors, the source and the destination, are 100% involved in communication, $i_s(e, L_1) = \varsigma(e, L_1)$, $i_r(e, L_l) = \varsigma(e, L_l)$.

Figure 8.16(a) displays the initial schedule after assigning A to P_1, with its leaving edges also scheduled on P_1. Next, node B is scheduled earliest on P_2 together with the corresponding entering edge e_{AB} on the link and P_2. The scheduling of C on P_1 is preceded by the removal of its entering edge e_{AC}, because the communication is local. In consequence, C can benefit from the freed space and starts at time unit 5 (Figure 8.16(b)). The scheduling of E on P_1 is straightforward including the removal of e_{CE}. After the scheduling of E on P_1, D starts earliest on P_2, however, its scheduling leaves a gap of 3 time units (Figure 8.16(c)). If the insertion technique were used, D could have been inserted into this gap, though this would separate D and its leaving edge e_{DF}. The final schedule is shown in Figure 8.16(d), after node F has been scheduled on P_1, preceded by the scheduling of its entering edges on L_1 and on P_1. Note that the schedule length (24) is above the sequential time (22), yet after the elimination of the gap between B and e_{BF} (see Section 8.3.1) it reduces to 22 time units. Nevertheless, the high involvement of the processors impedes an efficient parallelization.

Priority Schemes Due to the fact that the involvement–contention model is an extension of the contention model, it appears sensible to utilize the same priority schemes as under the contention model. In particular, it is proposed to employ the *bottom-level* order for list scheduling under the involvement–contention model.

Complexity In comparison to contention aware list scheduling, the time complexity under the involvement–contention model does not increase. The additional scheduling of the edge on the processors increases the effort to implement the algorithm, but it does not modify its time complexity. For example, the complexity of the second part of simple list scheduling (Algorithm 9) with start/finish time minimization and the end technique is $O(\mathbf{P}(\mathbf{V} + \mathbf{E}O(routing)))$ (see Section 7.5.1).

Also, the complexity does not increase for the insertion technique (Section 6.1), even though the individual scheduling of a node and an edge on a processor is more complex. In the involvement–contention model, there are more objects on the processors, namely, nodes *and* edges, which must be searched to find an idle time interval. On all processors there are at most $O(\mathbf{V} + \mathbf{E})$ nodes and edges; hence, the total complexity for the scheduling of the nodes increases from $O(\mathbf{V}^2)$, under the contention model, to $O(\mathbf{V}(\mathbf{V} + \mathbf{E}))$. The scheduling of the edges on the processors is $O(\mathbf{E}(\mathbf{V} + \mathbf{E}))$, compared to $O(\mathbf{PE})$ with the end technique, and on the links it remains $O(\mathbf{PE}^2O(routing))$. Hence, the total complexity of the second part of simple list scheduling with start/finish time minimization and the insertion technique is $O(\mathbf{V}^2 + \mathbf{VE} + \mathbf{PE}^2O(routing))$, that is, $O(\mathbf{V}^2 + \mathbf{PE}^2O(routing))$, as under the contention model (Section 7.5.1).

8.4.2 Two-Phase Heuristics

The generic form of a two-phase scheduling algorithm was already outlined in Algorithm 13. The first phase establishes the processor allocation and the second phase constructs the actual schedule with a simple list scheduling heuristic, as described in Section 5.2. The next paragraphs discuss how such an algorithm is applied under the involvement–contention model.

Phase 1—Processor Allocation The processor allocation can originate from any heuristic or can be extracted from a given schedule. For example, a schedule produced under the classic or contention model might serve as the input (see also Section 5.2).

Using the first two steps of a clustering based heuristic, that is, the clustering itself and the cluster-to-processor mapping (Algorithm 14, Section 5.3), is very promising, because clustering tries to minimize communication costs, which is even more important under the involvement–contention model. The clustering itself can be performed under the classic or, as described in Section 7.5.3, under the contention model. Even the involvement–contention model can be used with a straightforward extension of the considerations made in Section 7.5.3.

Genetic Algorithm In Sinnen [172] and Sinnen et al. [180], a genetic algorithm, called GICS (genetic involvement–contention scheduling), is proposed for the determination of the processor allocation. The simple idea is that the genetic algorithm searches for an efficient processor allocation, while the actual scheduling is performed with a list scheduling heuristic as discussed later for phase 2.

GICS follows the outline of Algorithm 20, Section 6.5. Naturally, the chromosome encodes the processor allocation, which is an indirect representation (Section 6.5.2). Hence, the construction of the schedule corresponding to each chromosome, which is necessary for the evaluation of its fitness, requires the application of a heuristic. As already mentioned, this is performed by a list scheduling heuristic under the involvement–contention model (see later discussion). Strictly speaking, this means that a two-phase heuristic is applied multiple times in GICS—once for each new chromosome.

To reduce the running time of the evaluation, the node order is determined only once at the beginning of the algorithm, namely, according to their bottom levels. Like most GAs, GICS starts with a random initial population, enhanced with a chromosome representing a sequential schedule (all nodes are allocated to a single processor). The pool is completed by one allocation extracted from a schedule produced with a list scheduling heuristic (Section 8.4.1). Most other components are fairly standard. In fact, the algorithm can even be employed for scheduling under the classic or the contention model, using a modified evaluation (i.e., scheduling heuristic).

Phase 2—List Scheduling   From Section 5.2 it is known that the second phase can be performed with a simple list scheduling algorithm. The processor choice of list scheduling is simply a lookup from the given mapping $\mathcal{A}$ established in the first phase. Even though list scheduling under the involvement–contention model was already studied in Section 8.4.1, it must be revisited as there are three alternatives for the scheduling of the edges when the processor allocation is already given (Section 8.3.2).

Scheduling of the Edges   The first alternative corresponds to direct scheduling, and the procedures for the scheduling of a node and the calculation of the DRT are presented in Algorithms 23 and 24. The scheduling of an edge on the source processor must be modified (line 16 in Algorithm 23), since it is no longer necessary to assume the worst case. Lines 1–5 (removing local edges) and line 9 (correcting the finish time) can be dropped completely, since the provisional scheduling of the edges is not necessary with the given processor allocation. The other two alternatives are only distinguished from this procedure through the place where the edges are scheduled on the links and the destination processor. So, for the second alternative, where the edge is scheduled on the links as a leaving edge, lines 8 and 10 of Algorithm 23 move to the `for` loop on lines 15–17. In the third alternative, the edge is also scheduled on the destination processor within this loop, and the `for` loop for the entering edges (lines 6–13) is completely dropped.

Of course, in the determination of the DRT the entering edges do not need to be tentatively scheduled, as the processor allocation is already given.

Complexity   The complexity of list scheduling with a given processor allocation decreases slightly in comparison to list scheduling with start/finish time minimization. It is not necessary to tentatively schedule a node and its edges on every processor. Thus, with the end technique the complexity is $O(V + EO(routing))$ (instead of $O(P(V + EO(routing))))$ and with the insertion technique it is $O(V^2 + E^2O(routing))$ (instead of $O(V^2 + PE^2O(routing)))$. Remember, the third alternative of edge scheduling, where the leaving edges of a node are scheduled on the processors and links, is only meaningful with the insertion technique (Section 8.3.2).

8.4.3 Experimental Results

As for the contention model (Section 7.5.4), the question arises as to how much the accuracy of scheduling and the execution times of the produced schedules benefit from the involvement–contention model. As argued in Section 7.5.4, only an experimental evaluation on real parallel systems can help answer this question.

Such an experimental evaluation is performed in Sinnen [172] and Sinnen et al. [180]. The employed methodology is based on the one described in Section 7.5.4: a large set of graphs is generated and scheduled by algorithms under the different models to several target systems. Because the focus is on the comparison of the different scheduling models, the same list scheduling algorithm is employed under each model. From the produced schedules, code is generated—using C and MPI—and executed on the real parallel systems.

Modeling of the target machines as topology graphs is relatively simple, as discussed in Chapter 7. For the involvement–contention model it is additionally necessary to specify for each target system the overhead and involvement of the processors in communication. Due to the lack of a deep insight into the target systems' communication mechanisms and their MPI implementations, 100% involvement is assumed, that is, the source and destination processors are involved during the entire communication time on the first and last link, respectively:

$$i_s(e, L_1) = \varsigma(e, L_1) \quad \text{and} \quad i_r(e, L_l) = \varsigma(e, L_l). \tag{8.20}$$

The overhead is intuitively set to an experimentally measured setup time:

$$o_s(e, P) = o_r(e, P) = setup\_time. \tag{8.21}$$

While it is clear that this definition of the overhead and the involvement is probably not an accurate description of the target systems' communication behavior, it is very simple. The idea is to demonstrate that accuracy and efficiency of scheduling can be improved even with a rough but simple estimate of the overhead and involvement functions.

Results The experiments demonstrated that the involvement–contention model achieves profoundly more accurate schedules and significantly shorter execution times.

Thus, considering processor involvement in communication further improves the already improved accuracy under the contention model. The length of a schedule is now in a region where it can be seriously considered an estimate of the real execution time. Under the classic model and for some topology graphs under the contention model, the schedule lengths are only small fractions of the real execution times, in particular, for medium ($CCR = 1$) and high communication ($CCR = 10$), which can hardly be considered execution time estimations. For example, on the Cray T3E-900 (Section 7.5.4), the estimation error under the involvement–contention model was on average below 20%, while the estimation error under the classic model goes up to 1850%.

Still, the scheduling accuracy under the involvement–contention model is not perfect, especially for low communication ($CCR = 0.1$). A possible explanation might be the blocking communication mechanisms used in MPI implementations (White and Bova [199]), which does not match the assumption of nonblocking communication made in the involvement–contention model. Furthermore, the employed overhead and involvement functions are very rough estimates; a better approximation of these functions has the potential to further increase the accuracy. In any case, it is in the nature of any model that there is a difference between prediction and reality.

The profoundly improved accuracy under the involvement–contention model allows more than just the reduction of execution times: it also allows one to evaluate algorithms and their schedules without execution of the schedules. Hence, new algorithms can be developed and evaluated without the large efforts connected with an evaluation on real parallel systems.

In the experiments conducted, the involvement–contention model also clearly demonstrated its ability to produce schedules with significantly reduced execution times. The benefit of the high accuracy is apparent in the significantly improved execution times with speedup improvements of up to 82%.

Despite the very good results, the efficiency improvement lags behind the accuracy improvement. A possible explanation lies in the heuristic employed in the experiments. List scheduling is a greedy algorithm, which tries to reduce the finish time of each node to be scheduled. Thus, it does not consider the leaving communications of a node, which may impede an early start of following nodes. The high importance of communication under the involvement–contention model seems to demand research of more sophisticated algorithms in order to exploit the full potential of this new model.

8.5   CONCLUDING REMARKS

This chapter investigated processor involvement in communication and its integration into task scheduling. The motivation originated from the unsatisfying accuracy results obtained under the classic and the contention model as referenced in Section 7.5.4, which indicated a shortcoming of these scheduling models.

Thus, the chapter began by thoroughly analyzing the various types of processor participation in communication and their characteristics. Based on this analysis, another scheduling model was introduced that abandons the idealizing assumption of a dedicated communication subsystem and instead integrates the modeling of all types of processor involvement in communication. This model, referred to as the involvement–contention model, is a generalization of the contention model analyzed in Chapter 7. It thereby adopts all properties of the contention model, for example, the contention awareness and the ability to reflect heterogeneous systems.

Scheduling under the new model requires some modifications of the scheduling techniques, since edges are now also scheduled on the processors to represent the processors' involvement in communication. This chapter discussed the general issue and its possible solutions. The necessary alterations to list scheduling were investigated in order to employ it, in all its forms, under the involvement–contention model. Another strategy for scheduling under the new model is based on the initial determination of the processor allocation.

It will depend on the target system and its communication mechanisms whether the benefits of the involvement–contention model outweigh its higher conceptual complexity and additional effort. The experimental results referenced in Section 8.4.3 indicate that for various parallel systems the accuracy for schedules produced under the involvement–contention model fundamentally improved. This allows more than just the reduction of execution times: it also allows one to evaluate algorithms and their schedules without execution of the schedules.

8.6 EXERCISES

8.1 Visit the online *Overview of Recent Supercomputers* by van der Steen and Dongarra [193] on the site of the TOP500 Supercomputer Sites [186]. Find examples for systems that use different types of processor involvement in communication. (*Hint*: The information provided in Ref. [186] might not be sufficient and you will need to aquire more information from the manufacturer's site.)

8.2 Describe in your own words what the essential challenges are of task scheduling when processor involvement is considered.

8.3 Schedule the following task graph

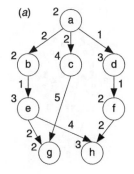

under the involvement–contention model on four processors connected via a central switch, employing half duplex links:

The nodes are preallocated to the processors as specified in the following table.

Node	a	b	c	d	e	f	g	h
Processor	1	3	2	4	1	4	1	4

Schedule the nodes in alphabetic order using the end technique. For the scheduling of the edges on the processors and links use the first alternative (Section 8.3.2), that is, edge e_{ij} is scheduled on P_{src} with n_i, on R, P_{dst} with n_j. The overhead is identical on the sending and receiving sides, namely, $o_{s,r}(e, P) = 0.5$, and the processors on both sides are involved 50% in communication, that is, $i_{s,r}(e, L) = 0.50\varsigma(e, L)$.

8.4 Perform Exercise 8.3 with the following modifications:

    (a) The processor involvement is one-sided (sending side), with 75% involvement, that is, $i_s(e, L) = 0.75\varsigma(e, L)$, $i_r(e, L) = 0$, and overhead $o_s(e, P) = 0.5$, $o_r(e, P) = 0$.

    (b) The processor involvement is one-sided (receiving side), with 75% involvement, that is, $i_s(e, L) = 0$, $i_r(e, L) = 0.75\varsigma(e, L)$, and $o_s(e, P) = 0$, $o_r(e, P) = 0.5$.

8.5 Perform Exercise 8.3 with the third alternative (Section 8.3.2) for the scheduling of the edges on the processors and links, that is, edge e_{ij} is scheduled on P_{src}, R, P_{dst} with n_i. Remember that this implies that the insertion technique must be used for the scheduling of the nodes and edges.

8.6 Use list scheduling with start time minimization and the end technique to schedule the following task graph

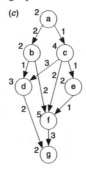

under the involvement–contention model on three processors connected via a central switch, employing half duplex links:

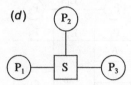

(d)

The overhead is negligible, hence $o_{s,r}(e, P) = 0$. The processors on both sides are 100% involved in communication, that is, $i_{s,r}(e, L) = \varsigma(e, L)$.

8.7 Perform Exercise 8.6 with the following modifications:

(a) The processor involvement is one-sided (sending side), with 50% involvement, that is, $i_s(e, L) = 0.50\varsigma(e, L)$, $i_r(e, L) = 0$.

(b) The processor involvement is one-sided (receiving side), with 50% involvement, that is, $i_s(e, L) = 0$, $i_r(e, L) = 0.50\varsigma(e, L)$.

(c) Why is one-sided involvement on the receiving side different from other involvement types in what relates to the scheduling approach?

8.8 In Section 8.2.3 the length of a path in a task graph was analyzed under the light of the overhead of interprocessor communication. Determine the bottom levels of the nodes of Exercise 8.6's task graph, assuming an overhead of $o_{s,r}(e, P) = 2$.

Is the decreasing bottom-level order of the nodes different without the consideration of the overhead?

8.9 To generate code from a given task graph and schedule assume that a node A is translated into `function(A)`. Remote communication is performed with the two nonblocking directives `send(to_node, p_dst)` and `receive(from_nodes, p_src)`. The parameters `from_node` and `to_node` stand for the origin and destination nodes, respectively, while `p_src` and `p_dst` stand for the source and destination processors, respectively. Local communication is performed through main memory and does not require explicit directives.

Using these simple statements, generate code for the following task graphs and schedules:

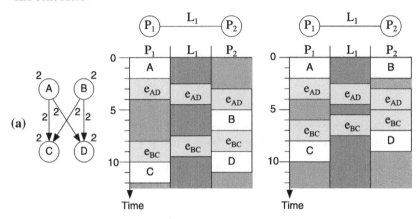

(a)

(b) The schedule obtained in Exercise 8.3.

How does this differ from code generation for schedules produced under the classic mode?

8.10 Research: The experimental results referenced in Section 8.4.3 suggest that list scheduling might not be the most adequate heuristic for scheduling under the involvement–contention model. As discussed in Section 8.4.2, an alternative is to use a two-phase scheduling heuristic, where the processor allocation is determined in the first phase. It was argued that using clustering in the first phase has the potential to produce good results, as it tries to reduce the cost of communication.

Implement a two-phase scheduling algorithm under the involvement–contention model, where clustering is used in the first phase to obtain a good processor allocation. Experimentally compare this algorithm to list scheduling.

■■■■■ BIBLIOGRAPHY

1. W. B. Ackerman. Data flow languages, February 1982.

2. T. L. Adam, K. M. Chandy, and J. R. Dickson. A comparison of list schedules for parallel processing systems. *Communications of the ACM*, 17:685–689, 1974.

3. I. Ahmad and M. K. Dhodhi. Multiprocessor scheduling in a genetic paradigm. *Parallel Computing*, 22:395–406, 1996.

4. I. Ahmad and Y.-K. Kwok. On exploiting task duplication in parallel program scheduling. *IEEE Transactions on Parallel and Distributed Systems*, 9(8):872–892, September 1998.

5. I. Ahmad, Y.-K. Kwok, and M.-Y. Wu. Performance comparison of algorithms for static scheduling of DAGs to multiprocessors. In *Proceedings of the 7th IEEE Symposium on Parallel and Distributed Processing (SPDP'95)*, pages 185–192, Fremantle, Western Australia, September 1995.

6. I. Ahmad, Y.-K. Kwok, and M.-Y. Wu. Analysis, evaluation, and comparison of algorithms for scheduling task graphs on parallel processors. In G.-J. Li, D. F. Hsu, S. Horiguchi, and B. Maggs, editors, *Second International Symposium on Parallel Architectures, Algorithms, and Networks, 1996*, pages 207–213, June 1996.

7. I. Ahmad, Y.-K. Kwok, M.-Y. Wu, and W. Shu. Automatic parallelization and scheduling of programs on multiprocessors using CASCH. In *Proceedings of the 1997 International Conference on Parallel Processing (ICPP'97)*, pages 288–291, Bloomingdale, Illinois, USA, August 1997.

8. A. Aiken and A. Nicolau. Perfect pipelining: a new loop parallelization technique. In *Proceedings of 1988 European Symposium on Programming*, pages 221–235, 1988.

9. A. Alexandrov, M. Ionescu, K. E. Schauser, and C. Scheimann. LogGP: incorporating long messages into the LogP-model—one step closer towards a realistic model for parallel computation. In *7th Annual Symposium on Parallel Algorithms and Architectures*, pages 95–105. ACM Press, 1995.

10. H. H. Ali and H. El-Rewini. The complexity of scheduling interval orders with communication is polynomial. *Parallel Processing Letters*, 3(1):53–58, 1993.

11. R. Allen and K. Kennedy. Conversion of control dependence to data dependence. In *Proceedings of the 10th ACM Symposium on Principles of Programming Languages*, January 1983.

12. R. Allen and K. Kennedy. *Optimizing Compilers for Modern Architectures*. Morgan Kaufmann/Academic Press, 2002.

13. P. Banerjee, E. W. Hodges IV, D. J. Palermo, J. A. Chandy, J. G. Holm, S. Ramaswamy, M. Gupta, A. Lain, and E. Su. An overview of the paradigm compiler for distributed-memory multicomputers. *IEEE Computer*, 28(10):37–47, October 1995.

14. U. Banerjee. *Speedup of Ordinary Programs*. PhD thesis, University of Illinois at Urbana-Champaign, Department of Computer Science, October 1979.

15. U. Banerjee. *Dependence Analysis for Supercomputing*. Kluwer Academic Publishers, 1988.

16. U. Banerjee. An introduction to a formal theory of dependence analysis. *The Journal of Supercomputing*, 2(2):133–149, 1988.

17. U. Banerjee, R. Eigenmann, A. Nicolau, and D. A. Padua. Automatic program parallelization. *Proceedings of the IEEE*, 81(2):211–243, February 1993.

18. O. Beaumont, V. Boudet, and Y. Robert. The iso-level scheduling heuristic for heterogeneous processors. In *PDP'2002, 10th Euromicro Workshop on Parallel, Distributed and Network-based Processing*. IEEE Computer Society Press, 2002.

19. O. Beaumont, V. Boudet, and Y. Robert. A realistic model and an efficient heuristic for scheduling with heterogeneous processors. In *HCW'2002, the 11th Heterogeneous Computing Workshop*. IEEE Computer Society Press, 2002.

20. M. S. T. Benten and S. M. Sait. Genetic scheduling of task graphs. *International Journal of Electronics*, 77(4):401–415, 1994.

21. C. Berge. *Graphs and Hypergraphs*, 2nd edition, North-Holland, 1976.

22. K. A. Berman and J. L. Paul. *Algorithms: Sequential, Parallel, and Distributed*. Thomson/Course Technology, 2005.

23. W. Blume and R. Eigenmann. The range test: a dependence test for symbolic, non-linear expressions. In *Proceedings Supercomputing '94*, pages 528–537, November 1994.

24. W. Blume, R. Eigenmann, J. Hoeflinger, D. A. Padua, P. M. Petersen, L. Rauchwerger, and P. Tu. *Automatic Detection of Parallelism: A Grand Challenge for High-Performance Computing*. Technical Report TR1348, Center for Supercomputing Research and Development, University of Illinois at Urbana-Champaign, 1994.

25. C. Boeres and V. E. F. Rebello. A versatile cost modelling approach for multicomputer task scheduling. *Parallel Computing*, 25(1):63–86, January 1999.

26. S. Booth, J. Fisher, and M. Bowers. *Introduction to the Cray T3E at EPCC*. Edinburgh Parallel Computing Centre, Scotland, UK, June 1999. http://www.epcc.ed.ac.uk/t3d/documents/t3e-intro.html.

27. P. Brucker. *Scheduling Algorithms*, 4th edition, Springer-Verlag, 2004.

28. P. Brucker, J. Hurink, and W. Kubiak. Scheduling identical jobs with chain precedence constraints on two uniform machines. *Mathematical Methods Operational Research*, 49(2):211–219, 1999.

29. P. Brucker and S. Knust. *Complexity Results for Scheduling Problems*. Technical Report, Mathematics Institute, University of Osnabrück, Germany, 2006-. http://www.mathematik.uni-osnabrueck.de/research/OR/class/.

30. P. Chrétienne. A polynomial algorithm to optimally schedule tasks over a virtual distributed system under tree-like precedence constraints. *European Journal of Operational Research*, 43:225–230, 1989.

31. P. Chrétienne. Task scheduling over distributed memory machines. In *Proceedings of the International Workshop on Parallel and Distributed Algorithms*. North-Holland, Amsterdam, 1989.

32. P. Chrétienne. Task scheduling with interprocessor communication delays. *European Journal of Operational Research*, 57:348–354, 1992.

33. P. Chrétienne. Tree scheduling with communication delays. *Discrete Applied Mathematics*, 49(1–3):129–141, 1994.

34. P. Chrétienne, E. G. Coffman, J. K. Lenstra, and Z. Liu, editors. *Scheduling Theory and Its Applications*. Wiley, 1995.

35. P. Chrétienne and C. Picouleau. Scheduling with communication delays: a survey. In P. Chrétienne, E. G. Coffman, J. K. Lenstra, and Z. Liu, editors, *Scheduling Theory and Its Applications*, pages 65–90. Wiley, 1995.

36. B. Cirou and E. Jeannot. Triplet: a clustering scheduling algorithm for heterogeneous systems. In *Proceedings of Workshop on Scheduling and Resource Management for Cluster Computing (ICPP 2001)*, pages 231–236, Valencia, Spain, September 2001. IEEE Press.

37. E. G. Coffman, editor. *Computer and Job-Scheduling Theory*. Wiley, 1976.

38. E. G. Coffman and R. L. Graham. Optimal scheduling for two-processor systems. *Acta Informatica*, 1:200–213, 1972.

39. M. Coli and P. Palazzari. Global execution time minimization by allocating tasks in parallel systems. In *Euromicro Workshop on Parallel and Distributed Processing, 1995*, pages 91–97, January 1995.

40. J.-Y. Colin and P. Chrétienne. CPM scheduling with small interprocessor communication delays and task duplication. *Operations Research*, 39(3):680–684, 1991.

41. S. A. Cook. The complexity of theorem proving procedures. In *Proceedings of 3rd ACM Symposium on Theory of Computing*, pages 151–158, 1971.

42. T. H. Cormen, C. E. Leiserson, R. L. Rivest, and C. Stein. *Introduction to Algorithms*, 2nd edition, MIT Press, 2001.

43. R. C. Correa, A. Ferreira, and P. Rebreyend. Integrating list heuristics into genetic algorithms for multiprocessor scheduling. In *IEEE Symposium on Parallel and Distributed Processing 1996*, pages 462–469, 1996.

44. R. C. Correa, A. Ferreira, and P. Rebreyend. Scheduling multiprocessor tasks with genetic algorithms. *IEEE Transactions on Parallel and Distributed Systems*, 10(8):825–837, August 1999.

45. M. Cosnard and D. Trystram. *Parallel Algorithms and Architectures*. Thomson Computer Press, London, UK, 1995.

46. D. E. Culler, R. M. Karp, D. A. Patterson, A. Sahay, E. E. Santos, K. E. Schauser, R. Subramonian, and T. von Eicken. LogP: a practical model of parallel computation. *Communications of the ACM*, 39(11):78–85, November 1996.

47. D. E. Culler, R. M. Karp, D. A. Patterson, A. Sahay, K. E. Schauser, E. E. Santos, R. Subramonian, and T. von Eicken. LogP: towards a realistic model of parallel computation. *ACM SIGPLAN Notices, Proceedings of the Symposium on Principles and Practice of Parallel Programming*, 28(7):1–12, July 1993.

48. D. E. Culler and J. P. Singh. *Parallel Computer Architecture*. Morgan Kaufmann Publishers, 1999.

49. S. Darbha and D. P. Agrawal. Scalable scheduling algorithm for distributed memory machines. In *Eighth IEEE Symposium on Parallel and Distributed Processing, 1996*, pages 84–91, October 1996.

50. S. Darbha and D. P. Agrawal. Optimal scheduling algorithm for distributed-memory machines. *IEEE Transactions on Parallel and Distributed Systems*, 9(1):87–95, January 1998.

51. S. Darbha and D. P. Agrawal. SDBS: a task duplication based optimal scheduling algorithm. In *Proceedings of the Scalable High-Performance Computing Conference, 1994*, pages 756–763, May 1994.

52. A. Darte, Y. Robert, and F. Vivien. *Scheduling and Automatic Parallelization*. Brinkhäuser, Boston, USA, 2000.

53. A. L. Davis and R. M. Keller. Data flow program graphs. *IEEE Computer*, 15:26–41, February 1982.

54. L. Davis. *Handbook of Genetic Algorithms*. Van Nostrand-Reinhold, New York, USA, 1991.

55. M. K. Dhodhi, I. Ahmad, A. Yatama, and I. Ahmad. An integrated technique for task matching and scheduling onto distributed heterogeneous computing systems. *Journal of Parallel and Distributed Computing*, 62(9):1338–1361, September 2002.

56. J. Du, J. Y.-T. Leung, and G. H. Young. Scheduling chain-structured tasks to minimize makespan and mean flow time. *Information and Computation*, 92(2):219–236, 1991.

57. P.-F. Dutot, O. Sinnen, and L. Sousa. *A Note on the Complexity of Task Scheduling with Communication Contention*. Technical Report, University of Auckland, New Zealand, February 2005.

58. C. Eisenbeis and J.-C. Sogno. A general algorithm for data dependence analysis. In *International Conference on Supercomputing, Washington DC, USA*, pages 292–302. ACM Press, August 1992.

59. J. Eisenbiegler, W. Löwe, and A. Wehrenpfennig. On the optimization by redundancy using an extended LogP model. In *International Conference on Advances in Parallel and Distributed Computing (APDC'97)*, pages 149–155. IEEE Computer Society Press, 1997.

60. H. El-Rewini and M. Abd-El-Barr. *Advanced Computer Architecture and Parallel Processing*. Wiley, 2005.

61. H. El-Rewini and H. H. Ali. On considering communication in scheduling task graphs on parallel processors. *Journal of Parallel Algorithms and Applications*, 3:177–191, 1994.

62. H. El-Rewini and H. H. Ali. Static scheduling of conditional branching in parallel programs. *Journal of Parallel and Distributed Computing*, 24(1):41–54, 1995.

63. H. El-Rewini and T. G. Lewis. Scheduling parallel program tasks onto arbitrary target machines. *Journal of Parallel and Distributed Computing*, 9(2):138–153, June 1990.

64. H. El-Rewini and T. G. Lewis. *Distributed and Parallel Computing*. Manning, 1998.

65. H. El-Rewini, T. G. Lewis, and H. H. Ali. *Task Scheduling in Parallel and Distributed Systems*. Prentice Hall, 1994.

66. B. Falsafi and D. A. Wood. Scheduling communication on an SMP node parallel machine. In *Proceedings of IEEE International Symposium on High Performance Computer Architecture*, pages 128–138, 1997.

67. M. J. Flynn. Very high-speed computing systems. *Proceedings of the IEEE*, 54:1901–1909, 1966.

68. S. Fortune and J. Wyllie. Parallelism in random access machines. In *Proceedings of the 10th Annual ACM Symposium on Theory of Computing*, pages 114–118, May 1978.

69. I. Foster. *Designing and Building Parallel Programs*. Addison-Wesley, 1995.

70. M. Fujii, T. Kasami, and K. Ninomiya. Optimal sequencing of two equivalent processors. *SIAM Journal of Applied Mathematics*, 17(3):784–789, 1969.

71. H. Gabow. An almost linear algorithm for two-processor scheduling. *Journal of the ACM*, 29(3):766–780, 1982.

72. M. R. Garey and D. S. Johnson. "Strong" NP-completeness results: motivation, examples and implications. *Journal of the ACM*, 25(3):499–508, 1978.

73. M. R. Garey and D. S. Johnson. *Computers and Intractability: A Guide to the Theory of NP-Completeness*. Freeman, 1979.

74. M. R. Garey, D. S. Johnson, R. Tarjan, and M. Yannakakis. Scheduling opposing forests. *SIAM Journal of Algebraic and Discrete Methods*, 4(1):72–93, 1983.

75. A. Gerasoulis, J. Jiao, and T. Yang. A multistage approach for scheduling task graphs on parallel machines. In *Workshop on Parallel Processing of Discrete Optimization Problems*, pages 81–103. American Mathematical Society, 1994.

76. A. Gerasoulis and T. Yang. A comparison of clustering heuristics for scheduling DAGs on multiprocessors. *Journal of Parallel and Distributed Computing*, 16(4):276–291, December 1992.

77. A. Gerasoulis and T. Yang. On the granularity and clustering of directed acyclic task graphs. *IEEE Transactions on Parallel and Distributed Systems*, 4(6):686–701, June 1993.

78. D. E. Goldberg. *Genetic Algorithms in Search, Optimization, and Machine Learning*. Addison-Wesley, 1989.

79. D. K. Goyal. *Scheduling Processor Bound Systems*. Technical Report CS-7-036, Computer Science Department, Washington State University, Pullman, 1996.

80. R. L. Graham. Bounds for multiprocessing timing anomalies. *SIAM Journal of Applied Mathematics*, 17(2):416–419, 1969.

81. R. L. Graham, E. L. Lawler, J. K. Lenstra, and A. H. G. Rinnooy Kan. Optimization and approximation in deterministic sequencing and scheduling: a survey. *Annals of Discrete Mathematics*, 5:287–326, 1979.

82. A. Grama, A. Gupta, G. Karypis, and V. Kumar. *Introduction to Parallel Computing*, 2nd edition, Pearson Addison Wesley, London, UK, 2003.

83. T. Hagras and J. Janeček. A high performance, low complexity algorithm for compile-time task scheduling in heterogeneous systems. *Parallel Computing*, 31(7):653–670, 2005.

84. C. Hamacher, Z. Vranesic, and S. Zaky. *Computer Organization*, 5th edition, McGraw-Hill, 2002.

85. C. Hanen and A. Munier Kordon. Minimizing the volume in scheduling an out-tree with communication delays and duplication. *Parallel Computing*, 28(11):1573–1585, November 2002.

86. C. Hanen and A. Munier. An approximation algorithm for scheduling dependent tasks on m processsors with small communication delays. In *ETFA 95: INRIA/IEEE Symposium on Emerging Technology and Factory Animation*, pages 167–189. IEEE Press, 1995.

87. J. L. Hennessy and D. A. Patterson. *Computer Organization and Design: The Hardware/Software Interface*, 2nd edition. Morgan Kaufmann Publishers, San Francisco, USA, 1998.

88. J. L. Hennessy and D. A. Patterson. *Computer Architecture, A Quantitative Approach*, 3rd edition. Morgan Kaufmann Publishers, San Francisco, USA, 2003.

89. J. H. Holland. *Adaptation in Natural and Artificial Systems*. University of Michigan Press, Ann Arbor, USA, 1975.

90. J. A. Hoogeveen, J. K. Lenstra, and B. Veltman. *Three, Four, Five, Six or the Complexity of Scheduling with Communication Delays*. Technical Report BS-R9229, ISSN 0924-0659, Centre for Mathematics and Computer Science, The Netherlands, October 1992.

91. J. A. Hoogeveen, S. L. van de Velde, and B. Veltman. Complexity of scheduling multiprocessor tasks with prespecified processor allocations. *Discrete Applied Mathematics*, 55(3):259–272, 1994.

92. E. S. H. Hou, N. Ansari, and H. Ren. Genetic algorithm for multiprocessor scheduling. *IEEE Transactions on Parallel and Distributed Systems*, 5(2):113–120, February 1994.

93. T. Hu. Parallel sequencing and assembly line problems. *Operations Research*, 9(6):841–848, 1961.

94. J. J. Hwang, Y. C. Chow, F. D. Anger, and C. Y. Lee. Scheduling precedence graphs in systems with interprocessor communication times. *SIAM Journal of Computing*, 18(2):244–257, April 1989.

95. K. Hwang and F. A. Briggs. *Computer Architecture and Parallel Processing*. McGraw-Hill, London, UK, 1984.

96. IBM. *SP Switch2 Technology and Architecture*, March 2001. http://www-1.ibm.com/servers/eserver/pseries/hardware/whitepapers/sp_switch2.pdf.

97. H. Jung, L. M. Kirousis, and P. Spirakis. Lower bounds and efficient algorithms for multiprocessor scheduling of directed acyclic graphs with communication delays. *Information and Computation*, 105(1):94–104, 1993.

98. T. Kalinowski, I. Kort, and D. Trystram. List scheduling of general task graphs under LogP. *Parallel Computing*, 26:1109–1128, 2000.

99. R. M. Karp. Reducibility among combinatorial problems. In R. E. Miller and J. W. Thatcher, editors, *Complexity of Computer Computation*, pages 85–104. Plenum Press, 1972.

100. R. M. Karp and R. E. Miller. Properties of a model for parallel computations: determinacy, termination, queueing. *SIAM Journal of Applied Mathematics*, 14(6):1390–1411, November 1966.

101. R. M. Karp, R. E. Miller, and S. Winogard. The organization of computations for uniform recurrence equations. *Journal of the ACM*, 14(3):563–590, July 1967.

102. H. Kasahara and S. Narita. Practical multiprocessor scheduling algorithms for efficient parallel processing. *IEEE Transactions on Computers*, C-33:1023–1029, November 1984.

103. A. A. Khan, C. L. McCreary, and M. S. Jones. A comparison of multiprocessor scheduling heuristics. In *Proceedings of International Conference on Parallel Processing*, Volume 2, pages 243–250, August 1994.

104. S. J. Kim and J. C. Browne. A general approach to mapping of parallel computation upon multiprocessor architectures. In *International Conference on Parallel Processing*, Volume 3, pages 1–8, 1988.

105. B. Kruatrachue. *Static Task Scheduling and Grain Packing in Parallel Processing Systems*. PhD thesis, Oregon State University, USA, 1987.

106. B. Kruatrachue and T. G. Lewis. Grain size determination for parallel processing. *IEEE Software*, 5(1):23–32, January 1988.

107. W. Kubiak. Exact and approximate algorithms for scheduling unit time tasks with tree-like precedence constraints. In *Abstracts EURO IX–TIMS XXVIII*, Paris, France, 1988.

108. V. Kumar, A. Grama, A. Gupta, and G. Karypis. *Introduction to Parallel Computing— Design and Analysis of Algorithms*. Benjamin/Cummings, 1994.

109. S. Y. Kung. *VLSI Array Processors*. Information and System Sciences Series. Prentice Hall, 1988.

110. Y.-K. Kwok and I. Ahmad. Efficient scheduling of arbitrary task graphs to multi-processors using a parallel genetic algorithm. *Journal of Parallel and Distributed Computing*, 47(1):58–77, November 1997.

111. Y.-K. Kwok and I. Ahmad. Benchmarking the task graph scheduling algorithms. In *Proceedings of International Parallel Processing Symposium/Symposium on Parallel and Distributed Processing (IPPS/SPDP-98)*, pages 531–537, Orlando, Florida, USA, April 1998.

112. Y.-K. Kwok and I. Ahmad. A comparison of parallel search-based algorithms for multi-processors scheduling. In *Proceedings of the 2nd European Conference on Parallel and Distributed Systems (EURO-PDS'98)*, Vienna, Austria, July 1998.

113. Y.-K. Kwok and I. Ahmad. Static scheduling algorithms for allocating directed task graphs to multiprocessors. *ACM Computing Surveys*, 31(4):406–471, December 1999.

114. Y.-K. Kwok and I. Ahmad. Link contention-constrained scheduling and mapping of tasks and messages to a network of heterogeneous processors. *Cluster Computing*, 3(2):113–124, 2000.

115. Y.-K. Kwok and I. Ahmad. On multiprocessor task scheduling using efficient state space approaches. *Journal of Parallel and Distributed Computing*, 65:1515–1532, 2005.

116. L. Lamport. Time, clocks, and the ordering of events in a distributed system. *Communications of the ACM*, 21(7):558–565, July 1978.

117. C. Y. Lee, J. J. Hwang, Y. C. Chow, and F. D. Anger. Multiprocessor scheduling with interprocessor communication delays. *Operations Research Letters*, 7(3):141–147, 1988.

118. F. T. Leighton. *Introduction to Parallel Algorithms and Architectures: Arrays, Trees, Hypercubes*. Morgan Kaufmann Publishers, 1992.

119. J. K. Lenstra, A. H. G. Rinnooy Kan, and P. Brucker. Complexity of machine scheduling problems. *Annals of Discrete Mathematics*, 1:343–362, 1977.

120. J. K. Lenstra, M. Veldhorst, and B. Veltman. The complexity of scheduling trees with communication delays. *Journal of Algorithms*, 20(1):157–173, 1996.

121. J. Y.-T. Leung, editor. *Handbook of Scheduling*. Chapman and Hall/CRC, 2004.

122. T. G. Lewis. *Foundations of Parallel Programming, A Machine-Independent Approach*. IEEE Press, 1993.

123. T. G. Lewis and H. El-Rewini. *Introduction to Parallel Computing*. Prentice Hall, 1992.

124. T. G. Lewis and H. El-Rewini. Parallax: a tool for parallel program scheduling. *IEEE Parallel and Distributed Technology: Systems and Applications*, 1(2):62–72, May 1993.

125. J.-C. Liou and M. A. Palis. A new heuristic for scheduling parallel programs on multi-processor. In *1998 International Conference on Parallel Architectures and Compilation Techniques*, pages 358–365, October 1998.

126. G. Q. Liu, K. L. Poh, and M. Xie. Iterative list scheduling for heterogenous computing. *Journal of Parallel and Distributed Computing*, 65(5):654–665, May 2005.

127. J. W. S. Liu. *Real-Time Systems*. Prentice Hall, 2000.

128. Z. Liu. A note on Graham's bound. *Information Processing Letters*, 36:1–5, October 1990.

129. V. M. Lo. Temporal communication graphs: Lamport's process-time graphs augmented for the purpose of mapping and scheduling. *Journal of Parallel and Distributed Computing*, 16(4): 378–384, December 1992.

130. V. M. Lo, S. Rajopadhye, S. Gupta, D. Keldsen, M. A. Mohamed, B. Nitzberg, J. A. Telle, and X. Zhong. OREGAMI: tools for mapping parallel computations to parallel architectures. *International Journal of Parallel Programming*, 20(3):237–270, June 1991.

131. W. Löwe, W. Zimmermann, and J. Eisenbiegler. On linear schedules of task graphs for generalized LogP-machines. In *Euro-Par '97*, Volume 1300 of *Lecture Notes in Computer Science*, pages 895–904. Springer, 1997.

132. B. S. Macey and A. Y. Zomaya. A performance evaluation of CP list scheduling heuristics for communication intensive task graphs. In *Parallel Processing Symposium, 1998. Proceedings of IPPS/SPDP 1998*, pages 538–541, 1998.

133. K. F. Man, K. S. Tang, and S. Kwong. *Genetic Algorithms: Concepts and Designs*. Springer Verlag, 1999.

134. S. Manoharan. Effect of task duplication on the assignment of dependency graphs. *Parallel Computing*, 27(3):257–268, February 2001.

135. C. L. McCreary and H. Gill. Automatic determination of grain size for efficient parallel processing. *Communications of the ACM*, 32(9):1073–1078, September 1989.

136. C. L. McCreary, A. A. Khan, J. J. Thompson, and M. E. McArdle. A comparison of heuristics for scheduling DAGs on multiprocessors. In *Eighth International Parallel Processing Symposium, 1994*, pages 446–451, April 1994.

137. D. A. Menascé, D. Saha, S. C. S. S. Porto, V. A. F. Almeida, and S. K. Tripathi. Static and dynamic processor scheduling disciplines in heterogeneous parallel architectures. *Journal of Parallel and Distributed Computing*, 28(1):1–18, July 1995.

138. Message Passing Interface Forum. *MPI:A Message-Passing Interface Standard*, June 1995. http://www.mpi-forum.org/docs/docs.html.

139. H. Oh and S. Ha. A static scheduling heuristic for heterogeneous processors. In *Proceedings of Europar'96*, Volume 1124 of *Lecture Notes in Computer Science*. Springer-Verlag, 1996.

140. M. A. Palis, J.-C. Liou, and D. S. L. Wei. Task clustering and scheduling for distributed memory parallel architectures. *IEEE Transactions on Parallel and Distributed Systems*, 7(1):46–55, January 1996.

141. C. H. Papadimitriou and M. Yannakakis. Scheduling interval ordered tasks. *SIAM Journal of Computing*, 8:405–409, 1979.

142. C. H. Papadimitriou and M. Yannakakis. Towards an architecture-independent analysis of parallel algorithms. *SIAM Journal of Computing*, 19(2):322–328, April 1990.

143. B. Parhami. *Introduction to Parallel Processing: Algorithms and Architectures*. Plenum Press, 1999.

144. K. K. Parhi. Algorithm transformation techniques for concurrent processors. *Proceedings of the IEEE*, 77(12):1879–1895, December 1989.

145. K. K. Parhi. *VLSI Digital Signal Processing*. Wiley, 1999.

146. K. K. Parhi and D. G. Messerschmitt. Static rate-optimal scheduling of iterative data-flow programs via optimum unfolding. *IEEE Transactions on Computers*, 40(2):178–195, December 1991.

147. D. A. Patterson. A case for NOW (networks of workstations). In *Proceedings of the 14th Annual ACM Symposium on Principles of Distributed Computing (PODC '95)*, pages 17–19, New York, August 1995. ACM.

148. P. M. Petersen and D. A. Padua. Static and dynamic evaluation of data dependence analysis techniques. *IEEE Transactions on Parallel and Distributed Systems*, 7(11):1121–1132, November 1996.

149. C. Picouleau. *Two New NP-Complete Scheduling Problems with Communication Delays and Unlimited Number of Processors*. Technical Report 91-94, IBP, Université Pierre et Marie Curie, France, April 1991.

150. C. Picouleau. New complexity results on scheduling with small communication delays. *Discrete Applied Mathematics*, 60(1–3):331–342, 1995.

151. M. Pinedo. *Scheduling: Theory, Algorithms, and Systems*. Prentice Hall, 2002.

152. C. D. Polychronopoulos. *Parallel Programming and Compilers*. Kluwer Academic Publishers, 1988.

153. C. D. Polychronopoulos, M. Girkar, M. Reza Haghighat, C.-L. Lee, B. Leung, and D. Schouten. Parafrase-2: a new generation parallelizing compiler. *International Journal of High Speed Computing*, 1(1):45–72, May 1989.

154. J. Protić, M. Tomašević, and V. Milutinvć. Distributed shared memory: concepts and systems. *IEEE Transactions on Parallel and Distributed Technology*, pages 63–79, 1996.

155. W. Pugh. The Omega test: a fast and practical integer programming algorithm for dependence analysis. *Communications of the ACM*, 8:102–114, August 1992.

156. M. J. Quinn. *Parallel Programming in C with MPI and OpenMP*. McGraw-Hill, 2004.

157. A. Radulescu and A. J. C. van Gemund. Low-cost task scheduling for distributed-memory machines. *IEEE Transactions on Parallel and Distributed Systems*, 13(6): 648–658, 2002.

158. V. J. Rayward-Smith. UET scheduling with unit interprocessor communication delays. *Discrete Applied Mathematics*, 18:55–71, 1987.

159. V. J. Rayward-Smith, F. W. Burton, and G. J. Janacek. Scheduling parallel programs assuming preallocation. In P. Chrétienne, E. G. Coffman, J. K. Lenstra, and Z. Liu, editors, *Scheduling Theory and Its Applications*, pages 145–165. Wiley, 1995.

160. P. Rebreyend, F. E. Sandnes, and G. M. Megson. *Static Multiprocessor Task Graph Scheduling in the Genetic Paradigm: A Comparison of Genotype Representations*. Research Report 98-25, Ecole Normale Superieure de Lyon, Laboratoire de Informatique du Parallelisme, Lyon, France, 1998.

161. C. R. Reeves and J. E. Rowe. *Genetic Algorithms: Principles and Perspectives: A Guide to GA Theory*. Kluwer Academic Publishers, 2003.

162. R. Reiter. Scheduling parallel computations. *Journal of the ACM*, 15(4):590–599, October 1968.

163. P. Sadayappan, F. Ercal, and J. Ramanujam. Cluster partitioning approaches to mapping parallel programs onto a hypercube. *Parallel Computing*, 13:1–16, 1990.

164. F. E. Sandnes and G. M. Megson. Improved static multiprocessor scheduling using cyclic task graphs: a genetic approach. *Proceedings of the International Conference on Parallel*

Computing: Fundamentals, Applications and New Directions (Parco'97), pages 703–710, Bonn, Germany, 1997.

165. F. E. Sandnes and G. M. Megson. An evolutionary approach to static taskgraph scheduling with task duplication for minimised interprocessor traffic. In *Proceedings of the International Conference on Parallel and Distributed Computing, Applications and Technologies (PDCAT 2001)*, pages 101–108, Taipei, Taiwan, July 2001. Tamkang University Press.

166. F. E. Sandnes and O. Sinnen. A new strategy for multiprocessor scheduling of cyclic task graphs. *International Journal of High Performance Computing and Networking*, 3(1):62–71, 2005.

167. V. Sarkar. *Partitionning and Scheduling Parallel Programs for Execution on Multiprocessors*. MIT Press, 1989.

168. R. Sethi. Scheduling graphs on two processors. *SIAM Journal of Computing*, 5(1):73–82, 1976.

169. G. C. Sih and E. A. Lee. A compile-time scheduling heuristic for interconnection-constrained heterogeneous processor architectures. *IEEE Transactions on Parallel and Distributed Systems*, 4(2):175–186, February 1993.

170. H. Singh and A. Youssef. Mapping and scheduling heterogeneous task graphs using genetic algorithms. In *Proceedings of the Heterogeneous Computing Workshop (HCW'96)*, pages 86–97, Honolulu, HI, April 1996. IEEE Computer Society.

171. O. Sinnen. *Experimental Evaluation of Task Scheduling Accuracy*. Tese de Mestrado (Master's thesis), Instituto Superior Técnico, Technical University of Lisbon, Portugal, December 2001.

172. O. Sinnen. *Accurate Task Scheduling for Parallel Systems*. PhD thesis, Instituto Superior Técnico, Technical University of Lisbon, Portugal, April 2003.

173. O. Sinnen and L. Sousa. *A Classification of Graph Theoretic Models for Parallel Computing*. Technical Report RT/005/99, INESC-ID, Instituto Superior Técnico, Technical University of Lisbon, Portugal, May 1999.

174. O. Sinnen and L. Sousa. A comparative analysis of graph models to develop parallelising tools. In *Proceedings of 8th IASTED International Conference on Applied Informatics (AI 2000)*, pages 832–838, Innsbruck, Austria, February 2000.

175. O. Sinnen and L. Sousa. A platform independent parallelising tool based on graph theoretic models. In *Vector and Parallel Processing—VECPAR 2000, Selected Papers*, Volume 1981 of *Lecture Notes in Computer Science*, pages 154–167. Springer-Verlag, 2001.

176. O. Sinnen and L. Sousa. Experimental evaluation of task scheduling accuracy: implications for the scheduling model. *IEICE Transactions on Information and Systems*, E86-D(9):1620–1627, September 2003.

177. O. Sinnen and L. Sousa. List scheduling: extension for contention awareness and evaluation of node priorities for heterogeneous cluster architectures. *Parallel Computing*, 30(1):81–101, January 2004.

178. O. Sinnen and L. Sousa. On task scheduling accuracy: evaluation methodology and results. *The Journal of Supercomputing*, 27(2):177–194, February 2004.

179. O. Sinnen and L. Sousa. Communication contention in task scheduling. *IEEE Transactions on Parallel and Distributed Systems*, 16(6):503–515, June 2005.

180. O. Sinnen, L. Sousa, and F. E. Sandnes. Toward a realistic task scheduling model. *IEEE Transactions on Parallel and Distributed Systems*, 17(3):263–275, 2006.

181. T. Sterling, D. Savarese, D. J. Becker, J. E. Dorband, U. A. Ranawake, and C. V. Packer. BEOWULF: a parallel workstation for scientific computation. In *International Conference on Parallel Processing, Volume 1: Architecture*, pages 11–14. Boca Raton, USA, August 1995. CRC Press.

182. H. S. Stone. Muliprocessor scheduling with the aid of network flow alogorithms. *IEEE Transactions on Software Engineering*, SE-3(1):85–93, January 1977.

183. A. Tam and C. L. Wang. Contention-aware communication schedule for high speed communication. *Cluster Computing*, 6(4):339–353, 2003.

184. S. Telford. *BOBCAT User Guide*. Edinburgh Parallel Computing Centre, Scotland, UK, May 2000. http://www.epcc.ed.ac.uk/sun/documents/introdoc.html.

185. S. Tongsima, E. H.-M. Sha, and N. L. Passos. Communication-sensitive loop scheduling for DSP applications. *IEEE Transactions on Signal Processing*, 45(5):1309–1322, May 1997.

186. The 500 most powerful computer systems. Web site, TOP500 Supercomputer Sites, http://www.top500.org/.

187. H. Topcuoglu, S. Hariri, and M.-Y. Wu. Task scheduling algorithms for heterogenous machines. In *Proceedings of Heterogeneous Computing Workshop*, pages 3–14, 1999.

188. H. Topcuoglu, S. Hariri, and M.-Y. Wu. Performance-effective and low complexity task scheduling for heterogeneous computing. *IEEE Transactions on Parallel and Distributed Systems*, 13(3):260–274, 2002.

189. R. A. Towle. *Control and Data Dependence for Program Transformations*. PhD thesis, University of Illinois, Urbana-Champaign, Department of Computer Science, March 1976.

190. E. V. Trichina and J. Oinonen. Parallel program design in visual environment. In *IEEE International Conference on High Performance Computing*, pages 198–203. Bangalore, India, December 1997.

191. T. Tsuchiya, T. Osada, and T. Kikuno. Genetic-based multiprocessor scheduling using task duplication. *Microprocessors and Microsystems*, 22:197–207, 1998.

192. J. D. Ullman. NP-complete scheduling problems. *Journal of Computing System Science*, 10:384–393, 1975.

193. A. J. van der Steen and J. J. Dongarra. *Overview of Recent Supercomputers*. Technical Report, TOP500 Supercomputer Sites, http://www.top500.org/ORSC/ 1996.

194. T. A. Varvarigou, V. P. Roychowdhury, and T. Kailath. Scheduling in and out forests in the presence of communication delays. In *Proceedings of 7th International Parallel Processing Symposium*, pages 222–229, 1993.

195. B. Veltman. *Multiprocessor Scheduling with Communication Delays*. PhD thesis, CWI, Amsterdam, The Netherlands, 1993.

196. B. Veltman, B. J. Lageweg, and J. K. Lenstra. Multiprocessor scheduling with communication delays. *Parallel Computing*, 16(2–3):173–182, 1990.

197. T. von Eicken, D. E. Culler, S. C. Goldstein, and K. E. Schauser. Active Messages: a mechanism for integrated communication and computation. In *Proceedings of the 19th Annual International Symposium on Computer Architecture*, pages 256–266. Gold Coast, Australia, May 1992.

198. L. Wang, H. J. Siegel, V. P. Roychowdhury, and A. A. Maciejewski. Task matching and scheduling in heterogeneous computing environments using a genetic-algorithm-based approach. *Journal of Parallel and Distributed Computing*, 47:8–22, November 1997.

199. J. White III and S. Bova. Where's the overlap? An analysis of popular MPI implementations. In *Proceedings of MPIDC 1999*, 1999.

200. B. Wilkinson and C. M. Allen. *Parallel Programming: Techniques and Applications Using Networked Workstations and Parallel Computers*, 2nd edition, Prentice Hall, 2005.

201. G. Wirtz. Developing parallel programs in a graph-based environment. In D. Trystram, editor, *Proceedings of Parallel Computing 93, Grenoble, France*, pages 345–352. Amsterdam, September 1993. Elsevier Science Publishing North Holland.

202. M. Wolfe. *Optimizing Supercompilers for Supercomputers*. MIT Press, 1989.

203. M. Wolfe. Data dependence and program restructuring. *The Journal of Supercomputing*, 4(4):321–344, January 1991.

204. M. Wolfe. *High Performance Compilers for Parallel Computing*. Addison-Wesley, 1996.

205. S.-H. Woo, S.-B. Yang, S.-D. Kim, and T.-D. Han. Task scheduling in distributed computing systems with a genetic algorithm. In *High Performance Computing on the Information Superhighway, 1997. HPC Asia '97*, pages 301–305, April 1997.

206. A. S. Wu, H. Yu, S. Jin, K. Lin, and G. Schiavone. An incremental genetic algorithm approach to multiprocessor scheduling. *IEEE Transactions on Parallel and Distributed Systems*, 15(9):824–834, September 2004.

207. M. Y. Wu and D. D. Gajski. Hypertool: a programming aid for message-passing systems. *IEEE Transactions on Parallel and Distributed Systems*, 1(3):330–343, July 1990.

208. T. Yang and C. Fu. Heuristic algorithms for scheduling iterative task computations on distributed memory machines. *IEEE Transactions on Parallel and Distributed Systems*, 8(6):608–622, June 1997.

209. T. Yang and A. Gerasoulis. PYRROS: static scheduling and code generation for message passing multiprocessors. In *Proceedings of 6th ACM International Conference on Supercomputing*, pages 428–437, Washington, DC, August 1992.

210. T. Yang and A. Gerasoulis. List scheduling with and without communication delays. *Parallel Computing*, 19(12):1321–1344, 1993.

211. T. Yang and A. Gerasoulis. DSC: scheduling parallel tasks on an unbounded number of processors. *IEEE Transactions on Parallel and Distributed Systems*, 5(9):951–967, September 1994.

212. A. Yazici and T. Terzioglu. A comparison of data dependence analysis tests. In M. Valero, E. Onate, M. Jane, J. L. Larriba, and B. Suarez, editors, *Parallel Computing and Transputer Applications*, pages 575–583. IOS Press, Amsterdam, 1992.

213. A. Y. Zomaya, C. Ward, and B. S. Macey. Genetic scheduling for parallel processor systems: comparative studies and performance issues. *IEEE Transactions on Parallel and Distributed Systems*, 10(8):795–812, August 1999.

■■■■■■ AUTHOR INDEX

■ SUBJECT INDEX

A*, 182
Accuracy, 222
Active wait, 243
Adjacency, 201
Adjacency list, 44
Adjacency matrix, 44
Agglomeration, 23, 120, 138
ALAP (as-late-as-possible), 98
Algorithm, 51
Allocated level, 99, 107
α field, 159, 214, 249
$\alpha|\beta|\gamma$ classification, 158
$\alpha|\beta|\gamma$ notation, 214, 249
AltiVec, 8, 154, 156
Ancestor, 43
Annotation, 183
Antidependent, 53
Application, 22, 24
    area, 27
    specification, 22
Architecture, *see* Parallel architecture
Array, 32
    dimension, 34
    element, 32
Array processors, 55, 61, 67, 127
    systolic, 62
    wavefront, 61
ASAP (as-soon-as-possible), 98
Assignment
    spatial, 23, 75
    temporal, 23, 75
Asymptotic notation, 44
Asynchronous, 60
Atomic instruction, 49
Attribution of start times, 118

Bandwidth, 12
β field, 159, 214, 249
BFS, *see* Breadth first search
Bin packing, 139

BINPACKING, 89, 165
Bisection width, 13
Blocking wait, 243
Bottom level, 95, 107
    computation, 96
Branching, 49
Breadth first search (BFS), 46
    algorithm, 46
    complexity, 46
Breeding, 170
Broadcast, 21
Bubble scheduling and allocation (BSA),
    149, 158, 191, 200
Buffer size, 202
Bus, 11, 22, 193

C-SCHED, 211
Cache, 76
Cache effect, 76
CASCH, 183
CCR, *see* Communication to computation
    ratio
CFG, *see* Control flow graph
CG, *see* Computation graph
Chess, 38
Child, 43
Chromosome, 171–177
    direct representation, 173–174
    encoding, 172
    fittest, 181
    indirect representation, 172–173
    node list, 172, 176
    pool, 171
    processor allocation, 172, 176
    representation type, 174–175
Clan, 138
Classic model, 76
Cluster, 120
    assignment, 139–140
    linear, definition, 124
    merge, 121

WILEY SERIES ON PARALLEL
AND DISTRIBUTED COMPUTING

Editor: Albert Y. Zomaya

A complete list of titles in this series appears at the end of this volume.

WILEY SERIES ON PARALLEL AND DISTRIBUTED COMPUTING
Series Editor: Albert Y. Zomaya

Printed and bound by CPI Group (UK) Ltd, Croydon, CR0 4YY

16/04/2025

14658587-0002